INSTITUTE OF ECONOMICS
AND STATISTICS

KU-511-240

QA 274 MAL

# STOCHASTIC METHODS IN ECONOMICS AND FINANCE

# ADVANCED TEXTBOOKS IN ECONOMICS

---

VOLUME 17

*Editors*:

C.J. BLISS

M.D. INTRILIGATOR

*Advisory Editors:*

L. JOHANSEN

D.W. JORGENSON

M.C. KEMP

J.-J. LAFFONT

J.-F. RICHARD

NORTH-HOLLAND PUBLISHING COMPANY
AMSTERDAM · NEW YORK · OXFORD

-7 MAY 1982

# STOCHASTIC METHODS IN ECONOMICS AND FINANCE

A.G. MALLIARIS
*Loyola University of Chicago*

with a Foreword and Contributions by

W.A. BROCK
*University of Wisconsin, Madison*

WITHDRAWN

ECONOMICS LIBRARY AND STATISTICS

NORTH-HOLLAND PUBLISHING COMPANY
AMSTERDAM · NEW YORK · OXFORD

© NORTH-HOLLAND PUBLISHING COMPANY — 1982

All rights reserved. No part of this publication may be reproduced, stored in a retrieval system, or transmitted, in any form or by any means, electronic, mechanical, photocopying, recording or otherwise, without the prior permission of the copyright owner.

*ISBN for this volume: 0 444 86201 3*

*Publishers*

NORTH-HOLLAND PUBLISHING COMPANY
AMSTERDAM · NEW YORK · OXFORD

*Sole distributors for the U.S.A. and Canada*

ELSEVIER/NORTH-HOLLAND, INC.
52 VANDERBILT AVENUE
NEW YORK, N.Y. 10017

PRINTED IN THE NETHERLANDS

## INTRODUCTION TO THE SERIES

The aim of the series is to cover topics in economics, mathematical economics and econometrics, at a level suitable for graduate students or final year under-graduates specializing in economics. There is at any time much material that has become well established in journal papers and discussion series which still awaits a clear, self-contained treatment that can easily be mastered by students without considerable preparation or extra reading. Leading specialists will be invited to contribute volumes to fill such gaps. Primary emphasis will be placed on clarity, comprehensive coverage of sensibly defined areas, and insight into fundamentals, but original ideas will not be excluded. Certain volumes will therefore add to existing knowledge, while others will serve as a means of communicating both known and new ideas in a way that will inspire and attract students not already familiar with the subject matter concerned.

The Editors

# CONTENTS

Contents

ix

6.  Portfolio jump processes                                 228
    7.  The demand for index bonds                               230
    8.  Term structure in an efficient market                    233
    9.  Market risk adjustment in project valuation              236
   10.  Demand for cash balances                                 238
  *11.  The price of systematic risk                             242
  *12.  An asset pricing model                                   246
  *13.  Existence of an asset pricing function                   254
  *14.  Certainty equivalance formulae                           256
  *15.  A testable formula                                       262
  *16.  An example                                               264
   17.  Miscellaneous applications and exercises                 268
   18.  Further remarks and references                           272


Selected bibliography                                            279

Author index                                                     295

Subject index                                                    299

*  Sections contributed by Professor W.A. Brock.

# FOREWORD

In *Stochastic Methods in Economics and Finance*, A.G. Malliaris has undertaken the extraordinarily difficult task of assembling the relevant literature on stochastic methods used in recent articles in economics and finance. Malliaris faces the following difficult tradeoffs.

On the one hand he must present the mathematical literature on stochastic calculus, stochastic differential equations, optimal stochastic control, and optimal stopping theory, among other topics, correctly and rigorously. On the other hand he must choose a low enough level of rigor so that the book is accessible to a wide audience. At the same time Malliaris must present an appropriate balance between the mathematical developments of the theory and a host of applications, so that the user is well prepared to read and contribute to frontier research in economics and finance.

Furthermore, since the mathematical background of the people who work in economics and finance covers the entire spectrum, from those having skills equivalent to PhD's from the best mathematics departments to those with a rudimentary understanding of first-year analysis, Malliaris' task of constructing a book to appeal to such a heterogeneous audience becomes almost impossible. Nor is this all. The book must be of finite size in order to keep the price finite. These are important pedagogical difficulties.

Dr. Malliaris has chosen the route of giving detailed citation to results that are further presented elsewhere. In short he has exposited what needs expositing and has told the reader where to go to find additional material. In this way he has prepared a "user's Guide" to an enormous mathematical—economic—financial literature.

The hope is that a reader equiped with Malliaris' book will find a much easier path to the cutting edge of ongoing research using stochastic methods in economics and finance over the alternative of hunting and pecking through the journals.

Anastasios G. Malliaris is well-equipped to write such a novel work. Even though he already possesses a PhD in economics, he took several years of courses in the University of Chicago's economics and mathematics departments. He is a patient English language craftsman with an encyclopedic mind.

Let me give an example of Malliaris' expository skills that appears in his book. He was able to take raw material from my lecture notes, problem sets, examinations, and oral delivery and machine this morass into readable material. Anyone who has attempted to write anything at all knows how much easier it is to communicate with the mouth than the pen.

In conclusion, A.G. Malliaris has presented the profession with a bold and novel concept in the area of textbook and reference book writing. My personal opinion is that he has solved a very hard constrained optimization problem admirably well. I believe that the market will think so too.

Chicago and Madison                                      W.A. Brock

# PREFACE

*Stochastic Methods in Economics and Finance* introduces the reader to certain mathematical techniques by presenting both their theoretical elements and their applications. Topics such as martingale methods, stochastic processes, optimal stopping, the modeling of uncertainty using a Wiener process, Itô's lemma as a tool of stochastic calculus, basic facts about stochastic differential equations, the notion of stochastic stability and the methods of stochastic control are discussed and their use in economic theory and finance is illustrated in numerous applications. Among these applications we mention futures pricing, job search, stochastic capital theory, stochastic economic growth, the rational expectations hypothesis, a stochastic macroeconomic model, competitive firm under price uncertainty, the Black—Scholes option pricing theory, optimum consumption and portfolio rules, demand for index bonds, term structure of interest rate, the market risk adjustment in project valuation, demand for cash balances and an asset pricing model.

Because the economics and finance professions have accepted the superiority of dynamic deterministic methods over static analysis techniques, the time has come to further encourage the trend towards the application of dynamic stochastic methods.

These methods are expected to capture the complexities, measurement errors and uncertainties associated with economic reality and to provide a way of modeling some of the researcher's pure ignorance about the future. Knowing the techniques or the methods presented here will enhance the ability to attack successfully some of the difficult applied problems arising in economics and finance. This book alone will not provide sufficient intellectual capital for the economic researchers to be fully equipped to solve a wide range of problems demanding stochastic methods. It is almost impossible to produce a book meeting such a standard. However, if used as an introductory survey of advanced stochastic methods, this book could serve as an efficient and useful guide to an enormous mathematics, economics and finance literature. Such, then is, the nature of this book: an introductory survey of advanced stochastic methods applied in economic analysis with detailed sections on further remarks and references to provide guidance for additional reading.

There are pedagogical challenges in writing a book where no close substitutes exist. For example, the content of the book, its mix between theory and application, and its level of exposition are three of the difficult questions to be dealt with. Concerning content, we chose to present a broader rather than a narrower coverage by describing several methods. We should point out that some important methods, including econometric methods, stochastic difference equations techniques and discrete time stochastic control, are not included, primarily because excellent sources on these subjects already exist. For the appropriate mix between theory and application, we chose to emphasize where possible each equally. Some theoretical topics have been included for which there is no immediate application. Such is the case with several measure theoretic probability concepts and the notion of stochastic integration. The justification for including these topics lies in their usefulness for understanding subsequent theoretical developments with immediate applications.

In illustrating the interplay between theory and applications we use two approaches. In Chapter 1, applications follow the presentation of martingale methods and optimal stopping and the use of theory is immediately illustrated. In the rest of the book, mathematical methods are contained in Chapter 2, while Chapters 3 and 4 contain economics and finance applications. Integrating theory and applications is natural in Chapter 1, but the separation of the two is more efficient elsewhere to maintain the flow of exposition. Some methods such as Itô's lemma have been intensively discussed regarding both theory and application. Other methods, despite completion of their theoretical development, have not yet been widely applied and therefore must await the work of future researchers. Examples of these methods include continuous optimal stopping, stochastic differential equations and stochastic control. Furthermore, certain areas of applied research in economics and finance, such as stochastic stability, have raised some difficult mathematical problems and appropriate mathematical methods are not yet fully developed.

It is very difficult to exposit the methods in this book without making use of some of their measure theoretic probability underpinnings. For that reason we chose to make use of various fundamental notions such as $\sigma$-field, probability space, measurable function, expectation and conditional probability. To make the book accessible to a wider audience, however, we kept the theoretical foundations to a skeletal minimum, and we supplied the interested reader with references where greater theoretical depth may be found.

The primary audience of this book will include Ph.D. students in economics and finance with aspirations of using stochastic methods as a tool of their research, but some MBA students specializing in financial theory or quantitative methods may find this book helpful in their coursework or as a supplementary source. In addition, economic and financial theory researchers may use this book to teach themselves stochastic methods. Finally, some applied mathematicians who specialize in the methods described here may find the book useful because of the numerous applications it contains.

An attempt has been made to keep mathematical prerequisites to a minimum. Many parts of this book could be understood by someone with a good background in analysis and basic probability theory. In the first five sections of Chapter 1 we have collected some results from courses beyond analysis and probability that are needed for Chapters 2, 3 and 4.

Most exercises fall in two categories: they either inform the reader of supplementary facts with an indication of where to find an answer or are designed to develop computational skills. There are, however, some exercises, such as nos. (5), (9) and (14) of Chapter 3, that move the reader into the area of model building. These are the hardest, and the reader is warned that they are suggested as possible illustrations rather than definitive contributions.

# ACKNOWLEDGMENTS

A book does not exist without a content and in our case the content relies heavily on the pioneering developments of the many mathematicians and economists cited in detail in the pages that follow. I acknowledge my intellectual debt to all authors mentioned in the pages that follow and hope that this book will help usher the reader to some of the works cited.

Professor W. Brock proposed the idea of writing this book about three years ago, and all during this period he generously supplied instruction, advice, encouragement and at the completion of the manuscript he offered to write the Foreword. He also gave permission to use his class notes, to reproduce several of his exam questions and he contributed several sections which are starred in the table of contents.

Insightful corrections and suggestions were offered by Jerry Bona (University of Chicago), F.R. Chang (University of Chicago), Patrick Brockett (University of California, Riverside), Mike Rothschild (University of Wisconsin, Madison), and anonymous referees. George Constantinides (University of Chicago) supplied valuable instruction and clarifications on the subject matter of Chapter 4. Mike Intriligator provided encouragement with his interest in this book, and through his advice and support the writing converged to its completion. Several individuals showed interest in this work or helped in various ways: Boyan Jovanovic (Bell Laboratories), George Kaufman (Loyola University of Chicago), Steve Magee (University of Texas, Austin), Mike Magill (University of Southern California), John McCall (University of California, Los Angeles), Donald Meyer (Loyola University of Chicago), Ronald Michener (University of Virginia), Samuel Ramenofsky (Loyola University of Chicago), Scott Richard (Carnegie-Mellon University), José Scheinkman (University of Chicago), and Sam Wu (University of Iowa). W. Brock and J. Ingersoll used portions of this book in courses they taught at the University of Chicago and I presented parts of the book at a finance seminar at the University of Texas at Austin. From such a use several errors have been corrected. Barbara Novy and Jean Shenoha helped with the bibliographical research and did extensive proofreading. Mary Beth Allen and John Schmadeke provided editorial assistance and Carmela Perno typed various versions of the manuscript accurately and efficiently. Dr. Ellen M. van Koten, economics editor, Mr. Leland K. Pierce, technical editor, and the staff at North-Holland Publishing Company were most cooperative and efficient during the production process. I am grateful to all, and for the errors that exist in the final product, almost surely, I take responsibility.

I happily dedicate this book to my wife, Mary Elaine, for her many contributions in my life.

Chicago

A.G. Malliaris

# CHAPTER 1

# RESULTS FROM PROBABILITY

> The evolution of probability theory is due
> precisely to the consideration of more
> and more complicated observables.
>
> M. Loève (1977, p. 7)

## 1. Introduction

In this chapter we present various ideas from modern probability theory in the form of definitions, theorems and examples. These ideas have been collected from a large number of sources and their presentation is intended to provide the reader with background material for the subsequent chapters of this book. Although some sections in this chapter develop ideas at a rapid rate, there are sections on martingales and on optimal stopping problems which are more detailed. Several examples illustrate the applicability of probabilistic martingale theory and the theory of optimal stopping rules in economic analysis and finance. Therefore, the reader may benefit from this chapter in at least two ways, firstly by developing an understanding of the theoretic underpinnings of the various applied probabilistic models, and secondly by being familiarized with martingale and optimal stopping theories which are actively used in applied research.

## 2. Probability spaces

The occurrence or nonoccurrence of an experiment or a trial in which chance intervenes is usually called an *outcome*. The totality of outcomes of an experiment are grouped together in a set denoted by $\Omega$. The elements of the *set* or *space* $\Omega$ are denoted by $\omega$ and called *sample points*. A subset of $\Omega$ is called an *event*. As an example, for the number of unemployed as a proportion of the total U.S. labor force during next month, the space $\Omega$ is all rational numbers in the interval $[0, 1]$, while $\omega = 0.065$ is an element or a sample point. Some events are: proportion of unemployed no more than 0.07, or, proportion of unemployed between 0.06 and 0.08. Intuitively, a probability is a valuation on some class of events.

Let $\Omega$ be an arbitrary space consisting of points $\omega$. Certain classes of subsets of $\Omega$ are important in the study of probability. We now define the class of subsets called *$\sigma$-field* or *$\sigma$-algebra*, denoted by $\mathscr{F}$ in this book. We say that $\mathscr{F}$ is a $\sigma$-field if the following three conditions are satisfied:

(1)  $\Omega \in \mathscr{F}$, i.e. $\mathscr{F}$ contains the space $\Omega$ itself;

(2)  $A \in \mathscr{F}$ implies $A^c \in \mathscr{F}$, i.e. if an event $A$ which is a subset of $\Omega$ belongs to $\mathscr{F}$, then the complement of $A$, denoted by $A^c$, also belongs to $\mathscr{F}$; and

(3)  $A_1, A_2, A_3, \ldots \in \mathscr{F}$ implies $A_1 \cup A_2 \cup A_3 \ldots \in \mathscr{F}$, i.e. if a countable sequence of subsets of $\Omega$ belongs to $\mathscr{F}$, then the countable union of these subsets also belongs to $\mathscr{F}$.

We call the elements of $\mathscr{F}$ *measurable sets*. Note that for a given set $\Omega$, the smallest $\sigma$-field consists of the empty set, $\varnothing$, and the set $\Omega$ itself, while the largest $\sigma$-field consists of the power set of $\Omega$, i.e. all the subsets of $\Omega$. The largest $\sigma$-field consists of $2^{\Omega}$ subsets of $\Omega$. The pair $(\Omega, \mathscr{F})$ is called a *measurable space*.

Now let $\mathscr{G}$ be a class of subsets of $\Omega$. The intersection of all $\sigma$-fields containing $\mathscr{G}$ is called the $\sigma$-field *generated* by $\mathscr{G}$ and is denoted by $\sigma(\mathscr{G})$. As an example we mention the $\sigma$-field generated by the class of intervals $(a, b]$ which are subsets of the real line $R^1$. This $\sigma$-field is denoted by $\mathscr{R}^1$ and its elements are called *Borel sets*. Note that $\mathscr{R}^1$ is the smallest $\sigma$-field containing the class of all intervals of $R$; in particular, it contains all open and all closed sets of $R$. Intuitively, $\mathscr{R}^1$ is obtained by starting with intervals of $R$ and forming repeated finite and countable set-theoretic operations (unions, intersections and complements) in all possible ways.

It was mentioned above that a probability is a valuation on a class of various events. We now make this notion precise. Let $(\Omega, \mathscr{F})$ be a measurable space. A set function $\mu$ defined on $\mathscr{F}$ is called a *measure* if it satisfies the following conditions:

(1)  $\mu(\varnothing) = 0$;

(2)  $A \in \mathscr{F}$ implies $0 \leqslant \mu(A) \leqslant \infty$; and

(3)  $A_1, A_2, \ldots \in \mathscr{F}$ and if $A_n$ are *pairwise disjoint*, i.e. $A_k \cap A_m = \varnothing$ for $k \neq m$, then

$$\mu\left(\bigcup_{n=1}^{\infty} A_n\right) = \sum_{n=1}^{\infty} \mu(A_n).$$

A measure of great importance is the *Lebesgue measure*, denoted by $\lambda$, and defined on the class of Borel sets $\mathscr{R}^1$ of the real line $R^1$. This measure assigns to every interval its length, i.e.

$$\lambda(a, b] = b - a.$$

The Lebesgue measure $\lambda$ can be extended in a straightforward manner to all Borel sets. For a detailed exposition see Ash (1972, ch. 1). Note that $\lambda(R^1) = \infty$ and also that the Lebesgue measure of every countable set is zero. The Lebesgue measure $\lambda$ on $\mathscr{R}^1$ can be generalized to be defined on the Borel sets of the $k$-space $R^k$ denoted by $\mathscr{R}^k$.

A *probability* is a special kind of a measure denoted by $P$, where $P(\Omega) = 1$. Thus, for $A \in \mathscr{F}$ we have $0 \leqslant P(A) \leqslant 1$. The triple $(\Omega, \mathscr{F}, P)$ is called a *probability space*, where $\Omega$ is a nonempty space of trials, $\mathscr{F}$ is a $\sigma$-field of subsets of $\Omega$ representing various events, and $P$ is a probability measure defined on $\mathscr{F}$.

As an example consider as $\Omega$ all rational numbers in $[0, 1]$ denoting the unemployed as a proportion of the total U.S. labor force in some future month. Let $\mathscr{F}$ be the $\sigma$-field of all subsets of this countable space $\Omega$ and let $\mu(\omega)$ be a nonnegative function defined on $\Omega$ such that $\Sigma_{\omega \in \Omega} \mu(\omega) = 1$. Next, define $P(A) = \Sigma_{\omega \in A} \mu(\omega)$. Then the triple $(\Omega, \mathscr{F}, P)$ is a probability space.

A special case of a probability space is that of a complete probability space. Let $(\Omega, \mathscr{F}, P)$ be a probability space and let $N$ be a subset of this space. We say that $N$ is *negligible* if there exists a set $A \in \mathscr{F}$ such that $N$ is contained in $A$, and $P(A) = 0$. The probability space is *complete* if $\mathscr{F}$ contains every negligible subset of $\Omega$ with respect to $P$. This concept is used in section 7 of this chapter.

Consider a sequence $\{A_n\}$ of events in a probability space $(\Omega, \mathscr{F}, P)$. Define

$$\lim_n \sup A_n = \bigcap_{n=1}^{\infty} \bigcup_{k=n}^{\infty} A_k$$

$$= \{\omega : \omega \in A_n \text{ for infinitely many } n\}$$

and also

$$\lim_n \inf A_n = \bigcup_{n=1}^{\infty} \bigcap_{k=n}^{\infty} A_k$$

$$= \{\omega : \omega \in A_n \text{ for all but finitely many } n\}.$$

If all $A_n \in \mathscr{F}$ then both $\lim \sup_n A_n$ and $\lim \inf_n A_n$ belong to $\mathscr{F}$.

Next we need the definition of independence before we close this section with the statement of an important lemma.

From elementary probability we recall that two events, $A$ and $B$, are *independent* if $P(A \cap B) = P(A) \cdot P(B)$. We generalize this concept by stating that a finite collection of events $A_1, A_2, ..., A_n$ is *independent* if

$$P(A_{k_1} \cap ... \cap A_{k_j}) = P(A_{k_1}) \cdots P(A_{k_j}) \qquad (2.1)$$

for each set $k_1, ..., k_j$ of distinct indices from $1, ..., n$. An infinite collection of events is *independent* if each of its finite subcollection is independent. The following lemma is useful and will be used later in this book.

**Lemma 2.1** (Borel–Cantelli). Let $\{A_n\}$ be a sequence of events in a probability space $(\Omega, \mathscr{F}, P)$. If $\Sigma_n \, P(A_n) < \infty$, then $P(\limsup_n A_n) = 0$. Also, if the sequence of events $\{A_n\}$ is independent and $\Sigma_n \, P(A_n) = \infty$, then $P(\limsup_n A_n) = 1$.

*Proof.* See Neveu (1965, pp. 128–129).

## 3. Random variables

Let $(\Omega, \mathscr{F})$ and $(\Omega', \mathscr{F}')$ denote two measurable spaces. A mapping $X: \Omega \to \Omega'$ is $(\mathscr{F}, \mathscr{F}')$-*measurable* if for each $A' \in \mathscr{F}'$

$$X^{-1}(A') = \{\omega: X(\omega) \in A'\} \in \mathscr{F}.$$

Intuitively, measurability of a mapping means that for every meaningful event in the image space $\Omega'$ there is a meaningful event in the domain space $\Omega$. It turns out that an event is meaningful if it belongs to an appropriate $\sigma$-field. The $\sigma$-field generated by $X$ and denoted by $\sigma(X)$ is the smallest $\sigma$-field with respect to which $X$ is measurable.

A special case of importance is when the image space $\Omega'$ is the real line $R^1$. In this case we use as a $\sigma$-field $\mathscr{R}^1$, i.e. the class of Borel sets. A function $f: \Omega \to R^1$ is measurable if for each $A' \in \mathscr{R}^1$ we have $f^{-1}(A') = \{\omega: f(\omega) \in A'\} \in \mathscr{F}$. A real function that is measurable in the sense just described is called a *random variable*.

Suppose that $f: \Omega \to R^k$ is measurable. Then it is called a *random vector*. Note that $f$ has the form

$$f(\omega) = (f_1(\omega), ..., f_k(\omega)),$$

where each component $f_i(\omega)$ is a real function. The mapping $f$ is measurable if and only if each component function $f_i$ is measurable. In many applications we assume that $f$ is continuous, which implies that $f$ is measurable.

Let $(\Omega, \mathscr{F}, P)$ be a probability space, and $(R^1, \mathscr{R}^1, \lambda)$ the measurable space with $R^1$ the real line, $\mathscr{R}^1$ the $\sigma$-field of Borel sets, and $\lambda$ the Lebesgue measure, and then suppose that $X$ is a random variable such that $X: \Omega \to R^1$. For all $A \in \mathscr{R}^1$ we define the *distribution* or *law* of $X$, denoted by $P_X$, as

$$P_X(A) = P(X^{-1}(A))$$

$$= P\{\omega: X(\omega) \in A\}. \tag{3.1}$$

The distribution of the random variable $X$ assigns to a set $A \in \mathscr{R}^1$ in the target space a probability measure depending of course on $X$. The *distribution function* of $X$, denoted by $F$, is defined by

$$F(x) = P\{\omega: X(\omega) \leqslant x\} \tag{3.2}$$

for $x \in R^1$. We note that $F$ is a nondecreasing function, right continuous such that

$$F(x) \to 0 \quad \text{as } x \to -\infty,$$
$$F(x) \to 1 \quad \text{as } x \to \infty.$$

There is a powerful theorem in probability theory that says that for a given distribution function $F$ having the stated properties, there can always be constructed a probability space $(\Omega, \mathscr{F}, P)$ and a random variable $X$ such that $F$ is the distribution function of $X$. See Billingsley (1979, p. 159).

The concept of a distribution function may be generalized for a random vector $X$ such that $X: \Omega \to R^k$. We then have for $x \in R^k$

$$F(x) = F(x_1, x_2, ..., x_k)$$

$$= P\{\omega: X_1(\omega) \leqslant x_1, ..., X_k(\omega) \leqslant x_k\}. \tag{3.3}$$

Here $X_1, ..., X_k$ are the $k$-components of $X$ and we specify $F$ as the *joint distribution function*. Let $F$ be a joint distribution function of a $k$-random vector $X = (X_1, ..., X_k)$. The *marginal distribution* function of a subcollection of $m$ components of $X$, where $m \leqslant k$, is obtained by replacing the unencountered arguments with $\infty$. As an illustration, a two-dimensional, $m = 2$, marginal distribution can be written as

$$F(x_1, x_2, \infty, ..., \infty)$$

$$= P\{\omega: X_1(\omega) \leqslant x_1, X_2(\omega) \leqslant x_2, X_3(\omega) \leqslant \infty, ..., X_k(\omega) \leqslant \infty\}. \qquad (3.4)$$

A random variable $X$ and its distribution $P_X$ have *density f* with respect to the Lebesgue measure if $f$ is a non-negative real function such that for each $A \in \mathscr{R}^1$,

$$P\{\omega: X(\omega) \in A\} = \int_A f(x) \, dx. \qquad (3.5)$$

Observe that the density function $f$ is determined only to within a set of Lebesgue measure zero. In the case where the random variable $X$ has density $f$, then the density $f$ and the distribution function $F$ are related by the equation

$$F(x) = \int_{-\infty}^{x} f(s) \, ds. \qquad (3.6)$$

In many applications the distribution function $F$ is continuously differentiable, in which case the derivative of $F$ can serve as the density $f$.

There are several examples of distributions that are well known from elementary probability, e.g. the *binomial distribution*

$$P_X(x) = \binom{n}{x} p^x (1-p)^{n-x} \qquad (3.7)$$

for $x = 0, 1, 2, ..., n$. In this example we may take $\Omega = \{0, 1, 2, ..., n\}$ with $\mathscr{F}$ consisting of all subsets of $\Omega$, and define the random variable $X$ as $X(x) = x$.

Another example is the *Poisson distribution* with positive parameter $m$ given by

$$P_X(x) = \frac{e^{-m} m^x}{x!} \qquad (3.8)$$

for $x = 0, 1, 2, ...$ . Finally, we say that the random variable $X$ has a *normal distribution* with parameters $\mu$ and $\sigma > 0$ if

$$P_X(A) = \frac{1}{\sqrt{2\pi}\sigma} \int_A \exp\left[\frac{-(x-\mu)^2}{2\sigma^2}\right] dx \qquad (3.9)$$

for $x \in A$ and $A \in \mathscr{R}^1$.

Suppose that $X_1, ..., X_n$ are random variables defined on a probability space $(\Omega, \mathscr{F}, P)$ such that $X_i: \Omega \to R^1$, $i = 1, 2, ..., n$. We say that $X_1, ..., X_n$ are *independent random variables* if for all Borel sets $A_1, ..., A_n$,

$$P\{\omega: X_1(\omega) \in A_1, ..., X_n(\omega) \in A_n\}$$
$$= P\{\omega: X_1(\omega) \in A_1\} \cdots P\{\omega: X_n(\omega) \in A_n\}. \tag{3.10}$$

There are two other equivalent definitions: $X_1, ..., X_n$ are independent random variables if for $x_1, ..., x_n$ real numbers

$$P[X_1 \leqslant x_1, ..., X_n \leqslant x_n] = P[X_1 \leqslant x_1] \cdots P[X_n \leqslant x_n]. \tag{3.11}$$

Note the notation in (3.11) where the argument $\omega$ has been suppressed for notational convenience. Alternatively, $X_1, ..., X_n$ are independent if for $x_1, ..., x_n$ real numbers

$$F(x_1, ..., x_n) = F_1(x_1) \cdots F_n(x_n), \tag{3.12}$$

where $F$ is the joint distribution function of $X_1, ..., X_n$ and $F_1, ..., F_n$ are the one-dimensional marginal distribution functions.

The definitions above may be generalized to random vectors very simply. In (3.10) we understand $X_i$ to be a random vector and $A_i$ to be a $k$-dimensional Borel set. Furthermore, the definition of independence can be extended to an infinite collection of random variables $X_1, X_2, ...$ simply by requiring that each finite subcollection is independent according to (3.10) or the other equivalent definitions.

For a sequence of random variables $\{X_n\}$ defined on $(\Omega, \mathscr{F}, P)$ and a random variable $X$, also defined on the same space, there are three useful convergence concepts. First, if there is a set $N \in \mathscr{F}$ having $P(N) = 0$ and such that for all $\omega \notin N$, the sequence of real numbers $\{X_n(\omega)\}$ converges in the usual sense to $X(\omega)$, then we say that $\{X_n\}$ *converges with probability* 1 to $X$ and we write

$$X_n \to X \text{ w.p.1 as } n \to \infty. \tag{3.13}$$

Secondly, we say that the sequence of random variables $\{X_n\}$ *converges in probability* to $X$ if for each positive $\epsilon$ we have

$$P\{\omega: |X_n(\omega) - X(\omega)| > \epsilon\} = P[|X_n - X| > \epsilon] \to 0 \tag{3.14}$$

as $n \to \infty$ and we write

$$X_n \xrightarrow{P} X \text{ as } n \to \infty. \tag{3.15}$$

Finally, let $\{F_n\}$ and $F$ denote the distributions of $\{X_n\}$ and $X$ and suppose that as $n \to \infty$

$$\int_R f(x)\,\mathrm{d}F_n(x) \to \int_R f(x)\,\mathrm{d}F(x) \tag{3.16}$$

for every real-valued continuous bounded function $f$ defined on $R$. Then we say that $\{X_n\}$ *converge in distribution* to $X$ and we write

$$X_n \Rightarrow X \quad \text{as } n \to \infty. \tag{3.17}$$

Observe that (3.13) implies (3.15), and that (3.15) implies (3.17).

## 4. Expectation

If $X$ is a random variable defined on a probability space $(\Omega, \mathscr{F}, P)$, the *expectation* of $X$ is defined by

$$E(X) = \int_\Omega X\,\mathrm{d}P, \tag{4.1}$$

provided the integral exists, i.e. $\int_\Omega X\,\mathrm{d}P < \infty$. A question naturally arises: What is the meaning of the right-hand side of (4.1)? We immediately proceed to answer this question by defining the integral

$$\int_\Omega X(\omega)\,\mathrm{d}P(\omega) = \int_\Omega X(\omega)P(\mathrm{d}\omega) = \int_\Omega X\,\mathrm{d}P. \tag{4.2}$$

First, assume that $X$ is a non-negative random variable. For each finite decomposition of the space $\Omega$ into sets $A_i$, i.e. a collection $\{A_i\}$ such that $A_i \in \mathscr{F}$, $A_i \cap A_j = \varnothing$ for $i \neq j$ and $\cup_i A_i = \Omega$, consider the sum

$$\sum_i [\inf_{\omega \in A_i} X(\omega)]P(A_i). \tag{4.3}$$

Using (4.3) we now define the meaning of (4.2) as follows:

$$\int_\Omega X\,\mathrm{d}P = \sup \sum_i [\inf_{\omega \in A_i} X(\omega)]P(A_i). \tag{4.4}$$

In (4.4) the supremum extends over all finite decompositions $\{A_i\}$ of $\Omega$ into $\mathscr{F}$-sets.

Secondly, observe that an arbitrary random variable $X$ may be written as

$$X = X^+ - X^-, \tag{4.5}$$

where $X^+$ denotes the *positive part of X* defined by

$$X^+(\omega) = \begin{cases} X(\omega) & \text{if } 0 \leqslant X(\omega) \leqslant \infty, \\ 0 & \text{otherwise,} \end{cases} \tag{4.6}$$

and where $X^-$ denotes the *negative part of X* defined by

$$X^-(\omega) = \begin{cases} -X(\omega) & \text{if } -\infty \leqslant X(\omega) \leqslant 0, \\ 0 & \text{otherwise.} \end{cases} \tag{4.7}$$

Note that $X^+$ and $X^-$ are non-negative and measurable and therefore (4.4) applies to both of them. Thus, the integral of an arbitrary random variable is defined by

$$\int_\Omega X \, \mathrm{d}P = \int_\Omega X^+ \, \mathrm{d}P - \int_\Omega X^- \, \mathrm{d}P. \tag{4.8}$$

If each term of the right-hand side of (4.8) is finite then we say that $X$ is *integrable*. Since $|X| = X^+ + X^-$, note that $X$ is integrable if and only if $\int_\Omega |X| \mathrm{d}P < \infty$.

The concept of expectation has several desirable properties. Before we collect these properties in the next theorem we explain a usual convention in measure and probability theory. In a probability space $(\Omega, \mathscr{F}, P)$, a statement or proposition regarding the elements $\omega$ of the space $\Omega$ is understood to hold *almost surely*, written as a.s., or equivalently *with probability* 1, written as w.p.1, if the statement or proposition is true for all $\omega \in \Omega$ except possibly for those $\omega$ in some set $N$, where $N \in \mathscr{F}$ and $P(N) = 0$. If instead of $P$ we have an arbitrary measure we replace w.p.1 by *almost everywhere*, written as a.e.

**Theorem 4.1.** Let $(\Omega, \mathscr{F}, P)$ be a probability space and assume that $f$ and $h$ are random variables defined on this space.

(1) If $f$ is integrable and $\alpha \in R$ then $\int_\Omega \alpha f \mathrm{d}P < \infty$ and also

$$\int_\Omega \alpha f \mathrm{d}P = \alpha \int_\Omega f \mathrm{d}P.$$

(2) If $f$ and $h$ are integrable and $\alpha$ and $\beta$ are finite real numbers, then $\alpha f + \beta h$ is integrable and also

$$\int_\Omega (\alpha f + \beta h)\, dP = \alpha \int_\Omega f\, dP + \beta \int_\Omega h\, dP.$$

(3)  If $f$ and $h$ are integrable and if $f \leqslant h$ w.p.1 then

$$\int_\Omega f\, dP \leqslant \int_\Omega h\, dP.$$

(4)  If $f$ is integrable then

$$\left| \int_\Omega f\, dP \right| \leqslant \int_\Omega |f|\, dP.$$

(5)  If $f$ is integrable then for $A \in \mathscr{F}, \int_A f\, dP < \infty$.

For a proof of this theorem see Ash (1972, pp. 41–42).

For sequences of random variables we state three basic theorems.

**Theorem 4.2.** (Monotone convergence).  Let $\{X_n\}$ be an increasing sequence of non-negative random variables on $(\Omega, \mathscr{F}, P)$ and let $X_n(\omega) \to X(\omega)$ w.p.1. Then

$$\int_\Omega X_n\, dP \to \int_\Omega X\, dP.$$

Observe that this theorem justifies interchanging the limit and expectation operations, i.e. $\lim_{n\to\infty} E(X_n) = E(\lim_{n\to\infty} X_n)$.

Next we define $\limsup_n X_n$ and $\liminf_n X_n$ for a sequence of real random variables as follows:

$$\left( \liminf_n X_n \right)(\omega) = \sup_n \inf_{k \geqslant n} X_k(\omega),$$

$$\left( \limsup_n X_n \right)(\omega) = \inf_n \sup_{k \geqslant n} X_k(\omega).$$

**Theorem 4.3.** (Fatou).  For a non-negative sequence of random variables $\{X_n\}$ on $(\Omega, \mathscr{F}, P)$,

$$\int_\Omega \liminf_n X_n\, dP \leqslant \liminf_n \int X_n\, dP.$$

**Theorem 4.4.** (Dominated convergence).  Let $\{X_n\}$ be a sequence of random variables and $Y$ an integrable random variable all on $(\Omega, \mathscr{F}, P)$ such that $|X_n| \leqslant Y$ w.p.1, for all $n$. If $X_n \to X$ w.p.1, then $X$ and $X_n$ are integrable and

$$\int_\Omega X_n\, dP \to \int_\Omega X\, dP.$$

Proofs of these theorems may be found in Ash (1972, pp. 44–50).

With the above background on integration we now state some fundamental definitions. Let $X$ be a random variable on $(\Omega, \mathscr{F}, P)$ and let $k > 0$. We say that $E(X^k)$ is the *kth moment* of $X$ and that $E((X - E(X))^k)$ is the *kth central moment*. When $k = 1$, $E(X)$ is usually called the *mean* of $X$ and when $k = 2$, the second central moment is called the *variance* of $X$, written as var $X$ and denoted by $\sigma^2$, i.e.

$$\sigma^2 = \text{var } X = E((X - E(X))^2), \tag{4.9}$$

provided $E(X) < \infty$. The positive square root $\sigma$ is called the standard deviation.

Note that if $k > 0$ and $E(X^\kappa) < \infty$, then $E(X^\ell) < \infty$ for $0 < \ell < k$. Also, if $X_1, ..., X_n$ are independent random variables on $(\Omega, \mathscr{F}, P)$ such that $E(X_i) < \infty$ for all $i = 1, 2, ..., n$, then $E(X_1, ..., X_n) < \infty$ and also

$$E(X_1, ..., X_n) = E(X_1) \cdots E(X_n); \tag{4.10}$$

furthermore

$$\text{var}(X_1 + ... + X_n) = \sum_{i=1}^{n} \text{var } X_i. \tag{4.11}$$

For two random variables, $X$ and $Y$, each having a finite expectation, the *covariance* of $X$ and $Y$ is defined by

$$\begin{aligned} \text{cov}(X, Y) &= E((X - E(X))(Y - E(Y))) \\ &= E(XY) - E(X)E(Y). \end{aligned} \tag{4.12}$$

From (4.10) observe that if $X$ and $Y$ are independent, then $\text{cov}(X, Y) = 0$; however, the converse is not true.

Consider the random variables $X$ and $Y$ and suppose that their variances, denoted by $\sigma_X^2$ and $\sigma_Y^2$, respectively, are nonzero and finite. We define the *correlation coefficient* between $X$ and $Y$, denoted by $\rho(X, Y)$ as

$$\rho(X, Y) = \text{cov}(X, Y)/\sigma_X \sigma_Y. \tag{4.13}$$

We close this section by stating a useful fact. Let $X$ be a random variable on $(\Omega, \mathscr{F}, P)$ with distribution function $F$. Assume that $g: R \to R$ is a measurable function and let $Y = g(X)$. Then

$$E(Y) = \int_R g(x)\, dF(x). \tag{4.14}$$

In various applications it becomes easier to compute the expectation of a random variable by integrating over $R$ instead of over $\Omega$. In such cases we may use (4.14) by letting $g(x) = x$.

## 5. Conditional probability

Recall from elementary probability that the conditional probability of a set $A$ given a set $B$, denoted by $P(A \mid B)$, is given by

$$P(A \mid B) = P(A \cap B)/P(B), \tag{5.1}$$

provided $P(B) \neq 0$. A conditional probability is associated with events in a given subset of the space $\Omega$. Intuitively, a *conditional probability* represents a re-evaluation of the probability of $A$ occurring in light of the information that $B$ has already occurred. In this section we study conditional probability in a general context for a space $(\Omega, \mathscr{F}, P)$, where the conditional probability of a set $A$ is defined with respect to a $\sigma$-field $\mathscr{G}$ in $\mathscr{F}$. The notation used is $P[A \mid \mathscr{G}]$ and the intuitive interpretation is analogous to (5.1), in the sense that the conditional probability of the set $A$ is being evaluated in light of the information available in the $\sigma$-field $\mathscr{G}$, with $\mathscr{G}$ contained in $\mathscr{F}$.

At this point a digression is necessary to present two definitions and the Radon–Nikodym theorem which will establish the existence of conditional probability.

Consider two measures $\nu$ and $P$ on a measurable space $(\Omega, \mathscr{F})$. We say that the measure $\nu$ is *absolutely continuous* with respect to the measure $P$, or equivalently that $\nu$ is *dominated* by $P$, if for each $A \in \mathscr{F}$, $P(A) = 0$ implies $\nu(A) = 0$ also. If $\Omega$ is the union of an at most countably infinite family of sets in $\mathscr{F}$, each with finite measure, then the measure $P$ is called *$\sigma$-finite*. We are now ready to state

**Theorem 5.1.** (Radon–Nikodym). Let $(\Omega, \mathscr{F})$ be a measurable space. Suppose that $\nu$ and $P$ are two $\sigma$-finite measures on $(\Omega, \mathscr{F})$ such that $\nu$ is dominated by $P$. Then, there is a non-negative measurable function $f$ such that

$$\nu(A) = \int_A f \, \mathrm{d}P \tag{5.2}$$

for all $A \in \mathscr{F}$.

For a proof see Ash (1972, pp. 63–65).

The function $f$ in (5.2) is called the *Radon–Nikodym derivative* of $\nu$ with respect to $P$ and is denoted as $\mathrm{d}\nu/\mathrm{d}P$.

With the above brief background we motivate the definition of conditional probability of a set $A$ given a $\sigma$-field $\mathcal{G}$ in $\mathcal{F}$ for an arbitrary probability space $(\Omega, \mathcal{F}, P)$. In this analysis we follow Billingsley (1979).

For a given set $A \in \mathcal{F}$, define the measure $\nu$ on the $\sigma$-field $\mathcal{G}$ by

$$\nu(G) = P(A \cap G) \tag{5.3}$$

for all $G \in \mathcal{G}$. Restrict the probability measure $P$ to the $\sigma$-field $\mathcal{G}$ and note that if $P(G) = 0$ then $\nu(G) = 0$, which means that the measure $\nu$, as defined in (5.3), is dominated by $P$. For the measurable space $(\Omega, \mathcal{G})$, the facts that $\nu$ and $P$ are $\sigma$-finite and also that $\nu$ is dominated by $P$ imply that by the Radon—Nikodym theorem there is a non-negative random variable $f$ such that (5.2) holds. Combining (5.2) and (5.3) we obtain that

$$\nu(G) = P(A \cap G) = \int_G f \, \mathrm{d}P \tag{5.4}$$

for all $G \in \mathcal{G}$. The random variable $f$ is precisely the *conditional probability* $P[A \mid \mathcal{G}]$ which has, by the preceding analysis, two properties:

(1)  $P[A \mid \mathcal{G}]$ is measurable with respect to $\mathcal{G}$ and integrable with respect to $P$; and

(2)  $P[A \mid \mathcal{G}]$ satisfies the equation for $G \in \mathcal{G}$,

$$\int_G P[A \mid \mathcal{G}] \, \mathrm{d}P = P(A \cap G). \tag{5.5}$$

There are many random variables $P[A \mid \mathcal{G}]$ which are equal w.p.1. For a given $P[A \mid \mathcal{G}]$ we say that this random variable is a *version* of conditional probability. When the conditional probability of $A$ is with respect to a $\sigma$-field generated by a random variable $X$ we write $P[A \mid \sigma(X)]$, or simply $P[A \mid X]$, which can be generalized to the conditional probability of $A$ given an arbitrary sequence of random variables $X_1, X_2, \ldots$ written as $P[A \mid \sigma(X_1, X_2, \ldots)]$ or $P[A \mid X_1, X_2, \ldots]$.

Some of the basic properties of conditional probability are summarized in the next theorem.

**Theorem 5.2.**  Let $(\Omega, \mathcal{F}, P)$ be a probability space and $\mathcal{G}$ a $\sigma$-field in $\mathcal{F}$.

(1)  If $A \in \mathcal{F}$, then $0 \leqslant P[A \mid \mathcal{G}] \leqslant 1$. In particular, if $P(A) = 1$, then $P[A \mid \mathcal{G}] = 1$ and if $P(A) = 0$, then $P[A \mid \mathcal{G}] = 0$.

(2)  If $\{A_n\}, n = 1, 2, \ldots$ is a sequence of disjoint sets, then

$$P[\bigcup_n A_n \mid \mathcal{G}] = \sum_n P[A_n \mid \mathcal{G}].$$

(3)  If $A \in \mathscr{F}$ and $B \in \mathscr{F}$ such that $A \subset B$, then

$$P[B - A \mid \mathscr{G}] = P[B \mid \mathscr{G}] - P[A \mid \mathscr{G}]$$

and also

$$P[A \mid \mathscr{G}] \leqslant P[B \mid \mathscr{G}].$$

(4)  If $\{A_n\}$ is an increasing (decreasing) sequence with $A$ as its limit, then $P[A_n \mid \mathscr{G}]$ increases (decreases) to $P[A \mid \mathscr{G}]$ as its limit as $n \to \infty$.

Having briefly discussed conditional probability we next present the notion of *conditional expectation.*

Let $X$ be a non-negative integrable random variable on $(\Omega, \mathscr{F}, P)$ and define the measure $\nu$ on $\mathscr{G}$ by

$$\nu(G) = \int_G X \, \mathrm{d}P \tag{5.6}$$

for $G \in \mathscr{G}$, where $\mathscr{G}$ is a $\sigma$-field in $\mathscr{F}$. As earlier in this section, if $P$ is restricted to $\mathscr{G}$ then $\nu$, as defined in (5.6), is dominated by $P$ and an application of the Radon–Nikodym theorem yields the existence of a random function $f$ such that

$$\nu(G) = \int_G f \, \mathrm{d}P. \tag{5.7}$$

This random variable is denoted by $E[X \mid \mathscr{G}]$ and called the *conditional expected value* of the random variable $X$ given the $\sigma$-field $\mathscr{G}$. Note that $E[X \mid \mathscr{G}]$ satisfies the two properties:

(1)  $E[X \mid \mathscr{G}]$ is integrable and $\mathscr{G}$ measurable.
(2)  From (5.6) and (5.7) we obtain that for $G \in \mathscr{G}$

$$\int_G E[X \mid \mathscr{G}] \, \mathrm{d}P = \int_G X \, \mathrm{d}P. \tag{5.8}$$

For a general random variable $X$, not necessarily non-negative, its conditional expected value can be obtained by using (4.5) and defining

$$E[X \mid \mathscr{G}] = E[X^+ \mid \mathscr{G}] - E[X^- \mid \mathscr{G}].$$

Since there are many $E[X \mid \mathscr{G}]$ which vary on a set of probability zero, one such given conditional expected value is called a *version* of the conditional expectation of the random variable $X$. It is worth mentioning that $E[X \mid \mathscr{G}]$ is a random variable whose value depends on $\omega$, i.e. $E[X \mid \mathscr{G}](\omega)$, and we interpret this to mean the expected value of $X$ given the information in the $\sigma$-field $\mathscr{G}$.

Conditional expectation satisfies various properties reported in the theorems below. In theorem 5.3 we assume that $X$ and $Y$ are integrable.

**Theorem 5.3.** Let $(\Omega, \mathscr{F}, P)$ be a probability space and $\mathscr{G}$ a $\sigma$-field in $\mathscr{F}$.

(1) If $X$ is a constant random variable, i.e. $X = c$, $c \in R$, then $E[X \mid \mathscr{G}] = c$ w.p.1.

(2) If $X$ and $Y$ are random variables and $a, b \in R$, then

$$E[aX \pm bY \mid \mathscr{G}] = a E[X \mid \mathscr{G}] \pm b E[Y \mid \mathscr{G}].$$

(3) If $X$ and $Y$ are random variables such that $X \leqslant Y$ w.p.1, then

$$E[X \mid \mathscr{G}] \leqslant E[Y \mid \mathscr{G}].$$

(4) For $X$ a random variable

$$|E[X \mid \mathscr{G}]| \leqslant E[|X| \mid \mathscr{G}].$$

For a proof see Billingsley (1979, p. 397).

**Theorem 5.4.** Let $X$ be a random variable on $(\Omega, \mathscr{F}, P)$ and suppose that it is integrable. Suppose that $\mathscr{G}_1$ and $\mathscr{G}_2$ are $\sigma$-fields in $\mathscr{F}$ such that $\mathscr{G}_1 \subset \mathscr{G}_2$. Then

$$E[E[X \mid \mathscr{G}_2] \mid \mathscr{G}_1] = E[X \mid \mathscr{G}_1] \quad \text{w.p.1.}$$

For a proof see Tucker (1967, p. 212).

**Theorem 5.5.** Let $X$ and $Y$ be random variables on $(\Omega, \mathscr{F}, P)$, and suppose $Y$ and $XY$ are integrable and $X$ is $\mathscr{G}$-measurable. Then

$$E[XY \mid \mathscr{G}] = X E[Y \mid \mathscr{G}] \quad \text{w.p.1}$$

for $\mathscr{G}$ a $\sigma$-field in $\mathscr{F}$.

For a proof see Tucker (1967, pp. 213–214).

Finally we state

**Theorem 5.6.** (Conditional form of Jensen's inequality). Let $h$ be a convex function on $R$ and $X$ a random variable on $(\Omega, \mathcal{F}, P)$ with $\mathcal{G}$ a $\sigma$-field in $\mathcal{F}$. Suppose that $X$ and $h(X)$ are both integrable. Then

$$h(\mathrm{E}[X \mid \mathcal{G}]) \leqslant \mathrm{E}[h(X) \mid \mathcal{G}] \quad \text{w.p.1.}$$

For a proof see Tucker (1967, p. 217).

We close this section by stating two important facts.

Suppose that $X$ is a random variable on $(\Omega, \mathcal{F}, P)$ and let $\mathcal{G}$ be a $\sigma$-field in $\mathcal{F}$. In our discussion thus far we have explained the meaning of conditional probability which readily extends to the case of a random variable since $P[X \in A' \mid \mathcal{G}]$ $= P[A \mid \mathcal{G}]$, with $A = \{\omega: X(\omega) \in A'\}$ for $A' \in \mathcal{R}^1$. We also explained the meaning of conditional expectation, written as $\mathrm{E}[X \mid \mathcal{G}]$. For both concepts of conditional probability and conditional expectation we would like to associate a probability distribution function. In this analysis we follow Billingsley (1979, p. 390) and we state

**Fact 1.** For the random variable $X$ on $(\Omega, \mathcal{F}, P)$ with conditional probability

$$P[A \mid \mathcal{G}] = P[X \in A' \mid \mathcal{G}] = P[\omega: X(\omega) \in A' \mid \mathcal{G}]$$

for $A' \in \mathcal{R}^1$, there exists a function $p(A', \omega)$ defined for $A' \in \mathcal{R}^1$ and $\omega \in \Omega$ satisfying two properties: (1) for each $\omega \in \Omega, p(A', \omega)$ is, as a function of $A'$, a probability measure on $\mathcal{R}^1$, and (2) for each $A' \in \mathcal{R}^1, p(A', \omega)$ is, as a function of $\omega$, a version of $P[X \in A \mid \mathcal{G}]$. The probability measure $p(\cdot, \omega)$ is called the *conditional distribution* of $X$ given $\mathcal{G}$. We also have

**Fact 2.** Let $p(\cdot, \omega)$ be the conditional distribution of $X$ given $\mathcal{G}$, as described in fact 1, and suppose that $X$ is integrable. Then

$$\int_R x p(\mathrm{d}x, \omega)$$

is a version of $\mathrm{E}[X \mid \mathcal{G}]$. See Billingsley (1979, p. 399).

## 6. Martingales and applications

Having presented the elements of conditional probability we now study martingale theory and some of its applications in economics and finance. Martingale theory uses conditional probability extensively so the usefulness of the previous section will shortly become apparent. Let us also mention that martingale theory, like various other areas of probability theory, has its origins in notions of gambling in the following sense: let $X_1$, $X_2$, ... be a sequence of random variables defined on a common probability space $(\Omega, \mathscr{F}, P)$. Within the gambling context these random variables denote the gambler's total winnings after $1, 2, ..., n, ...$ trials in a succession of games. The gambler's expected fortune after trial $n + 1$, given that he has completed $n$ trials already, is denoted as $E[X_{n+1} \mid X_1, ..., X_n]$. As in the previous section $E[X_{n+1} \mid X_1, ..., X_n]$ denotes the expectation of $X_{n+1}$ conditioned upon the $\sigma$-field generated by the random variables $X_1, ..., X_n$. Intuitively, the $\sigma$-field generated by $X_1, ..., X_n$ contains all the past information of the gambler's fortune up to and including the $n$th trial. If a game is *fair* then the gambler after the $(n + 1)$st trial will expect on the average to be neither wealthier nor poorer than he was before this trial, i.e.

$$E[X_{n+1} \mid X_1, ..., X_n] = X_n. \tag{6.1}$$

Stated differently, eq. (6.1), called the *martingale property*, states that in a fair game the gambler's fortune on the next play is on the average his current fortune and is not otherwise affected by the previous history.

If instead of $=$ in (6.1) we put $\geqslant$ or $\leqslant$, then the game is favorable or unfavorable, respectively. With this motivation we state the definition of a martingale.

## 6.1 Definitions

Let $X_1$, $X_2$, ... be a sequence of random variables defined on a common probability space $(\Omega, \mathscr{F}, P)$ and let $\mathscr{F}_1$, $\mathscr{F}_2$, ... be a sequence of $\sigma$-fields all belonging to $\mathscr{F}$. The sequence $\{(X_n, \mathscr{F}_n), n = 1, 2, ...\}$ is a *martingale* if for each $n$ it satisfies the four conditions below:

(1) $\mathscr{F}_n \subset \mathscr{F}_{n+1}$;
(2) $X_n$ is measurable $\mathscr{F}_n$;
(3) $E(|X_n|) < \infty$; and
(4) $E[X_{n+1} \mid \mathscr{F}_n] = X_n$ w.p.1.

Condition (1) states that $\{\mathscr{F}_n\}$, $n = 1, 2, ...$ is an increasing sequence of $\sigma$-fields in $\mathscr{F}$. Intuitively, the requirement of an increasing $\{\mathscr{F}_n\}$ implies that the amount of information contained in the sequence of $\sigma$-fields $\{\mathscr{F}_n\}$ is increasing. This is called the *monotoneity property* of the $\sigma$-fields $\{\mathscr{F}_n\}$ in $\mathscr{F}$ and it attempts to capture the practical idea that the past to time $n + 1$ includes more events, information or history than the past to time $n$. The overall informational structure represented by the monotonically increasing sequence $\{\mathscr{F}_n\}$ captures the concept of *learning without forgetting*. In some applications $\mathscr{F}_n$ is the σ-field generated by the random variables $X_1, ..., X_n$, or perhaps some other sequence of random variables, say $Y_1, ..., Y_n$. In such applications, instead of $E[X_{n+1} | \mathscr{F}_n]$ we write $E[X_{n+1} | X_1, ..., X_n]$ or $E[X_{n+1} | Y_1, ..., Y_n]$ as the case may be.

The *measurability property* stated in condition (2) means that for each $n$, $X_n$ has as its domain the measurable space $(\Omega, \mathscr{F}_n)$ and as its target space $(R, \mathscr{R})$. Some authors, such as Meyer (1966, p. 77), use the terminology that $X_n$ is adapted to the σ-field $\mathscr{F}_n$. As was remarked in the previous paragraph, when $\{\mathscr{F}_n\}$ are taken to be the σ-fields generated by $X_1, X_2, ..., X_n$, i.e. $\mathscr{F}_n = \sigma(X_1, ..., X_n)$, then the measurability condition is automatically satisfied. Recall that $\sigma(X_1, ..., X_n)$, the σ-field generated by $X_1, ..., X_n$, is the smallest σ-field making $X_1, ..., X_n$ measurable. Suppose that $Y$ is a random variable defined on the same space and consider the σ-field generated by $Y, X_1, ..., X_n$ and denoted by $\sigma(Y, X_1, ..., X_n)$. We have that

$$\sigma(X_1, ..., X_n) \subset \sigma(Y, X_1, ..., X_n)$$

and $X_1, ..., X_n$ continue to be measurable with respect to the new σ-field $\sigma(Y, X_1, ..., X_n)$. It is theoretically natural and in many applications appropriate to allow σ-fields $\{\mathscr{F}_n\}$ to be larger than the minimal ones $\sigma(X_1, ..., X_n)$. Fama (1970) has used various sizes of σ-fields to denote various degrees of information.

Condition (3), called the *integrability property* says simply that $X_n$ is integrable, i.e. the expectation of $X_n$ is finite. Finally, the martingale property expressed in condition (4) says that $X_n$ is a version of $E[X_{n+1} | \mathscr{F}_n]$ and in a gambling context this condition indicates that the game is fair. Note that condition (4) is equivalent to

$$\int_A E[X_{n+1} | \mathscr{F}_n] \, dP = \int_A X_n \, dP \qquad (6.2)$$

for $A \in \mathscr{F}_n$, $n = 1, 2, ...$ . Observe, however, that by eq. (5.8)

$$\int_A E[X_{n+1} | \mathscr{F}_n] \, dP = \int_A X_{n+1} \, dP \qquad (6.3)$$

for $A \in \mathscr{F}_n$. Using (6.2) and (6.3) we conclude that

$$\int_A X_{n+1} \, dP = \int_A X_n \, dP \qquad (6.4)$$

for $A \in \mathscr{F}_n$. Therefore condition (4) and (6.4) are equivalent. The same reasoning used inductively yields

$$\int_A X_n \, dP = \int_A X_{n+1} \, dP = \dots = \int_A X_{n+k} \, dP$$

for $A \in \mathscr{F}_n \subset \mathscr{F}_{n+1} \subset \dots \subset \mathscr{F}_{n+k}$ and $k \geq 1$. This means that $X_n$ is a version of $E[X_{n+k} | \mathscr{F}_n]$ and therefore the martingale property in condition (4) can also be written as

$$E[X_{n+k} | \mathscr{F}_n] = X_n.$$

Some additional definitions are appropriate. Condition (4) still makes sense for non-negative $X_n$ which does not satisfy condition (3). For such non-negative $X_n$ which satisfy conditions (1), (2) and (4) but not necessarily condition (3), we say that $\{(X_n, \mathscr{F}_n), n = 1, 2, \dots\}$ is a *generalized martingale*. The sequence $\{(X_n, \mathscr{F}_n), n = 1, 2, \dots\}$ is a *submartingale* if it satisfies conditions (1), (2), (3) and

(4\*) $\quad E[X_{n+1} | \mathscr{F}_n] \geq X_n$ w.p.1.

If instead of $\geq$ in (4\*) we put $\leq$, then $\{(X_n, \mathscr{F}_n), n = 1, 2, \dots\}$ is called a *supermartingale*. Using (6.4) we state that $\{(X_n, \mathscr{F}_n), n = 1, 2, \dots\}$ is a submartingale if and only if

$$\int_A X_{n+1} \, dP \geq \int_A X_n \, dP \tag{6.5}$$

for $A \in \mathscr{F}_n$ and $\{(X_n, \mathscr{F}_n), n = 1, 2, \dots\}$ is a supermartingale if and only if for $A \in \mathscr{F}_n$

$$\int_A X_{n+1} \, dP \leq \int_A X_n \, dP. \tag{6.6}$$

## 6.2. Examples

With the notions stated above we now give some examples of a probabilistic nature following Billingsley (1979, p. 408) and Doob (1971), before we illustrate the application of the martingale concept in economics and finance.

(1) Let $(\Omega, \mathscr{F}, P)$ be a probability space, let $v$ be a finite measure on $\mathscr{F}$, and let $\mathscr{F}_1, \mathscr{F}_2, \dots$ be a nondecreasing sequence of $\sigma$-fields in $\mathscr{F}$. Suppose that $v$ is dominated by $P$ when both are restricted to $\mathscr{F}_n$, i.e. suppose that $A \in \mathscr{F}_n$ and

$P(A) = 0$ imply $v(A) = 0$. By theorem 5.1 there exists a Radon–Nikodym derivative of $v$ with respect to $P$, when both are restricted to $\mathscr{F}_n$, and denote it by $X_n$.
Note that $X_n$ is measurable $\mathscr{F}_n$, integrable with respect to $P$ and by (5.2) it satisfies for $A \in \mathscr{F}_n$

$$v(A) = \int_A X_n \, dP. \tag{6.7}$$

But $A \in \mathscr{F}_n$ implies $A \in \mathscr{F}_{n+1}$ so that $v(A) = \int_A X_{n+1} \, dP$. Therefore (6.4) holds
and $\{(X_n, \mathscr{F}_n), n = 1, 2, \ldots\}$ is a martingale.

(2)  Another example is this: suppose $Y$ is an integrable random variable on
$(\Omega, \mathscr{F}, P)$ and $\{\mathscr{F}_n\}$ is a nondecreasing sequence of $\sigma$-fields in $\mathscr{F}$. Define $X_n$ as

$$X_n = E[Y \mid \mathscr{F}_n],$$

i.e. $X_n$ is the conditional expectation of $Y$ conditioned on the $\sigma$-field $\mathscr{F}_n$. From
the very definition of conditional expectation in the previous section $X_n$ is both
measurable and integrable. Furthermore, condition (4) holds since

$$\begin{aligned}
E[X_{n+1} \mid \mathscr{F}_n] &= E[E[Y \mid \mathscr{F}_{n+1}] \mid \mathscr{F}_n] \\
&= E[Y \mid \mathscr{F}_n] \\
&= X_n
\end{aligned}$$

by theorem 5.4. Thus $\{(X_n, \mathscr{F}_n), n = 1, 2, \ldots\}$ is a martingale, obtained from successive conditional expectations of $Y$ as we know more and more.

(3)  This example supposes that $Y_0, Y_1, Y_2, \ldots$ are independent random variables with $Y_0 = 0$, $E(\mid Y_n \mid) < \infty$ and $E(Y_n) = 0$ for all $n \geq 1$. Define $X_0 = Y_0$
and $X_n = Y_1 + \ldots + Y_n$ for $n \geq 1$. Then $X_n$ is a martingale with respect to the
$\sigma$-field generated by $Y_n$. Condition (1) holds because the $\sigma$-field generated by
$Y_1, \ldots, Y_n$, denoted by $\sigma(Y_1, \ldots, Y_n)$, is contained in the $\sigma$-field generated by
$Y_1, \ldots, Y_{n+1}$, denoted by $\sigma(Y_1, \ldots, Y_{n+1})$, i.e.

$$\sigma(Y_1, \ldots, Y_n) \subset \sigma(Y_1, \ldots, Y_{n+1}).$$

Condition (2) holds since measurability is preserved by addition. Condition (3)
follows from

$$E(\mid X_n \mid) \leqslant E(\mid Y_1 \mid) + \ldots + E(\mid Y_n \mid) < \infty.$$

Finally, condition (4) holds since

$$E[X_{n+1} \mid Y_0, ..., Y_n] = E[X_n + Y_{n+1} \mid Y_0, ..., Y_n]$$
$$= E[X_n \mid Y_0, ..., Y_n] + E[Y_{n+1} \mid Y_0, ..., Y_n]$$
$$= X_n + E(Y_{n+1})$$
$$= X_n.$$

In establishing condition (4) note that theorem 5.3 is used. Also note that $E[Y_{n+1} \mid Y_0, ..., Y_n] = E(Y_{n+1})$ follows from the assumption that $Y_0, Y_1, ...$ are independent random variables.

(4) As an example of a submartingale assume that $\{(X_n, \mathscr{F}_n), n = 1, 2, ...\}$ is a martingale. We claim that $\{(\mid X_n \mid, \mathscr{F}_n), n = 1, 2, ...\}$ is a submartingale. Condition (1) holds trivially since we consider the same family of $\sigma$-fields, while conditions (2) and (3) hold because measurability and integrability of $X_n$ imply the same for $\mid X_n \mid$. Condition (4*) follows from theorem 5.3 because

$$E[\mid X_{n+1} \mid \mid \mathscr{F}_n] \geqslant \mid E[X_{n+1} \mid \mathscr{F}_n] \mid = \mid X_n \mid.$$

## 6.3. *Applications in economics and finance*

The concept of a martingale has found several applications in economics and finance. Below we present some such applications.

(1) In this application we follow Samuelson (1965), with minor modifications, to show that *futures pricing*, under certain conditions, is a martingale.

Let $..., X_t, X_{t+1}, ..., X_{t+T}, ...$ represent a sequence of bounded random variables defined on a probability space $(\Omega, \mathscr{F}, P)$. For example, this sequence may represent a time sequence of prices, say spot prices of wheat or gold. $X_t$ denotes the present spot price and $X_{t+T}$ denotes the spot price that is to prevail $T$ units of time from now. The assumption of boundedness of the random variables is not too restrictive because commodity prices are always finite. An economic agent is assumed to know at least today's price as well as the past prices. In the language of probability we assume that the economic agent knows all the available information generated by the process which is in the $\sigma$-fields $\mathscr{F}_t$, where

$$\mathscr{F}_t = \sigma(X_0, X_1, ..., X_t). \tag{6.8}$$

Note that $\mathscr{F}_t$ contains in particular the past price realizations of the process. These prices, denoted by $x_0, x_1, ..., x_{t-1}, x_t$, are specific values of the process for a specific $\omega \in \Omega$, where,

$$X_0(\omega) = x_0, ..., X_{t-1}(\omega) = x_{t-1}, X_t(\omega) = x_t.$$

The economic agent cannot know with certainty tomorrow's price, $X_{t+1}$, or any future price, $X_{t+T}$. However, as time goes on his information increases because he observes additional price realizations. Needless to say $\mathscr{F}_t$ is a monotone increasing sequence by the definition in (6.8). Parenthetically we mention that Fama (1970) introduced the terminology *weak*, *semi-strong*, and *strong information sets* to describe information sets $\mathscr{F}_t$ which include past values of the process, all publicly available past information, and all past information both public and internal to an economic agent, respectively. In this present example $\mathscr{F}_t$ contains at least weak information which guarantees the measurability of $X_t$.

Next consider today's futures price quotation for the actual spot price that will prevail $T$ periods from now. We use the notation $Y(T, t)$ to write the futures price that will prevail $T$ periods from period $t$ and quoted at $t$. Let another period pass; then the new quotation for the same futures price is written as $Y(T-1, t+1)$. Thus, we have a sequence

$$Y(T, t), Y(T-1, t+1), ..., Y(T-n, t+n), ..., Y(1, t+T-1). \qquad (6.9)$$

Samuelson's (1965) fundamental assumption is that a futures price is to be set by competitive bidding at the *now-expected level of the terminal spot price*. This is similar to Muth's (1961) *rational expectations* hypothesis and can be written as

$$Y(T, t) = E[X_{t+T} | \mathscr{F}_t] \qquad (6.10)$$

for $T = 1, 2, ...$ . Note that (6.10) makes sense because the conditional expectation of $X_{t+T}$ with respect to $\mathscr{F}_t$ exists and this is so because $X_{t+T}$ is integrable. The integrability of $X_{t+T}$ follows from the assumption of the boundedness of $X_{t+T}$ made at the beginning of this application.

Now we are ready to establish that (6.9) is a martingale. The $\sigma$-fields associated with (6.9) are $\mathscr{F}_t, \mathscr{F}_{t+1}, ..., \mathscr{F}_{t+n}, ..., \mathscr{F}_{t+T-1}$. Such $\sigma$-fields defined in (6.8) form a monotone increasing sequence which satisfies the first condition of a martingale. The measurability and integrability conditions of (6.9) follow from eq. (6.10) and the properties of conditional expectation of $X_{t+T}$. More precisely, $E[X_{t+T} | \mathscr{F}_t]$ is measurable $\mathscr{F}_t$ and by (6.10) so is $Y(T, t)$. Furthermore, $E[X_{t+T} | \mathscr{F}_t]$ is integrable and

$$\int_A E[X_{t+T} | \mathscr{F}_t] \, dP = \int_A X_{t+T} \, dP < \infty \qquad (6.11)$$

for $A \in \mathscr{F}_t$ and so is $Y(T, t)$ by (6.10). The right-hand side of (6.11) follows from the boundedness assumption of $X_{t+T}$. Thus, we only need to establish the martingale property,

$$E[Y(T-1, t+1) \mid \mathscr{F}_t] = Y(T, t). \tag{6.12}$$

Eq. (6.12) is easy to establish by using (6.10) and theorem 5.5. More specifically,

$$
\begin{aligned}
E[Y(T-1, t+1) \mid \mathscr{F}_t] &= E[E[X_{t+T} \mid \mathscr{F}_{t+1}] \mid \mathscr{F}_t] \\
&= E[X_{t+T} \mid \mathscr{F}_t] \\
&= Y(T, t).
\end{aligned}
$$

Thus, futures pricing under the assumptions stated is a martingale.

The theoretical paper of Samuelson (1965) summarized in this application and the work of Mandelbrot (1966) generated great interest in econometric testing of the properties of stock prices. Although Samuelson's paper establishes the martingale property for futures pricing rather than for an equity asset, a share of a stock may be regarded as a sequence of futures claims due to mature at successive intervals. Thus, the martingale property properly applied to stock prices may be used as a measure of capital market efficiency. A capital market is *efficient* if the prices of the securities incorporate all the available information. Here we briefly report some findings of Jensen (1969) and refer the interested reader to Fama (1970) for a detailed survey of the theory and empirical findings on efficient capital markets. Jensen distinguishes between weak and strong information and he proceeds to test the strong martingale hypothesis where the expectation of prices $T$ periods from $t$, denoted by $X_{t+T}$, is conditioned upon all information available at $t$. As stated earlier such information includes the past values of the process and information both public and internal to the firm. Notationally, Jensen (1969) writes $\Theta_t$ for $\sigma$-fields that contain all such information. Then, the strong martingale hypothesis is written as

$$E[X_{t+T} \mid \Theta_t] = f(T) X_t \tag{6.13}$$

where, according to Jensen (1969), $f(T)$ represents the *normal accumulation rate* which depends on the length of the period $T$. Note that (6.13) actually represents a submartingale for the sequence $X_t, X_{t+1}, ..., X_{t+T}$ since,

$$E[X_{t+T} \mid \Theta_t] \geqslant X_t$$

holds for $f(T) \geqslant 1$. Jensen's (1969) empirical analysis of 115 mutual funds shows that current prices of securities incorporate all available information and therefore the best forecast of future prices is the present prices plus a normal expected return over the period $T$.

We close this application with a remark on Samuelson's (1965) sequences ..., $X_t$, ..., $X_{t+T}$, ..., and their unconditional expectations in (6.9). Note that these sequences are assumed to be given exogenously. LeRoy (1973) has attempted to derive the martingale property when the assumption of exogenously given sequences is relaxed. He analyzes the relation between the riskiness of a stock and the risk aversion of investors to formally derive endogenously probability distributions of rates of return. LeRoy (1973, p. 437) concludes "not that any particular systematic departure from the martingale property is to be expected, but only that under risk aversion no rigorous theoretical justification for an exact martingale property is available". Ohlson (1977), in commenting on LeRoy's (1973) paper, shows that the martingale property holds when investors have constant relative risk aversion and the percentage change in dividends is stationary.

(2)  The application of futures' pricing as a martingale can be generalized following Samuelson (1965). Let $\alpha = (1 + r)^{-1}$, where $r$ is a measure of forgone safe interest and postulate

$$Y(T, t) = \alpha^T E[X_{t+T} \mid \mathscr{F}_t] \tag{6.14}$$

instead of (6.10), i.e. we allow the conditional expectation of the spot price, $T$ periods from now, to be appropriately discounted. Then the sequence $Y(T, t)$, ..., $Y(T - n, t + n)$, ..., which satisfies the axiom of expected present discounted value in (6.14), is a submartingale with respect to the $\sigma$-fields $\mathscr{F}_t$, ..., $\mathscr{F}_{t+n}$, ... . Here we assume that the $\sigma$-fields include at least weak information.

To establish that the sequence is a submartingale we only check the submartingale property. Conditions (1), (2) and (3) are easily established, as in the previous application. Note that in the economics and finance literature conditions (1), (2) and (3) are seldom checked carefully. The reader, however, may use the analysis in the previous example as a model in establishing the monotoneity, measurability and integrability conditions.

The submartingale condition is obtained by using theorem 5.5 and eq. (6.14) as follows:

$$E[Y(T - 1), t + 1) \mid \mathscr{F}_t]$$
$$= E[\alpha^{T-1} E[X_{t+T} \mid \mathscr{F}_{t+1}] \mid \mathscr{F}_t]$$
$$= \alpha^{T-1} E[E[X_{t+T} \mid \mathscr{F}_{t+1}] \mid \mathscr{F}_t]$$

$$= \alpha^{T-1} E[X_{t+T} | \mathscr{F}_t]$$

$$= \alpha^{T-1} \alpha^{-T} Y(T, t)$$

$$= \alpha^{-1} Y(T, t)$$

$$= (1 + r) Y(T, t)$$

$$\geqslant Y(T, t).$$

In this application the discounted futures price will rise in each period by a percentage equal to $r$. Samuelson (1965) uses this application to provide a rational explanation of the doctrine of *normal backwardation*. The moral is this: the martingale property of futures pricing establishes that all methods used to read out of the past sequence of known prices any profitable pattern of prediction are doomed to failure.

(3) In this application we follow Samuelson (1973) to show that under certain conditions stocks that are capitalized at their expected present discounted values satisfy the martingale property. Although our purpose is to establish the martingale property, before we do so in this application we generalize the familiar *present discounted-value rule of capitalization*.

Let $x_t, ..., x_{t+T}$ be a sequence of dividends of a given stock paid out at time $t, ..., t + T$, on each dollar invested at time $t, ..., t + T$. Suppose that the discount rate is $r$ and remains constant. Allowing $r$ to change per unit of time does not add any new insights into the analysis; it just complicates the notation. Initially, we assume that $x_t, ..., x_{t+T}$ are nonrandom and we write the familiar equation

$$V_t = \sum_{T=1}^{\infty} \frac{x_{t+T}}{(1+r)^T}, \tag{6.15}$$

where $V_t$ is the value of the stock at time $t$ obtained from the present discounted-value rule of capitalization. Using (6.15) we can obtain the value of the stock next period, $V_{t+1}$, as follows:

$$V_{t+1} = \sum_{T=1}^{\infty} \frac{x_{t+1+T}}{(1+r)^T} = \sum_{T=2}^{\infty} \frac{x_{t+T}}{(1+r)^{T-1}}. \tag{6.16}$$

Using (6.15) and (6.16) it follows that

$$V_t - \frac{V_{t+1}}{1+r} = \frac{x_{t+1}}{1+r}$$

which can be written as

$$V_{t+1} = (1 + r) V_t - x_{t+1}.$$ (6.17)

Eq. (6.17) is useful because it expresses the value of the stock next period as a function of its current price, $V_t$, the discount rate $r$, and the stock's dividend at the end of next period, $x_{t+1}$. As a special case, if $rV_t = x_{t+1}$, then $V_{t+1} = V_t$, which means that if the discount rate is equal to the rate of return, $x_{t+1}/V_t$, then the value of the stock will remain the same.

With the above review we now generalize eqs. (6.15) and (6.17) to make them stochastic. Let $(\Omega, \mathscr{F}, P)$ be a probability space and $x_t(\omega), ..., x_{t+T}(\omega)$, be a sequence of random variables which denotes stock dividends. In what follows we delete the $\omega$'s to simplify the notation. For each random variable $x_{t+T}$, $T = 1$, 2, ..., we assume the existence of a distribution and of a conditional expectation. The sequence of $\sigma$-fields in $\mathscr{F}$ is denoted by $\mathscr{F}_{t+T}$ and it contains at least weak information, i.e. $\sigma(x_t, ..., x_{t+T}) \subset \mathscr{F}_{t+T}$. This means that the investor knows at least the dividend history of the stock.

At this point we can immediately generalize (6.15) by writing it as

$$v_t \equiv E[V_t \mid \mathscr{F}_t]$$

$$= E\left[ \sum_{T=1}^{\infty} \frac{x_{t+T}}{(1+r)^T} \;\middle|\; \mathscr{F}_t \right]$$

$$= \sum_{T=1}^{\infty} \left( E\left[ \frac{x_{t+T}}{(1+r)^T} \;\middle|\; \mathscr{F}_t \right] \right).$$ (6.18)

Note that the stochastic generalization of (6.17) is also easy. As a first step, compute $v_{t+1}$ using (6.16) and (6.17)

$$v_{t+1} \equiv E[V_{t+1} \mid \mathscr{F}_{t+1}]$$

$$= E\left[ \sum_{T=2}^{\infty} \frac{x_{t+T}}{(1+r)^{T-1}} \;\middle|\; \mathscr{F}_{t+1} \right].$$ (6.19)

Next, observe that from (6.18), (6.19) and theorem 5.5 we get

$$E[v_{t+1} \mid \mathscr{F}_t]$$

$$= E\left[ E\left[ \sum_{T=2}^{\infty} \frac{x_{t+T}}{(1+r)^{T-1}} \,\middle|\, \mathscr{F}_{t+1} \right] \,\middle|\, \mathscr{F}_t \right]$$

$$= E\left[ \sum_{T=2}^{\infty} \frac{x_{t+T}}{(1+r)^{T-1}} \,\middle|\, \mathscr{F}_t \right]$$

$$= E\left[ \left( \pm x_{t+1} + \sum_{T=2}^{\infty} \frac{x_{t+T}}{(1+r)^{T-1}} \right) \,\middle|\, \mathscr{F}_t \right]. \tag{6.20}$$

Finally, multiplying both sides by $1/(1 + r)$ and using the definition of $v_t$ in (6.18), we conclude

$$E[v_{t+1} \mid \mathscr{F}_t] = (1+r)E\left[ \sum_{T=1}^{\infty} \frac{x_{t+T}}{(1+r)^{T}} \,\middle|\, \mathscr{F}_t \right] - E[x_{t+1} \mid \mathscr{F}_t]$$

$$= (1+r)v_t - E[x_{t+1} \mid \mathscr{F}_t]. \tag{6.21}$$

Eq. (6.21) is the stochastic generalization of (6.17) and it can be used to help us decide under what conditions the sequence $v_t, ..., v_{t+T}$ is a martingale. In other words we ask: Under what conditions does $E[v_{t+1} \mid \mathscr{F}_t] = v_t$? The answer is when $rv_t = E[x_{t+1} \mid \mathscr{F}_t]$. This means that the martingale condition holds when the discount rate is equal to the conditional expected return of the stock.

We conclude therefore that the sequence $v_t, v_{t+T}, T = 1, 2, ...$ of discounted expected values of a stock is a martingale provided

$$r = E[x_{t+1+T} \mid \mathscr{F}_{t+T}]/v_{t+T}. \tag{6.22}$$

From this last equation we at once decide that $v_t, v_{t+T}, T = 1, 2, ...,$ is a submartingale if in (6.22) instead of = we have $\geqslant$. This is so because (6.21) with (6.22) having $\geqslant$ instead of = yields

$$E[v_{t+1+T} \mid \mathscr{F}_{t+T}] = v_{t+T} + \left( r - \frac{E[x_{t+1+T} \mid \mathscr{F}_{t+T}]}{v_{t+T}} \right) v_{t+T}$$

$$\geqslant v_{t+T}.$$

The submartingale property says that the conditional expected value of the stock next period is greater or equal to its current value. This is so because the conditional expected rate of return is smaller than or equal to the discount rate.

We close this application with an important remark. Recall that in this application we assumed the existence of conditional expectation for the sequence $x_t, ..., x_{t+T}$ but we did not explain the way the individual investor forms his expectations. A more complete analysis would require an individual investor to form his subjective expectations $y_t, ..., y_{t+T}$ as a sequence of random variables where $y_{t+T}$, $T = 1, 2, ...$, denotes the investor's expected rate of return $T$ periods from now. Having done so, a relationship needs to be established between $y_{t+T}$ and $x_{t+T}$. At this point, Samuelson's (1965) *axiom of expectation formation* or Muth's (1961) *rational expectation hypothesis* can be used to postulate

$$y_{t+T} = E[x_{t+T} \mid \mathscr{F}_t]. \tag{6.23}$$

Eq. (6.23) links the investor's subjective expectations to the markets' objective conditional expectations. Thus, we may start with an investor and his subjective expectations $y_{t+T}$, use (6.23) to link the subjective expectations to the objective ones and proceed from there using $x_{t+T}$ as was done above.

(4) In this application we follow Hall (1978) to study a simple model of *intertemporal stochastic optimilization* to obtain a martingale property for the marginal utility of consumption.

Consider an individual with a strictly concave one-period utility function $u(\cdot)$, whose life-cycle consumption problem under uncertainty is given by

$$\max E\left[ \sum_{T=0}^{N} \frac{u(c_{t+T})}{(1+\delta)^T} \, \mathscr{F}_t \right] \tag{6.24}$$

subject to the condition

$$\sum_{T=0}^{N} \frac{c_{t+T} - w_{t+T}}{(1+r)^T} = A_t. \tag{6.25}$$

In (6.24) and (6.25) the notation used is: $c_{t+T}$, $T = 0, 1, ..., N$ is consumption, $\delta$ is the rate of subjective time preference, $r$ is the real rate of interest, $w_{t+T}$ is earnings and $A_t$ is the value of assets at time $t$. The individual considers the problem today, at time $t$, and his length of economic life is $t + N$ periods. The $\sigma$-fields $\mathscr{F}_t, \mathscr{F}_{t+1}, ..., \mathscr{F}_{t+N}$ each include information, at least, about the sequence of consumption and the sequence of marginal utility of such consumption, respectively, at $t, ..., t + N$.

Hall (1978) derives a necessary condition for this maximization problem of the form

$$E[u'(c_{t+1}) \mid \mathcal{F}_t] = \frac{1+\delta}{1+r} \, u'(c_t),$$ (6.26)

where $u'(\cdot)$ denotes marginal utility. Assuming $\delta = r$, (6.26) says that the sequence of marginal utilities satisfies the martingale property. If we assume that $r < \delta$, then (6.26) says that the sequence of marginal utilities is a submartingale.

Hall (1978) also shows that if the stochastic change in marginal utility from one period to the next is small, then consumption is a submartingale, written as

$$E[c_{t+1} \mid \mathcal{F}_t] = \lambda_t c_t \geqslant c_t,$$ (6.27)

where $\lambda_t$ is given by

$$\lambda_t = \left( \frac{1+\delta}{1+r} \right)^{u'(c_t)/c_t u''(c_t)}.$$ (6.28)

From (6.28) we obtain that $\lambda_t \geqslant 1$ because $u'' < 0$ and $r$ is assumed to be greater than $\delta$.

Before we close this application we mention that Foldes (1978) has obtained martingale conditions for a dynamic discrete time model of stochastic optimal saving.

## 6.4. Basic theorems on martingales

We conclude the section on martingales by stating some useful facts in the form of theorems. These facts have not yet been widely used in the economics and finance literature on martingales and their presentation is motivated by probabilistic interest. There are several excellent sources where these and additional results on martingales may be found, such as Ash (1972), Billingsley (1979), Doob (1953), Meyer (1966), Neveu (1975) and Tucker (1967). Here we follow Ash (1972) and Billingsley (1979).

**Theorem 6.1.**

   (1) Let $X_1, X_2, \ldots$ be a martingale relative to $\mathcal{F}_1, \mathcal{F}_2, \ldots$ on $(\Omega, \mathcal{F}, P)$. If $\phi$ is a convex function and if $\phi(X_n)$ are integrable, then, $\phi(X_1), \phi(X_2), \ldots$ is a submartingale relative to $\mathcal{F}_1, \mathcal{F}_2, \ldots$.

   (2) Let $X_1, X_2, \ldots$ be a submartingale relative to $\mathcal{F}_1, \mathcal{F}_2, \ldots$ on $(\Omega, \mathcal{F}, P)$. If $\phi$ is a nondecreasing and convex function, and if the $\phi(X_n)$ are integrable, then $\phi(X_1), \phi(X_2), \ldots$ is a submartingale relative to $\mathcal{F}_1, \mathcal{F}_2, \ldots$.

*Proof.* To prove (1) we only need to show that $E[\phi(X_{n+1}) \mid \mathcal{F}_n] \geqslant \phi(X_n)$. Since $X_n$ is a martingale we have that $E[X_{n+1} \mid \mathcal{F}_n] = X_n$. So, $\phi\{E[X_{n+1} \mid \mathcal{F}_n]\} = \phi(X_n)$.

Since $\phi$ is convex, $X_n$ and $\phi(X_n)$ are integrable and we may apply Jensen's inequality for conditional expectations in theorem 5.6 to obtain $E[\phi(X_{n+1}) \mid \mathcal{F}_n] \geqslant \phi\{E[X_{n+1} \mid \mathcal{F}_n]\} = \phi(X_n)$.

To prove (2) we need only to show again that $E[\phi(X_{n+1}) \mid \mathcal{F}_n] \geqslant \phi(X_n)$. Since $X_n$ is a submartingale $E[X_{n+1} \mid \mathcal{F}_n] \geqslant X_n$. By hypothesis $\phi$ is nondecreasing so $\phi\{E[X_{n+1} \mid \mathcal{F}_n]\} \geqslant \phi(X_n)$. Now apply again Jensen's inequality to conclude

$$E[\phi(X_{n+1}) \mid \mathcal{F}_n] \geqslant \phi\{E[X_{n+1} \mid \mathcal{F}_n]\} \geqslant \phi(X_n).$$

**Theorem 6.2.** (Kolmogorov's inequality).  Let $X_1, X_2, ..., X_n$ be a submartingale on $(\Omega, \mathcal{F}, P)$ and let $\lambda > 0$. Then,

$$P\left[ \max_{i \leqslant n} X_i \geqslant \lambda \right] \leqslant \frac{1}{\lambda} \, E[\, | X_n | \,].$$

*Proof.* Let $\lambda > 0$ be given and define $A_1, ..., A_n, A$ as follows:

$$A_i = [\omega: X_1(\omega) \geqslant \lambda]$$
$$\vdots$$
$$A_k = \left[ \omega: \max_{i < k} X_i(\omega) < \lambda \leqslant X_k(\omega) \right]$$
$$A = \bigcup_{k=1}^{n} A_k = \left[ \omega: \max_{i \leqslant k} X_i(\omega) \geqslant \lambda \right].$$

Note that the $A_k$'s as defined above are disjoint. Also note that since $X_1, X_2, ..., X_n$ is a submartingale one inductively may obtain $E[X_{n+k} \mid \mathcal{F}_n] \geqslant X_n$ for $k \geqslant 1$. Then

$$\int_A X_n \, dP = \sum_{k=1}^{n} \int_{A_k} X_n \, dP$$

$$= \sum_{k=1}^{n} \int_{A_k} E[X_n \mid \mathcal{F}_k] \, dP$$

$$\geqslant \sum_{k=1}^{n} \int_{A_k} X_k \, dP$$

$$\geqslant \lambda \sum_{k=1}^{n} P(A_k) = \lambda P(A),$$

with $A_k \in \mathscr{F}_k = \sigma(X_1, ..., X_k)$. Therefore,

$$\lambda P(A) = \lambda P\left[\omega: \max_{i \leq n} X_i(\omega) \geq \lambda\right]$$

$$\leq \int_A X_n \, dP \leq \int_\Omega X_n^+ \, dP \leq E(|X_n|).$$

This concludes the proof.

Next we present the notion of an *upcrossing* which is fundamental for the Up-crossing Theorem. This upcrossing theorem is used by probabilists in the proof of the important result known as the martingale convergence theorem. However, we present the notion of upcrossing here because it may be useful to researchers in finance and economics also.

Let $[\alpha, \beta]$ be an interval with $\alpha < \beta$, and let $X_1, ..., X_n$ be random variables. The number of upcrossings of $[\alpha, \beta]$ by $X_1(\omega), ..., X_n(\omega)$ is the number of times the sequence passes from below $\alpha$ to above $\beta$ (see fig. 6.1).

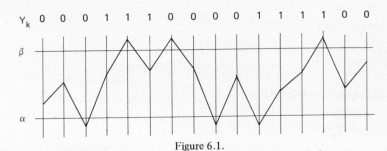

Figure 6.1.

In the figure above there are two upcrossings, for $n = 16$. These correspond to the strings of consecutive 1's above the graph for the variable $Y$ defined as follows:

$$Y_1 = 0, \quad 2 \leq k \leq n + 1$$

$$Y_k = \begin{cases} 0 \text{ if } Y_{k-1} = 1 \text{ and } X_{k-1} \geq \beta, \\ 0 \text{ if } Y_{k-1} = 0 \text{ and } X_{k-1} > \alpha, \\ 1 \text{ if } Y_{k-1} = 1 \text{ and } X_{k-1} < \beta, \\ 1 \text{ if } Y_{k-1} = 0 \text{ and } X_{k-1} \leq \alpha. \end{cases}$$

According to the definition an upcrossing corresponds in $Y_2$, $Y_3$, ..., $Y_n$ to an unbroken string of 1's with a 0 on either side. Now define

$$
Z_k = \begin{cases} 1 & \text{if } Y_k = 1 \text{ and } Y_{k+1} = 0, \\ \\ 0 & \text{otherwise,} \end{cases}
$$

then the number of upcrossings is

$$
U = \sum_{k=2}^{n} Z_k.
$$

Let $\mathscr{F}_0 = \{0, \Omega\}$ and $\mathscr{F}_k = \sigma(X_1, ..., X_k)$. Then $Y_k$ is measurable $\mathscr{F}_{k-1}$, $k = 1$, 2, ..., $n + 1$ and $U$ is measurable.

**Theorem 6.3.** (Martingale upcrossings). Let $X_1$, ..., $X_n$ be a submartingale on $(\Omega, \mathscr{F}, P)$. Then the number of upcrossings, $U$ of $[\alpha, \beta]$, satisfies

$$
E(U) \leqslant \frac{E(|X_n|) + |\alpha|}{\beta - \alpha}.
$$

For a proof see Billingsley (1979, p. 415) or Ash (1972, p. 291).

The final result establishes convergence of martingales.

**Theorem 6.4.** (Martingale convergence). Let $X_1, X_2, ...$ be a submartingale on $(\Omega, \mathscr{F}, P)$ and assume that $K = \sup_n E(|X_n|) < \infty$. Then $X_n \to X$ w.p.1, where $X$ is a random variable such that $E(|X|) \leqslant K$.

For a proof see Billingsley (1979, p. 416).

## 7. Stochastic processes

A *stochastic process* is a collection of random variables $\{X_t, t \in T\}$ on the same probability space $(\Omega, \mathscr{F}, P)$. Note that $X_t(\omega) \equiv X(t, \omega)$ has as its domain the product space $T \times \Omega$ and as its target space $R$ or $R^k$. The points of the *index* or *parameter* set $T$ are thought of as representing time. If $T$ is countable, especially if $T = \{0, 1, 2, 3, ...\} \equiv N$, i.e. the set of non-negative integers, then the process is called a *discrete parameter process*. If $T = R$ or $T = [a, b]$ for $a$ and $b$ real numbers or $T = [0, \infty)$, i.e. if $T$ is uncountable, then we have a *continuous parameter pro-*

*cess.* Although the index set $T$ can be rather arbitrary, in this section and the rest of this book the most often used index set is $T = [0, \infty)$. $\Omega$ denotes the random or sample space and for fixed $\omega \in \Omega$, $X_t(\omega) = X(\cdot, \omega)$ for $t \in T$, is called a *sample path* or *sample function* corresponding to $\omega$. Other terms used in various texts to describe $X_t(\omega)$ for a fixed $\omega$ are *realization* or *trajectory* of the process. For fixed $\omega \in \Omega$ the usual notation of the process is $X_t$; however, $X(t)$ is also used in this text as well as in the literature. Obviously for a fixed $t \in T$, $X_t(\omega) = X(t, \cdot)$ is a random variable. The space in which all the possible values of $X_t$ lie is called the *state space*. Usually the state space is the real line $R$ and $X_t$ is called a real-valued stochastic process or just a stochastic process. It is also possible for the state space to be $R^k$, in which case we say that $X_t$ is a $k$-vector stochastic process. The various martingales presented in the previous sections are good examples of stochastic processes. Additional examples will be discussed in this section; in particular we briefly plan to discuss the *Brownian motion* or *Wiener process*, the *Markov process*, and the *Poisson process*. However, before we do so we state some basic notions in the form of definitions and theorems of stochastic processes.

## 7.1. Basic notions

An important feature of a stochastic process $\{X_t, t \in T\}$ is the relationship among the random variables of the process, say $X_{t_1}, ..., X_{t_n}$ for $t_1, ..., t_n \in T$. This relationship is specified by the joint distribution function of these variables given by

$$P_{X_{t_1}, ..., X_{t_n}}(H) = P[\omega: (X_{t_1}(\omega), ..., X_{t_n}(\omega)) \in H] \tag{7.1}$$

for $H \in \mathscr{R}^n$. It must be pointed out at the outset that a system of *finite-dimensional distributions* of the form of (7.1) does not completely determine the properties of the process in the case of an arbitrary index set $T$. However, the first step in the general theory of stochastic processes is to construct processes for given finite-dimensional distributions.

Suppose that we are given a stochastic process having (7.1) as a finite-dimensional system. Note that (7.1) necessarily satisfies two *consistency properties*. The first property is the *condition of symmetry*. Let $p$ be a permutation of $(1, 2, ..., n)$ and define $f_p: R^n \to R^n$ by

$$f_p(x_{p1}, ..., x_{pn}) = (x_1, ..., x_n). \tag{7.2}$$

The random vector

$$(X_{t_1}, ..., X_{t_n}) = f_p(X_{t_{p1}}, ..., X_{t_{pn}}) \tag{7.3}$$

must have distribution $P_{X_{t_1}, \ldots, X_{t_n}}$ from the left-hand side of (7.3) and (7.1) and also $P_{X_{t_{p1}}, \ldots, X_{t_{pn}}} f_p^{-1}$ from the right-hand side of (7.3). This leads to the condition of symmetry written as

$$P_{X_{t_1}, \ldots, X_{t_n}} = P_{X_{t_{p1}}, \ldots, X_{t_{pn}}} f_p^{-1}. \tag{7.4}$$

The second consistency property is called the *condition of compatability*; it is written as

$$P_{X_{t_1}, \ldots, X_{t_n}}(H) = P_{X_{t_1}, \ldots, X_{t_n}, X_{t_{n+1}}}(H \times R^1) \tag{7.5}$$

for $H \in \mathscr{R}^n$.

The conclusion of the above analysis is this: given a stochastic process $\{X_t, t \in T\}$ then its finite-dimensional distributions satisfy properties (7.4) and (7.5). Naturally, the mathematical question arises: Does the converse hold true? That is to say, given finite-dimensional distributions having properties (7.4) and (7.5), does there exist a stochastic process having these finite-dimensional distributions? The question is answered affirmatively by the famous Kolmogorov theorem.

**Theorem 7.1.** (Kolmogorov). Given a system of finite-dimensional distributions satisfying the symmetry and compatability consistency conditions, then there exists a probability space $(\Omega, \mathscr{F}, P)$ and a stochastic process $\{X_t, t \in T\}$ defined on this space, such that the process has the given finite-dimensional distributions as its distributions.

For a proof see Billingsley (1979, section 36).

Kolmogorov's existence theorem puts the theory of stochastic processes on a firm foundation. Next, we proceed to state some useful definitions.

Consider two stochastic processes $\{X_t, t \in T\}$ and $\{Y_t, t \in T\}$, both defined on the same probability space $(\Omega, \mathscr{F}, P)$. These two processes are said to be *stochastically equivalent* if for every $t \in T$, $P[\omega: X_t(\omega) \neq Y_t(\omega)] = 0$. Alternatively, the two processes are equivalent if for every $t \in T$, $X_t(\omega) = Y_t(\omega)$ w.p.1. If two processes are equivalent we say that one is a version of the other and we conclude that their finite-dimensional distributions coincide. However, it is not always the case that equivalent processes have sample paths with the same properties. This can be illustrated with a simple example which also substantiates the remark made earlier that finite-dimensional distributions do not completely determine all the properties of the process. The example considers two processes

$\{X_t, t \geqslant 0\}$ and $\{Y_t, t \geqslant 0\}$ on the same probability space $(\Omega, \mathcal{F}, P)$. Define first $X_t$ for $t \geqslant 0$ as

$$X_t(\omega) = 0 \quad \text{for all } \omega \in \Omega, \tag{7.6}$$

and secondly define $Y_t$ for $t \geqslant 0$ as

$$Y_t(\omega) = \begin{cases} 1 & \text{if } V(\omega) = t, \\ \\ 0 & \text{if } V(\omega) \neq t, \end{cases} \tag{7.7}$$

where $V$ is a positive random variable on $(\Omega, \mathcal{F}, P)$ with a continuous distribution given by $P[\omega: V(\omega) = x] = 0$ for each $x > 0$. For $X_t$ and $Y_t$ defined as above, $P[\omega: X_t(\omega) \neq Y_t(\omega)] = 0$ for each $t \geqslant 0$ and therefore they are stochastically equivalent. Also, $X_t$ and $Y_t$ have the same finite-dimensional distribution which, for $t_1, ..., t_n$ say, are given by

$$P_{X_{t_1}, ..., X_{t_n}}(H) = P_{Y_{t_1}, ..., Y_{t_n}}(H) = \begin{cases} 1 & \text{if } H \text{ contains the origin of } R^n, \\ \\ 0 & \text{otherwise,} \end{cases}$$

for $H \in \mathcal{R}^n$. However, note that the equivalence of the two processes is not sufficient to guarantee the same sample paths for $X_t$ and $Y_t$. For $\omega \in \Omega$, from (7.6) we know that $X_t(\omega) = 0$ while from (7.7) $Y_t(\omega) = 0$ with a discontinuity at $t = V(\omega)$, where it obtains the value 1. Thus, the sample paths of these two processes are not the same. To correct irregularities of this form probabilists have introduced the concept of a *separable process*.

Consider a process $\{X_t, t \in [0, \infty)\}$ defined on a complete probability space $(\Omega, \mathcal{F}, P)$. This process is separable if there is a countable, dense subset of $T = [0, \infty)$ denoted by $S = \{t_1, t_2, ...\}$ such that for every interval $(a, b) \subset [0, \infty)$ and every closed set $A \subset R$ it holds that

$$P[\omega: X_t(\omega) \in A \quad \text{for all } t \in (a, b) \cap S]$$

$$= P[\omega: X_t(\omega) \in A \quad \text{for all } t \in (a, b)]. \tag{7.8}$$

Observe that the definition requires that the probability of the set where $[X_t \in A$ for all $t \in (a, b) \cap S]$ and $[X_t \in A$ for all $t \in (a, b)]$ differ is zero. The motivation of this definition is to make a countable set of time points serve to characterize the sample paths of a process. The important question is: Given a stochastic process having a system of consistent finite-dimensional distributions does there exist a separable version with the same distributions? Fortunately the answer is yes and we are therefore allowed to consider separable versions of a given process. For a detailed statement of the existence theorem and its proof see Billingsley (1979, section 38) or Tucker (1967, section 8.2).

We next present some simple facts about three important processes.

## 7.2. The Wiener or Brownian motion process

A *Wiener process* or a *Brownian motion process* $\{z_t, t \in [0, \infty)\}$ is a stochastic process on a probability space $(\Omega, \mathscr{F}, P)$ with the following properties:

(1) $z_0(\omega) = 0$ w.p.1, i.e. by convention we assume that the process starts at 0.

(2) If $0 \leqslant t_0 \leqslant t_1 \leqslant ... \leqslant t_n$ are time points then for $H \in \mathscr{R}^1$,

$$P[z_{t_i} - z_{t_{i-1}} \in H_i \text{ for } i \leqslant n] = \prod_{i \leqslant n} P[z_{t_i} - z_{t_{i-1}} \in H_i].$$

This means that the increments of the process $z_{t_i} - z_{t_{i-1}}, i \leqslant k$, are independent random variables.

(3) For $0 \leqslant s < t$, the increment $z_t - z_s$ has distribution

$$P[z_t - z_s \in H] = \frac{1}{\sqrt{2\pi(t-s)}} \int_H \exp\left[ -\frac{x^2}{2(t-s)} \right] dx$$

This means that every increment $z_t - z_s$ is normally distributed with mean 0 and variance $\sigma^2 (t-s)$; here we assume that $\sigma = 1$, i.e. we standardize it.

(4) For each $\omega \in \Omega, z_t(\omega)$ is continuous in $t$, for $t \geqslant 0$.

Note that condition (2) reflects a kind of lack of memory. This means that the displacements $z_{t_1} - z_{t_0}, ..., z_{t_{n-1}} - z_{t_{n-2}}$ of the process during the intervals $[t_0, t_1], ..., [t_{n-2}, t_{n-1}]$ in no way influence the displacement $z_{t_n} - z_{t_{n-1}}$ of the process during $[t_{n-1}, t_n]$. The past history of the process does not influence its future position. The future behavior of the process depends on its present position but it does not depend on how the process got there. Formally, if $0 \leqslant t_0 < t_1 < t_2 < ... < t_n < t$ then for real $x, x_0, ..., x_n$

$$P[z_t \leqslant x \mid z_{t_0} = x_0, ..., z_{t_n} = x_n] = P[z_t \leqslant x \mid z_{t_n} = x_n]. \tag{7.9}$$

Eq. (7.9) is called the *Markov property* and it will play an important role in defining Markov processes. Here we need to emphasize that condition (2) of the

Wiener process requires independent increments which is actually more restrictive than the Markov property.

To understand condition (3) suppose that $z_t$ denotes the height at time $t$ of a particle above a fixed horizontal plane. Then the fact that $z_t - z_s$ is assumed to have mean 0 says that the particle is as likely to go up as to go down, i.e. there is no *drift*. The fact that we assume that the variance grows as the length of the interval $[s, t]$ means that the particle tends to wander away from its position at time $s$ and having done so no force exists to restore it to its original position.

The increments of a Wiener process are *stationary* in the sense that the distribution of $z_t - z_s$ depends only on the difference $t - s$. From property (1) we have that $z_0 = 0$ which enables us to describe the behavior of increments by saying that $z_t$ is normally distributed with $E(z_t) = 0$ and $E(z_t^2) = t$. To compute its covariance note for $0 \leqslant s < t$ that

$$
\begin{aligned}
\mathrm{cov}(z_s z_t) &= E(z_s z_t) = E(z_s z_t - z_s z_s + z_s z_s) \\
&= E(z_s [z_t - z_s] + z_s^2) \\
&= E(z_s [z_t - z_s]) + E(z_s^2) \\
&= E(z_s) E(z_t - z_s) + E(z_s^2) = E(z_s^2) = s \\
&= \text{minimum } \{t, s\}.
\end{aligned}
$$

Finally, condition (4) is added because in many applications such continuity is essential. However, we immediately state an important theorem about the differentiability properties of the Wiener sample paths.

**Theorem 7.2.** (Nondifferentiability of the Wiener process). Let $\{z_t, t \geqslant 0\}$ be a Wiener process in $(\Omega, \mathscr{F}, P)$. Then for $\omega$ outside some set of probability 0, the sample path $z_t(\omega)$, $t \geqslant 0$, is nowhere differentiable.

For a proof see Billingsley (1979, section 37).

Intuitively, a nowhere differentiable sample path represents the motion of a particle which at no time has a velocity. Thus, although the sample paths are continuous, theorem 7.2 suggests that they are very kinky, and their derivatives exist nowhere. A complete analysis of the structure of the Brownian motion sample paths is found in Itô and McKean (1974).

Intuitively, a simple way to illustrate the nondifferentiability of the Wiener process is to observe from property (3) that for $0 \leqslant s < t$,

$$
E \left( \frac{z_t - z_s}{t - s} \right)^2 = \frac{\sigma^2}{t - s} = \frac{1}{t - s}
$$

assuming $\sigma = 1$; taking limits,

$$\lim_{t \to s} \text{E} \left( \frac{z_t - z_s}{t - s} \right)^2 = \infty. \tag{7.10}$$

Suppose, however, that the Wiener process were differentiable; then, if we denote its derivative by $z'_s$ we have

$$\lim_{t \to s} \text{E} \left( \frac{z_t - z_s}{t - s} - z'_s \right)^2 = 0.$$

This last equation implies

$$\lim_{t \to s} \text{E} \left( \frac{z_t - z_s}{t - s} \right)^2 = \text{E}(z'_s)^2. \tag{7.11}$$

Note that (7.11) contradicts (7.10).

In the engineering literature, the derivatives of a Wiener process is called *white noise*. Further remarks will be made about it in the next chapter.

Let $\{z_t, \, t \geqslant 0\}$ be a Wiener process and use it to construct a new process $\{w_t, t \geqslant 0\}$ defined by

$$w_t = z_t + \mu t, \quad t \geqslant 0, \tag{7.12}$$

where $\mu$ is a constant. Then we say that $\{w_t, t \geqslant 0\}$ is a Wiener process or Brownian motion process with drift and $\mu$ is called the *drift parameter*. In this case the only modification that occurs in the definition of a Wiener process is in property (3) where $w_t - w_s$ is normally distributed with mean $\mu(t - s)$ and variance $\sigma^2 (t - s)$, assuming $\sigma = 1$.

Finally, let $w_t$ be a Wiener process with drift as defined in (7.12). Consider the new process given by

$$y_t = \exp(w_t), \quad t \geqslant 0. \tag{7.13}$$

Then $\{y_t, t \geqslant 0\}$ is called a *geometric Brownian motion* or *geometric Wiener process*.

## 7.3. The Markov chain and the Markov process

A *Markov chain* is a stochastic process $\{X_t, t = 0, 1, 2, ...\}$ with a countable state space $E$ defined on a probability space $(\Omega, \mathcal{F}, P)$ provided it satisfies the property: for all $j \in E$ and $t \in T = N \equiv \{0, 1, 2, ...\}$,

$$P[X_{t+1} = j \mid X_0, ..., X_t] = P[X_{t+1} = j \mid X_t]. \tag{7.14}$$

Eq. (7.14) is called the *Markov property* and it says that the probability that the random variable $X_{t+1}$ will be at state $j$, conditioned on the past behavior of the random variables $X_0, ..., X_t$ (i.e. conditioned on the $\sigma$-field generated by $X_0$, ..., $X_t$ and denoted by $\sigma(X_0, ..., X_t) \equiv \sigma(X_n, n \leqslant t))$, is equal to the probability that $X_{t+1}$ will be at state $j$, conditioned only on the current or present information supplied by $X_t$ (i.e. conditioned on the $\sigma$-field generated by $X_t$ and denoted by $\sigma(X_t)$). Put differently, the Markov property says that the past and future are statistically independent when the present is known. All that matters in determining the next state $X_{t+1}$ of the process is its current state $X_t$ and it does not matter how the process got to $X_t$.

The probability of $X_{t+1}$ being in state $j$, given that $X_t$ is in state $i$, is called the *one-step transition probability* and is written as

$$p_{ij}^{t; t+1} \equiv P[X_{t+1} = j \mid X_t = i] \tag{7.15}$$

for $i, j \in E$.

Observe that (7.15) denotes the dependence of the probability on the initial and final state as well as the time. If the one-step transition probabilities are independent of time, i.e. if independent of $t \in N$,

$$P_{ij} \equiv P[X_{t+1} = j \mid X_t = i] \tag{7.16}$$

holds, then we say that the Markov chain is *time-homogeneous* or that the Markov chain has *stationary transition probabilities*. Such probabilities can be placed in a matrix such as

$$P \equiv \begin{bmatrix} P_{11} & P_{12} & P_{13} & \cdots \\ P_{21} & P_{22} & P_{23} & \cdots \\ \cdot & \cdot & \cdot & \\ \cdot & \cdot & \cdot & \\ \cdot & \cdot & \cdot & \end{bmatrix}. \tag{7.17}$$

The matrix $P$ in (7.17) is called the *transition matrix* of the Markov chain. The matrix in (7.17) satisfies, first, the condition $P_{ij} \geq 0$ for $i, j \in E$ and, secondly, the condition $\Sigma_{j \in E} P_{ij} = 1$ for $i \in E$.

The probability that the Markov chain moves from state $i$ to state $j$ in $s$ steps or periods of time is the $(i, j)$ th entry of the $s$th power of the transition matrix $P$. This can be written as

$$P[X_{t+s} = j \mid X_t = i] = P_{ij}^s. \tag{7.18}$$

In (7.18) $P_{ij}^s$ denotes the $(i, j)$ th entry of the $s$th power of $P$. Using the fact from matrix algebra that $P^{s+r} = P^s P^r$ we obtain for $i, j \in E$,

$$P_{ij}^{s+r} = \sum_{k \in E} P_{ik}^s P_{kj}^r, \tag{7.19}$$

which is called the *Chapman–Kolmogorov equation*. Eq. (7.19) says that if the Markov chain starts at state $i$, in order for it to be in state $j$ after $s + r$ periods or steps, it must be in some intermediate state $k$ after the $s$th step and then move from there into state $j$ during the remaining $r$ periods.

The function $\pi_0(i)$ for $i \in E$, defined by

$$\pi_0(i) = P[X_0 = i], \tag{7.20}$$

is called the *initial distribution* of the Markov chain and it satisfies, first, the condition $\pi_0(i) \geq 0$ for $i \in E$ and, secondly, the condition $\Sigma_{i \in E} \pi_0(i) = 1$. Finally, for a Markov chain $X_t$, $t \in N$, with transition probability $P_{ij}$ we say that the function $\pi(i)$, $i \in E$, is a *stationary distribution* if $\pi(i) \geq 0$ for $i \in E$, and $\Sigma_{i \in E} \pi(i) = 1$ and also

$$\sum_{i \in E} \pi(i) P_{ij} = \pi(j), \quad j \in E. \tag{7.21}$$

Suppose that the stationary distribution $\pi$ exists and that

$$\lim P^t(i, j) \to \pi(j), \quad j \in E, \tag{7.22}$$

as $t \to \infty$. Then, it may be proved that, regardless of the initial distribution of the chain $\pi_0$, the distribution of $X_t$ approaches $\pi$ as $t \to \infty$. If this occurs we say that $\pi$ is a *steady state distribution*. For details see Hoel, Port and Stone (1972, ch. 2).

Having briefly discussed some notions about Markov chains, we move on to present various definitions about Markov processes.

A stochastic process $\{X_t, t \in [0, \infty)\}$ defined on a probability space $(\Omega, \mathcal{F}, P)$

with state space the real line $R$, is called a *Markov process* if for $0 \leqslant s \leqslant t < \infty$ and $A \in \mathcal{R}^1$

$$P[X_t \in A \mid \sigma(X_u, u \leqslant s)] = P[X_t \in A \mid \sigma(X_s)] \tag{7.23}$$

holds w.p.1.

Equation (7.23) is a generalization of eq. (7.14) and it is also called the *Markov property*. In (7.23) the left-hand side is a conditional probability with conditioning on the $\sigma$-field generated by all the random variables $X_u$ for $u \in [0, s]$, while the right-hand side indicates the conditioning on the $\sigma$-field generated just by the random variable $X_s$. As before, what matters in a Markov process in determining its future behavior is not its past but only its current position. A condition that is equivalent to (7.23) is this: for $n \geqslant 1$, $A \in \mathcal{R}^1$ and $0 \leqslant t_0 < t_1 < ... < t_n < t < \infty$,

$$P[X_t \in A \mid X_{t_0}, ..., X_{t_n}] = P[X_t \in A \mid X_{t_n}] \tag{7.24}$$

holds w.p.1.

For a Markov process $X_t$, $t \geqslant 0$, we define the *transition probability*, denoted by $P(s, x, t, A)$, as follows: for $0 \leqslant s < t$, $A \in \mathcal{R}^1$ and $x \in R$,

$$P(s, x, t, A) = P[X_t \in A \mid X_s = x] \tag{7.25}$$

w.p.1. The transition probability exists by fact 1 of section 5 of this chapter and has two properties. First, for $0 \leqslant s < t$ and $A \in \mathcal{R}^1$, $P(s, \cdot, t, A)$ is a measurable function with respect to $\mathcal{R}^1$, where $s$, $t$ and $A$ are fixed, and secondly, $P(s, x, t, \cdot)$ is a probability measure on $\mathcal{R}^1$ for fixed $s$, $t$ and $x \in R^1$. In a manner analogous to (7.19) the *Chapman–Kolmogorov equation*, for a Markov process for $0 \leqslant s \leqslant u \leqslant t < \infty$ and $A \in \mathcal{R}^1$, and for all $x \in R^1$, with the possible exception of $x \in N$, where $P[\omega: X_s(\omega) = x \in N] = 0$, we have

$$P(s, x, t, A) = \int_R P(u, y, t, A) P(s, x, u, \mathrm{d}y). \tag{7.26}$$

The Markov process is *time-homogeneous* if its transition probability $P(s, x, t, A)$ is time independent or stationary, i.e. for any $u > 0$,

$$P(s + u, x, t + u, A) = P(s, x, t, A), \tag{7.27}$$

which means that the transition probability is a function of only three arguments, $x$, $t - s$, and $A$. In such a case the Chapman–Kolmogorov equation is written as

$$P(s+t, x, A) = \int_R P(s, y, A) P(t, x, \mathrm{d}y). \tag{7.28}$$

A standard example of a homogeneous Markov process is the Wiener process $z_t, t \geqslant 0$, with stationary transition probability for $t \geqslant 0$ and $A \in \mathscr{R}^1$,

$$P(t, x, A) = P[z_{s+t} \in A \mid z_s = x]$$

$$= \frac{1}{\sqrt{2\pi t}} \int_A \exp\left(-\frac{(y-x)^2}{2t}\right) \mathrm{d}y.$$

We conclude this section by discussing the Poisson process.

### 7.4  The Poisson process

A *Poisson process*, with parameter $\lambda$, is a collection of random variables $\{X_t, t \in [0, \infty)\}$ defined on $(\Omega, \mathscr{F}, P)$ having as state space $N = \{0, 1, 2, \dots\}$ and satisfying the following three properties:

(1)  $X_0 = 0$ w.p.1.

(2)  For each $0 < t_1 < t_2 < \dots < t_n$, the increments $X_{t_1}, X_{t_3} - X_{t_2}, \dots, X_{t_n} - X_{t_{n-1}}$ are independent.

(3)  For $0 \leqslant s < t < \infty$ the increment $X_t - X_s$ has a Poisson distribution with parameter $\lambda(t - s)$, i.e. the distribution of the increments is given by

$$P[X_t - X_s = k] = \frac{[\lambda(t-s)]^k}{k!} \exp[-\lambda(t-s)]$$

for $k \in N$.

Note that property (1) is just a convention while property (2) simply states that the number of events in disjoint time intervals are independent. Property (3) may be derived from two postulates as is done in Karlin and Taylor (1975, pp. 23–26). These two postulates are stated below.

First, the probability of at least one event happening in a time period of duration $\Delta t$ is

$$P(\Delta t) = \lambda \Delta t + o(\Delta t), \tag{7.29}$$

with $\lambda > 0$ and $\Delta t \to 0$. The notation $o(\Delta t)$ means that as $\Delta t$ approaches $0$, $o(\Delta t)$ approaches $0$ also, but at a rate faster than that of $\Delta t$.

Secondly, the probability of two or more events happening in $\Delta t$ is $o(\Delta t)$. This probability is very small and it essentially says that the simultaneous occurrence of more than one event during a time interval of length $\Delta t$ is almost 0.

In Chapter 2, section 12, we will have an occasion to use (7.29).

## 8. Optimal stopping

Suppose that a fair coin is tossed repeatedly and after each toss we have to make the decision of stopping or going on to the next toss. Let $Y_1, Y_2, \ldots$ be independent random variables denoting successive tosses with common probability distribution $P(Y_i = 1) = P(Y_i = -1) = \frac{1}{2}$, where $Y_i = 1$ represents heads on the $i$th toss and $Y_i = -1$ represents tails. If we stop after the $n$th toss we are to receive a *reward* denoted by $X_n$ which is a function of the first $n$ tosses, i.e. $X_n = f_n(Y_1, Y_2, \ldots, Y_n)$. The mathematical question is when we should stop so as to maximize our expected reward. A *stopping rule* or *stopping time* or *stopping variable* is a random variable $\tau$ with values in $\{1, 2, 3, \ldots\}$ and such that the event $\{\tau = n\}$ depends only on the past values of $Y_1, \ldots, Y_n$ and not on the future values $Y_{n+1}$, $Y_{n+2}, \ldots$ . If $\tau$ is a stopping rule then $\mathrm{E}(X_\tau)$ measures the average reward associated with the stopping rule $\tau$. Denote by $C$ the class of all stopping rules for which $\mathrm{E}(X_\tau) < \infty$. The $V$, where

$$V = \sup_{\tau \in C} \{\mathrm{E}(X_\tau)\}, \tag{8.1}$$

is called the *value of the sequence* $\{X_n\}$ and if a stopping rule $\tau$ exists such that

$$\mathrm{E}(X_\tau) = V, \tag{8.2}$$

then $\tau$ is said to be an *optimal stopping rule*.

To illustrate the concepts presented we discuss two examples from Robbins (1970, pp. 334–336). In the first example we consider the reward function given by

$$X_n = \min\{1, Y_1 + \ldots + Y_n\} - n/(n+1) \tag{8.3}$$

for $n \geqslant 1$, where min denotes the minimum or smallest value between 1 and $Y_1 + \ldots + Y_n$. As a stopping rule we consider

$$\tau = \text{first integer } n \geqslant 1 \text{ such that } Y_1 + \ldots + Y_n = 1. \tag{8.4}$$

For a stopping rule as in (8.4) it is difficult to compute $E(X_\tau)$ exactly, but we know that $E(X_\tau) > 0$, since

$$E(X_\tau) = E(\min\{1, Y_1, ..., Y_\tau\}) - E\left(\frac{\tau}{\tau+1}\right) = 1 - E\left(\frac{\tau}{\tau+1}\right) > 0. \quad (8.5)$$

To show that the stopping rule in (8.4) is optimal we next demonstrate that for any stopping rule other than the one in (8.4), its associated $E(X_\tau)$ is negative. To do this we need a famous result.

**Lemma 8.1.** (Wald). Suppose that $Y_1$, $Y_2$, ... are independent and identically distributed random variables such that

$$E(Y_i) \equiv \mu < \infty.$$

Suppose also that $\tau$ is any stopping time of the sequence $Y_1$, $Y_2$, ... such that $E(\tau) < \infty$. Then $E(\Sigma_{i=1}^\tau Y_i)$ always exists and

$$E\left(\sum_{i=1}^\tau Y_i\right) = \mu \cdot E(\tau). \quad (8.6)$$

For a proof see Wald (1944) or Shiryayev (1978, p. 175).

To apply Wald's lemma in our case note that $Y_1$, $Y_2$, ..., representing successive tosses of a fair coin, are independent identically distributed random variables with

$$\mu \equiv E(Y_i) = 1 \cdot \tfrac{1}{2} + (-1) \cdot \tfrac{1}{2} = 0.$$

Furthermore, suppose that $\tau$ is any stopping rule for which $E(\tau) < \infty$ and hence, in particular, if $\tau \in C$, then $E(Y_1 + ... + Y_\tau) = 0$. This implies that

$$E(X_\tau) \leqslant E(Y_1 + ... + Y_\tau) - E\left(\frac{\tau}{\tau+1}\right) \leqslant -\tfrac{1}{2}. \quad (8.7)$$

Thus, for all stopping rules other than (8.4), eq. (8.7) shows that $E(X_\tau) < 0$. The supremum of all $E(X_\tau)$ for $\tau \in C$ is therefore (8.5) and by (8.1) for $\tau$ as in (8.4)

$$V = E(X_\tau) > 0.$$

Robbins (1970, p. 335) concludes that (8.4) is optimal.

As a second example suppose that instead of (8.3) the reward function is given by

$$X_n = \frac{n2^n}{n+1} \prod_{i=1}^{n} \left( \frac{Y_i + 1}{2} \right) \tag{8.8}$$

for $n = 1, 2, \ldots$ . Note that (8.8) says that if we stop after the $n$th toss with all heads we receive an award of $n2^n/(n+1)$, while otherwise the award is 0. Suppose that we are at a stage $n$ in which all heads have appeared and our reward is $n2^n/(n+1)$. What is the conditional expected reward of making one more tossing before we stop? The answer is

$$E\left[ X_{n+1} \mid X_n = \frac{n2^n}{n+1} \right] = \frac{1}{2} \cdot \frac{(n+1)2^{n+1}}{n+2} = \frac{2^n(n+1)}{n+2} > X_n. \tag{8.9}$$

Eq. (8.9) suggests that we should not stop because our expected reward after one more toss, $X_{n+1}$, conditioned on $X_n$, is greater than our reward without any further tossing. But, suppose that we act wisely and have another tossing. Eventually, a tail will occur and our final reward will be 0. Thus, acting wisely at each stage does not imply the best long-run policy. This suggests that for the reward function (8.8) no optimal stopping rule exists. However, stopping rules do exist. Consider, for example, the class of stopping rules $\{\tau_k\}$, $k = 1, 2, \ldots$, where $\tau_k$ stops after the $k$th toss no matter what sequence of heads and tails has appeared. For such stopping rules we have

$$E(X_{\tau_k}) = \frac{1}{2^k} \cdot \frac{k2^k}{k+1} + \left( 1 - \frac{1}{2^k} \right) \cdot 0 = \frac{k}{k+1}. \tag{8.10}$$

From (8.10) as $k \to \infty$ we obtain using (8.1) that $V = 1$. Thus, stopping rules exist but no optimal rule exists.

## 8.1. *Mathematical results*

Having motivated the concept of optimal stopping we proceed to establish some basic mathematical results and then give some illustrations from the theory of job search and stochastic capital theory. Our analysis draws heavily from the two classic books of Chow, Robbins and Siegmund (1971), and DeGroot (1970).

Let $(\Omega, \mathscr{F}, P)$ be a probability space and let $\{\mathscr{F}_n, n = 1, 2, \ldots\}$ be a sequence of increasing $\sigma$-fields belonging to $\mathscr{F}$. Let $Y_1, Y_2, \ldots$ be random variables having a known joint probability distribution function $F$ and defined on $(\Omega, \mathscr{F}, P)$. We assume that we can observe sequentially $Y_1$, $Y_2$, ... and we denote by $X_1, X_2, ...$ the sequence of rewards. If we stop at the $n$th stage, $X_n = f_n(Y_1, Y_2, ..., Y_n)$. It is assumed that $X_1$, $X_2$, ... are measurable with respect to $\mathscr{F}_1, \mathscr{F}_2, ...$, i.e. $X_n$ is measurable $\mathscr{F}_n$ for $n = 1, 2, ...$ . A *stopping rule* or *stopping variable* or *stopping time* is a random variable $\tau = \tau(\omega)$ defined on $(\Omega, \mathscr{F}, P)$ with target space the positive integers $1, 2, ...$ which satisfies two conditions. First,

$$P[\omega: \tau(\omega) < \infty] = 1, \tag{8.11}$$

and secondly

$$\{\omega: \tau(\omega) = n\} \in \mathscr{F}_n \quad \text{for each } n. \tag{8.12}$$

Eq. (8.11) says that the stopping rule takes a finite value w.p.1 and (8.12) says that the decision to stop at time $n$ depends only on past information included in the $\sigma$-field $\mathscr{F}_n$. Put differently, (8.12) indicates that no future information is available to influence the decision to stop at time $n$.

In general $\mathscr{F}_n$ is quite arbitrary, although in some applications we may take $\mathscr{F}_n = \sigma(Y_1, Y_2, ..., Y_n)$. The collection of all sets $A \in \mathscr{F}$ such that $A \cap \{\tau = n\} \in \mathscr{F}_n$ for all integers $n$ is a $\sigma$-field in $\mathscr{F}$ and is denoted by $\mathscr{F}_\tau$. We note that $\tau$ and $X_\tau$ are measurable $\mathscr{F}_\tau$.

For any stopping rule $\tau$, the reward at time $\tau$ is denoted by $X_\tau$ which is a random variable of the form

$$X_\tau = \sum_{n=1}^{\infty} X_n I_{\{\tau=n\}} = \begin{cases} X_n & \text{on } \{\tau = n\}, \\ \\ 0 & \text{otherwise,} \end{cases} \tag{8.13}$$

for $n = 1, 2, ...$ . As in (8.1) the value of the reward sequence, $V$, is the supremum of $E(X_\tau)$ for $\tau \in C$. Note that $I_{\{\tau=n\}}$ is the *indicator function* taking the value 1 on the set $\{\omega: \tau(\omega) = n\}$ and zero otherwise.

For a given stopping rule $\tau$, the expectation of the reward $X_\tau$ exists, i.e. $E(X_\tau) < \infty$, if and only if

$$E(|X_\tau|) = \sum_{n=1}^{\infty} E[|X_n| \,|\, \{\tau = n\}] P[\tau = n] < \infty. \tag{8.14}$$

With the above notions available to us we now ask the question: Under what conditions does an optimal rule exist? The answer is given by the next theorem.

**Theorem 8.1.** (Existence of optimal rule). Suppose that $X_1, X_2, ...$ is a sequence of rewards on $(\Omega, \mathscr{F}, P)$ as described above such that w.p.1

$$E(|\sup_n X_n|) < \infty, \tag{8.15}$$

and also as $n \to \infty$

$$\lim X_n \to -\infty \text{ w.p.1.} \tag{8.16}$$

Then there exists an optimal stopping rule.

For a proof see DeGroot (1970, pp. 347–348).

The two sufficient conditions of theorem 8.1 have an intuitive explanation. Condition (8.15) says that even if we could observe the entire sequence of the random variables $Y_1, Y_2, ...$ and then select to stop so as to maximize the reward, the expected reward would still be finite. In other words, even with perfect foresight the payoff is limited. But even with a finite expected reward as in (8.15), it might be beneficial not to stop. Condition (8.16) makes certain that w.p.1 we stop at a finite time.

Although theorem 8.1 is very useful because it gives us sufficient conditions for the existence of an optimal rule, it does not, however, tell us anything about the nature of the optimal rule. Mathematical research has obtained results on the nature of optimal rule in some special cases. We follow DeGroot (1970) to discuss such an important case. Consider the reward function which has the form

$$X_n = \max \{Y_1, Y_2, ..., Y_n\} - nc \tag{8.17}$$

for $n = 1, 2, ...$, where $c > 0$ denotes the fixed cost of every observation. Eq. (8.17) says that our reward at period $n$ is the difference between the largest value observed among the random variables $Y_1, ..., Y_n$ and the cost of such observations or sampling. We then have the following theorem.

**Theorem 8.2.** Suppose that $Y_1, Y_2, ...$ is a sequence of independent and identically distributed random variables on $(\Omega, \mathscr{F}, P)$ with a distribution function $F$ and let $X_n, n = 1, 2, ...$ be a sequence of rewards as in (8.17). If

$$E(Y_n^2) < \infty \tag{8.18}$$

for $n = 1, 2, ...$, then there exists a stopping rule which maximizes $E(X_\tau)$ and which has the form: stop as soon as some observed value $Y \geqslant V$ and continue if

$Y < V$, where $V$ is the value of the reward sequence obtained as a unique solution of the equation

$$\int_V^\infty (Y - V)\,\mathrm{d}F(Y) = c. \tag{8.19}$$

For a proof see DeGroot (1970, p. 352).

## 8.2. Job search

Before we state additional mathematical theorems on optimal stopping we move to apply the results stated so far to the theory of job search. In this application we follow Lippman and McCall (1976a).

Consider an individual who is seeking employment and who searches daily, until he accepts a job and who receives exactly one job offer every day. The cost of generating each offer is $c > 0$. There are two possibilities: *sampling with recall*, i.e. when all offers are retained, and *sampling without recall*, i.e. when offers are made and not accepted are lost. Notationally, the random variables $Y_n$ represent the job offers at periods $n = 1, 2, \ldots$ and we assume that the job searcher knows the parameters of the wage distribution $F$ from which his wage offers $Y_n$ are randomly generated. To keep the analysis simple we assume the participant in the job search to be risk neutral and seeking to maximize his expected net benefit. The searcher's decision is when to stop searching and accept an offer. Note that his sequence of rewards has the simple form

$$X_n = \max \{Y_1, \ldots, Y_n\} - nc$$

as in (8.17), for $n = 1, 2, \ldots$, in the case of sampling with recall, and the form

$$\bar{X}_n = Y_n - nc$$

in the case of sampling without recall. This last equation simply says that in the case of sampling without recall the searcher's reward is the difference between the current offer and his total cost of the search. Below we discuss the case of sampling with recall for independent and identically distributed $Y_1, \ldots, Y_n$.

This job search problem can be analyzed using the results from the theory of optimal stopping. We proceed to discuss the existence and nature of optimal stopping and then we obtain some additional insights from the analysis.

By theorem 8.1, to establish the existence of an optimal stopping rule for the job search problem, we need to establish the two conditions in (8.15) and (8.16). This is accomplished in the following lemma in which independence of $Y_1$, $Y_2$, ... is not needed.

**Lemma 8.2.** Let $Y_1$, $Y_2$, ... be a sequence of identically distributed random variables having a distribution function $F$. Let $c > 0$ be a given number and define

$$Z = \sup_n X_n, \qquad (8.20)$$

where $X_n$ is as in eq. (8.17).

If the mean of $F$ exists, then $\lim X_n \to -\infty$ as $n \to \infty$, w.p.1.

If the variance of $F$ is finite then $E(|Z|) < \infty$.

For a proof see DeGroot (1970, pp. 350–352).

This lemma is useful because it helps us establish the existence of an optimal rule for the job search problem. Note that this lemma is a purely mathematical result which states sufficient conditions for (8.15) and (8.16) to hold. More specifically, if the mean and variance of the common distribution $F$ exist then (8.15) and (8.16) hold. Thus, assuming that $E(Y_n^2) < \infty$, $n = 1, 2, ...$, we can use lemma 8.2 and theorem 8.1 to conclude the existence of an optimal rule.

The nature of the optimal rule in the job search model is described by theorem 8.2. More specifically, for any wage offer $Y$, the optimal stopping rule for the job searcher is of the nature or form

$$\begin{aligned} &\text{accept job if} &&Y \geqslant V, \\ &\text{continue search if} &&Y < V. \end{aligned} \qquad (8.21)$$

In the job search literature the critical number $V$ in (8.21) is called the *reservation wage* and any policy of the form of (8.21) is said to possess the *reservation wage property*.

Consider the first observation $Y_1$ from a sequence of independent and identically distributed random variables $Y_1$, $Y_2$, ... . The expected return from following the optimal policy in (8.21) is $E(\max\{V, Y_1\}) - c$. From the definition of $V$, the optimal expected return from the optimal stopping rule satisfies

$$V = E \max (V, Y_1) - c. \qquad (8.22)$$

Note that

$$\begin{aligned}
E \max (V, Y_1) &= V \int_0^V dF(Y) + \int_V^\infty Y dF(Y) \\
&= V \int_0^V dF(Y) \pm V \int_V^\infty dF(Y) + \int_V^\infty Y dF(Y) \\
&= V \int_0^V dF(Y) + V \int_V^\infty dF(Y) + \int_V^\infty Y dF(Y) - \\
&\quad - V \int_V^\infty dF(Y) \\
&= V \int_0^\infty dF(Y) + \int_V^\infty (Y - V) dF(Y) \\
&= V + \int_V^\infty (Y - V) dF(Y).
\end{aligned} \tag{8.23}$$

Putting the result of (8.23) into (8.22) we obtain eq. (8.19) of theorem 8.2. Let us go a step further in analyzing eq. (8.19) which has just been derived. Define

$$H(V) \equiv \int_V^\infty (Y - V) dF(Y).$$

The function $H$ is convex, non-negative, strictly decreasing and satisfies the following properties: $\lim H(V) \to 0$ as $V \to \infty$, $\lim H(V) \to E(Y_1)$ as $V \to 0$, $dH(V)/dV = -[1 - F(V)]$, $d^2 H(V)/dV^2 \geqslant 0$. Graphically, we may illustrate $H$ as in fig. 8.1.

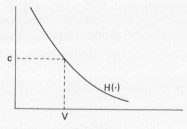

Figure 8.1.

From fig. 8.1 we see that the lower the cost of search $c > 0$ is, the higher the reservation wage $V$ and the longer the duration of search will be. Studying the equation $H(V) = c$, which is (8.19), we obtain a simple economic interpretation, i.e. the value $V$ is chosen so that equality holds between the expected marginal return from one more observation, $H(V)$, and the marginal cost of obtaining one more job offer, $c$. In this application the job searcher behaves *myopically* by comparing his wage from accepting a job with the expected wage from exactly one more job offer.

We conclude by remarking that in the infinite time horizon case with $F(\cdot)$ known there is no difference in the analysis between sampling with recall and sampling without recall. In the latter case search always continues until the reservation wage is exceeded by the last offer. Note, however, that if the time horizon is finite, or if $F(\cdot)$ is not known, these two assumptions cause the results to be different, as is shown in Lippman and McCall (1976a).

## 8.3. *Additional mathematical results*

We continue the analysis on optimal stopping by illustrating the role of martingale theory in problems of optimal stopping.

Let $X_1, X_2, \dots$ be a sequence of random variables on $(\Omega, \mathscr{F}, P)$, denoting rewards and let $\mathscr{F}_1, \mathscr{F}_2, \dots$ be an increasing sequence of $\sigma$-fields in $\mathscr{F}$. We assume, as earlier, that $X_n$ is measurable with respect to $\mathscr{F}_n$, $n = 1, 2, \dots$, and that $X_n = f_n(Y_1, \dots, Y_n)$. The pair $\{(X_n, \mathscr{F}_n), n = 1, 2, \dots\}$ is called a *stochastic sequence*. If $E(|X_n|) < \infty$, $n = 1, 2, \dots$, then we say that $\{(X_n, \mathscr{F}_n), n = 1, 2, \dots\}$ is an *integrable stochastic sequence*. If we interpret a stochastic sequence as a martingale it is natural to inquire whether $E(X_1) = E(X_\tau)$, with $\tau$ a stopping time. The reader will recall that the martingale property may be viewed as representing the notion of a fair gamble and therefore asking whether $E(X_1) = E(X_\tau)$ means that we are inquiring whether the property of fairness is preserved under any stopping time $\tau$. The results that are stated next explore this question and are obtained from Chow, Robbins and Siegmund (1971) and DeGroot (1970).

**Theorem 8.3.** Suppose that $\{(X_n, \mathscr{F}_n), n = 1, 2, \dots\}$ is a submartingale on $(\Omega, \mathscr{F}, P)$, $\tau$ is a stopping time and $n$ is a positive integer.

(1) If $P[\tau \leqslant n] = 1$, then $E[X_n | \mathscr{F}_\tau] \geqslant X_\tau$. \hfill (8.24)

(2) If $P[\tau < \infty] = 1$ and $E(X_\tau) < \infty$ and also

$$\liminf_{n} \int_{\{\tau > n\}} X_n^+ \, dP = 0, \tag{8.25}$$

then for each $n$,

$$E[X_\tau | \mathscr{F}_n] \geqslant X_n \quad \text{on } \{\tau \geqslant n\}. \tag{8.26}$$

For a proof of this theorem see Chow, Robbins and Siegmund (1971, p. 21).

From eq. (8.24) we conclude that the unconditional expectations satisfy the relation

$$E(X_n) \geqslant E(X_\tau) \geqslant E(X_1) \tag{8.27}$$

provided $P[\tau \leqslant n] = 1$. For supermartingales the conclusion is as in (8.27) with the inequalities reversed and for martingales we have (8.27) with equality signs.

Concerning the condition in (8.25), we note that it is satisfied for a sequence of random variables $X_1, X_2, \ldots$ which are uniformly integrable. The definition is this: a sequence of random variables $\{X_n, n = 1, 2, \ldots\}$ on $(\Omega, \mathscr{F}, P)$ is *uniformly integrable* if, as $\alpha \to \infty$, then

$$\limsup_{n} \int_{\{|X_n| > \alpha\}} |X_n| \, dP \to 0. \tag{8.28}$$

Note that (8.28) implies that $\sup_n E(|X_n|) < \infty$, which allows us to conclude in particular that $E(X_\tau) < \infty$ for $\tau$ a stopping time, and therefore (8.26) holds.

The last result of this subsection is the monotone case theorem. For a stochastic sequence $\{(X_n, \mathscr{F}_n), n = 1, 2, \ldots\}$ assumed to be integrable let

$$A_n = \{E[X_{n+1} | \mathscr{F}_n] \leqslant X_n\} \tag{8.29}$$

for $n = 1, 2, \ldots$ . If

$$A_1 \subset A_2 \subset \ldots \subset A_n \subset \ldots \tag{8.30}$$

and

$$\bigcup_{n=1}^{\infty} A_n = \Omega \tag{8.31}$$

both hold, we say that the *monotone case* holds. In this case the next theorem tells us the nature of the optimal stopping rule.

**Theorem 8.4.** (Monotone case).  Suppose that $\{(X_n, \mathscr{F}_n), n = 1, 2, ...\}$ is a stochastic sequence which satisfies the monotone case. Let the stopping variable $s$ be defined by

$$s = \text{first } n \geq 1 \text{ such that } X_n \geq E[X_{n+1} | \mathscr{F}_n],$$

provided $E(X_s^-) < \infty$. Then if

$$\liminf_n \int_{\{s>n\}} X_n^+ = 0$$

holds, we obtain that

$$E(X_s) \geq E(X_\tau)$$

for all $\tau$ such that $E(X_\tau) < \infty$ and

$$\liminf_n \int_{\{\tau \geq n\}} X_n^- = 0.$$

For a proof see Chow, Robbins and Siegmund (1971, p. 55).

## 8.4. Stochastic capital theory

We conclude this section by presenting a brief analysis of stochastic capital theory following Brock, Rothschild and Stiglitz (1979).

   Introductory lectures on capital theory often begin by analyzing the following problem: you have a tree which will be worth $X(t)$ if cut down at time $t$, where $t = 0, 1, 2, ...$ . If the discount rate is $r$, when should the tree be cut down? Also, what is the present value of such a tree? The answers to these questions are straightforward. Suppose we choose the cutting date $\tau$ to maximize $e^{-rt} X(\tau)$. Note that at $t < \tau$ a tree is worth $e^{rt} e^{-r\tau} X(\tau)$. Most of capital theory can be built on this simple foundation. It is our purpose to analyze how these simple questions of timing and evaluation change when the tree's growth is stochastic rather than deterministic. Suppose a tree will be worth $X(t, \omega)$ if cut down at time $t$, where $X(t, \omega)$ is a stochastic process. When should it be cut down? What is its present value? We ask these questions because, as in the certainty case, one can use such an analysis to answer many other questions of valuation and timing.

   Before we can analyze the problem of when to cut down a tree which grows stochastically, we must specify both the stochastic process which governs the tree's growth and the valuation principle used. Here we analyze a discrete time model because for such models analysis of some problems can be done both more easily and in greater generality. We assume the tree's value follows a discrete time real-valued Markov process which we write as $X_1, X_2, ..., X_t$. To complete the specification of our problem, we must describe what the person who owns the tree is trying to maximize. The simplest assumption is that he is maximizing expected present discounted value. Thus, if we let $C$ be the set of stopping times for $X_t$, then the problem is to find $\tau \in C$ to maximize $EX(\tau) e^{-r\tau}$. An apparently more general approach would involve the maximization of expected utility, $E(U[X(\tau)] e^{-r\tau})$. This is only apparently more general because if $U(X)$ is a strictly increasing function, $U(X(t, \omega))$ is a Markov process with essentially the same properties as $X(t, \omega)$ and there is no analytical difference between maximizing $E(e^{-r\tau} X(\tau))$ and maximizing $E(U[X(\tau)] e^{-r\tau}$. Of course, the interpretation of the stochastic properties of $X(t)$ depends on whether $X(t)$ is thought of as being measured in dollars or in utiles.

If the discount rate is

$$\beta = \frac{1}{1+r},$$ (8.32)

our problem is to choose a stopping time $\tau$ to maximize $E(\beta^\tau X_\tau)$. Let us choose the simplest possible specification for the $X_t$ process, i.e. suppose that

$$X_{t+1} = X_t + \epsilon_t,$$ (8.33)

where $\epsilon_t$ are independent and identically distributed random variables with expected value $\mu$. It is clear that in this case the optimal rule will be of a particularly simple form: pick a tree size $\hat{X}$ and cut the tree down the first time that $X_t \geqslant \hat{X}$.

If the process $X_t$ is deterministic, $X_{t+1} = X_t + \mu$, it is easy to find the optimal cutting size, which we denote by $\bar{X}$. $\bar{X}$ must satisfy

$$X = \beta(X + \mu),$$ (8.34)

since the l.h.s. of (8.34) is the value of cutting a tree down now and the r.h.s. is the present discounted value of the tree next period. If $\bar{X}$ satisfies (8.34), then the tree owner is indifferent as to whether he sells the tree now or keeps it for a period. For small $X$, the value of the r.h.s. (keeping the tree for a period) exceeds the value of the l.h.s. (harvesting the tree now). Note that if $\bar{X}$ is a solution to (8.34), then $\mu/\bar{X} = (1 - \beta)/\beta = r$, or $\bar{X} = \mu/r$, i.e. the growth rate equals the interest rate. How is this analysis changed if the tree's growth is uncertain? The answer depends on whether the stochastic process $X_t$ is strictly increasing or not. Let $\hat{X}$ be the optimal cutting size for the random process in (8.33). Then if $\epsilon$ is positive, which implies that $X_t$ is an increasing process, uncertainty has no effect on the cutting size.

**Theorem 8.5.** If

$$P[X_{t+1} > X_t] = 1, \tag{8.35}$$

then $\bar{X} = \hat{X}$.

We give a heuristic argument rather than a rigorous proof. This result is implied by the more general theorem 8.7 below. Let $V(X)$ be the value of having a tree of size $X$ assuming it will be cut down when it reaches the optimal size. Then if $\hat{X}$ is the cutting size it must be that

$$V(X) = X \quad \text{for } X \geqslant \hat{X} \tag{8.36}$$

and

$$V(X) > X \quad \text{for } X < \hat{X}. \tag{8.37}$$

Also at $\hat{X}$, the tree owner must be indifferent between cutting the tree down now and letting it grow for a period. That is to say, $\hat{X}$ must satisfy

$$\hat{X} = \beta \, \mathrm{E}(\max \{\hat{X} + \epsilon, \ V(\hat{X} + \epsilon)\}). \tag{8.38}$$

However, since $\epsilon$ is non-negative, $V(\hat{X} + \epsilon) = \hat{X} + \epsilon$ and (8.38) becomes $\hat{X} = \beta \, \mathrm{E}(\hat{X} + \epsilon) = \beta(\hat{X} + \mu)$ which is the same as (8.34). Since the solution to (8.34) is unique, $\hat{X} = \bar{X}$.

If $\epsilon$ is negative, this argument does not go through.

**Theorem 8.6.** If $P[\epsilon < 0] > 0$, then $\hat{X} > \bar{X}$.

This theorem says that trees will be harvested when they are larger under uncertainty rather than under certainty. Again we give a heuristic rather than a rigorous proof. Suppose for simplicity that $\epsilon$ has a density function $f(\cdot)$ with support on $[-1, +1]$. Then (8.38) becomes

$$\hat{X} = \beta \left( \int_{-1}^{0} V(\hat{X} + \epsilon) f(\epsilon) \, d\epsilon + \int_{0}^{1} (\hat{X} + \epsilon) f(\epsilon) \, d\epsilon \right)$$

$$> \beta \left( \int_{-1}^{0} (\hat{X} + \epsilon) f(\epsilon) \, d\epsilon + \int_{0}^{1} (\hat{X} + \epsilon) f(\epsilon) \, d\epsilon \right)$$

$$= \beta(\hat{X} + \mu).$$

Thus,

$$\hat{X} > \frac{\mu\beta}{1 - \beta} = \frac{\mu}{r} = \bar{X}.$$

We note here two implications of these simple propositions: that uncertainty can increase the value of a tree and that strictly increasing processes behave differently from processes which can decrease.

Theorem 8.6 implies that uncertainty can in some cases increase the value of a tree. Suppose you have a tree of size $\bar{X}$. Then if its future growth is certain, its value is just $\bar{X}$. If its evolution is uncertain and if its size may decrease, then you will not cut it down; its value exceeds $\bar{X}$. Writing $V^d(X)$ as the value of a tree of size $X$ in the deterministic case and $V^\epsilon(X)$ as the value of a tree when the increments in the tree's growth are the (nondegenerate) random variable $\epsilon$, we have shown that if $\epsilon$ were negative

$$V^\epsilon(X) > V^d(X) \tag{8.39}$$

for some $X$. (Continuity implies that (8.39) holds for $X \in [\bar{X} - \delta, \bar{X})$ for some $\delta$.) It is natural to ask whether (8.39) holds under more general circumstances.

Another implication of theorems 8.5 and 8.6 is that strictly increasing processes are different from processes which may decrease. We show that this is generally true by showing in the next theorem below that theorem 8.5 holds for a very wide class of increasing processes.

**Theorem 8.7.** Let $X_t$ be a Markov process such that

$$P[X_{t+1} \geq X_t] = 1$$

and that

$$E[\beta X_{t+1} - X_t \mid X_t \geq \bar{X}_t] \leq 0,$$
$$E[\beta X_{t+i} - X_t \mid X_t \leq \bar{X}_t] \geq 0,$$

where $\bar{X}_t$ is a nonincreasing sequence. Let $Y_t = \beta^t X_t$. If $X_t \geqslant \bar{X}_t$, then $Y_t, Y_{t+1}, Y_{t+2}, \dots$ is a supermartingale.

This proposition implies that uncertainty does not affect the time at which the tree is cut down. By theorem 8.3 we know that if $\tau$ is any stopping time, then $E[Y_\tau \mid X_t \geqslant \bar{X}_t] \leqslant Y_t$; it is optimal to stop when the tree's height first exceeds $\bar{X}_t$.

To establish this theorem we must show that for all $t$

$$E[Y_{t+\tau+1} - Y_{t+\tau} \mid X_t \geqslant \bar{X}_t] \leqslant 0.$$

However, since

$$Y_{t+\tau+1} - Y_{t+\tau} = \beta^{t+\tau}(\beta X_{t+\tau+1} - X_{t+\tau}),$$

it suffices to observe that

$$
\begin{aligned}
E[\beta X_{t+\tau+1} - X_{t+\tau} \mid X_t \geqslant \bar{X}_t] &= E(E[\beta X_{t+\tau+1} - X_{t+\tau} \mid X_{t+\tau}] \mid X_t \geqslant \bar{X}_t) \\
&= E(E[\beta X_{t+\tau+1} - X_{t+\tau} \mid X_{t+\tau} \\
&\geqslant \bar{X}_{t+\tau}] \mid X_t \geqslant \bar{X}_t) \leqslant 0.
\end{aligned}
$$

## 9. Miscellaneous applications and exercises

(1) Suppose that $\{A_n\}$ is a countable sequence of sets belonging to the $\sigma$-field $\mathscr{F}$. Show that $\cap_n A_n \in \mathscr{F}$. This fact establishes that a $\sigma$-field is closed under the formation of countable intersections. Next, suppose that $A \in \mathscr{F}$ and $B \in \mathscr{F}$, where $\mathscr{F}$ is a $\sigma$-field. Show that $A - B \in \mathscr{F}$.

(2) Let $Z$ be the set of all positive integers and let $\mathscr{A}$ be the class of subsets $A$ of $Z$ such that $A$ or $A^c$ is finite. Is $\mathscr{A}$ a $\sigma$-field? Explain.

(3) Let $R^1$ be the set of real numbers and let $\mathscr{R}^1$ be the $\sigma$-field of Borel sets. The class $\mathscr{R}^1$ contains all the open sets and all the closed sets on the real line. This shows that the class $\mathscr{R}^1$ of Borel sets is sufficiently large. However, the reader must be warned: there do exist sets in $R^1$ not belonging to $\mathscr{R}^1$. For such an example see Billingsley (1979, section 3).

(4) Suppose that $\Omega = \{1, 2, 3, \dots\}$, i.e. $\Omega$ is the class of positive integers and $\mathscr{F}$ is the $\sigma$-field consisting of all subsets of $\Omega$. Define $\mu(A)$ as the number of points, i.e. integers, in $A$ and $\mu(A) = \infty$ if $A$ is an infinite set. Then $\mu$ is a measure on $\mathscr{F}$; it is called the *counting measure*. Next, consider $(\Omega, \mathscr{F})$ as above and let $p_1, p_2, \dots$ be non-negative numbers corresponding to the set of positive integers such that $\Sigma_i p_i = 1, i = 1, 2, \dots$. Define

$$\mu(A) = \sum_{x_i \in A} p_i.$$

Then $\mu$ is a probability measure.

(5) Probability as a special kind of a measure satisfies several useful properties. Here are some such properties. Let $(\Omega, \mathcal{F}, P)$ be a probability space.

(a) If $A \in \mathcal{F}$ and $B \in \mathcal{F}$ and $A \subset B$ then $P(A) \leqslant P(B)$. This is the *monotonicity property*.

(b) If $A \in \mathcal{F}$ then $P(A^c) = 1 - P(A)$.

(c) If $\{A_n\}$ is a countable sequence of sets in $\mathcal{F}$ then

$$P\left( \underset{n}{\cup} A_n \right) \leqslant \sum_n P(A_n).$$

This is called the countable *subadditivity property* or *Boole's inequality*.

(d) If $\{A_n\}$ is an increasing sequence of sets in $\mathcal{F}$ having $A$ as its limit then $P(A_n)$ increases to $P(A)$. For proofs of these statements see Billingsley (1979, section 2) and Tucker (1967, pp. 6–7).

(6) Suppose $X$ and $Y$ are random variables defined on the same space $(\Omega, \mathcal{F}, P)$. Then

(a)  $cX$ is a random variable for $c \in R$;

(b)  $X + Y$ is a random variable provided $X(\omega) + Y(\omega) \neq \infty - \infty$ for each $\omega$;

(c)  $XY$ is a random variable provided $X(\omega) Y(\omega) \neq 0 \cdot \infty$ for each $\omega$; and

(d)  $X/Y$ is a random variable provided $X(\omega)/Y(\omega) \neq \infty / \infty$ for each $\omega$.

(7) Let $X$ be a random variable on $(\Omega, \mathcal{F}, P)$ with finite expectation $E(X)$ and finite variance var $X$. Then for $k$ such that $0 < k < \infty$ we have

$$P[\omega : |X(\omega) - E(X(\omega))| \geqslant k] \leqslant \frac{\text{var } X}{k^2}.$$

This is called *Chebyshev's inequality* and it indicates that a random variable with small variance is likely to take its values close to its mean.

(8) Let $X$ be a random variable on $(\Omega, \mathcal{F}, P)$ with finite expectation and suppose that $\phi$ is a convex real-valued function such that $E(\phi(X)) < \infty$. Then $\phi(E(X)) \leqslant E(\phi(X))$. This is called *Jensen's inequality* and it has found several applications in economic theory. If a consumer prefers risk then his utility function is convex, and Jensen's inequality says that the expected value of his random utility function is greater or equal to the utility of the expected random variable $X$. In this case the consumer will pay to participate in a fair gamble. Suppose that $\phi$ is a concave real-valued function with $E(X) < \infty$ and $E(\phi(X)) < \infty$. Then $\phi(E(X)) \geqslant E(\phi(X))$, which is also called Jensen's inequality, and in economics it may be

used to describe a risk-averse consumer who will avoid a fair gamble by purchasing insurance. Friedman and Savage (1948) have suggested a utility function composed of both concave and convex segments. For further applications of Jensen's inequality in economics see Rothschild and Stiglitz (1970).

(9)  Let $(\Omega, \mathscr{F}, P)$ be a probability space with $A \in \mathscr{F}$ such that $P(A) > 0$. Let $\{B_n\}$ be a finite or countable sequence of disjoint events such that $P(\cup_n B_n) = 1$ and $P(B_n) > 0$ for all $n$. Then for every $k$,

$$P[B_k \mid A] = P[A \mid B_k] P[B_k] / \sum_n P[A \mid B_n] P[B_n].$$

This is called *Bayes' theorem*. Using conditional probability this theorem may be stated as follows:

$$P[B \mid A] = \int_B P[A \mid \mathscr{B}] \, dP / \int_\Omega P[A \mid \mathscr{B}] \, dP$$

for $B \in \mathscr{B}$, where $\mathscr{B}$ is a $\sigma$-field in $\mathscr{F}$.

(10)  Theorems 4.2, 4.3 and 4.4 can be extended for conditional expectation. They are as follows.

(a)  **Conditional form of monotone convergence theorem.**  Let $\{X_n\}$ be an increasing sequence of non-negative random variables on $(\Omega, \mathscr{F}, P)$ and let $X_n(\omega) \to X(\omega)$ w.p.1 and assume that $E(X) < \infty$. Then for $\mathscr{G}$ a $\sigma$-field in $\mathscr{F}$,

$$E[X_n \mid \mathscr{G}] \to E[X \mid \mathscr{G}]$$

w.p.1 as $n \to \infty$.

(b)  **Conditional form of Fatou's lemma.**  If $\{X_n\}$ is a sequence of non-negative random variables on $(\Omega, \mathscr{F}, P)$ with finite expectations and if $E(\liminf_n X_n) < \infty$ as $n \to \infty$, then

$$E[\liminf_n X_n \mid \mathscr{G}] \leqslant \liminf_n E[X_n \mid \mathscr{G}]$$

w.p.1 as $n \to \infty$, for $\mathscr{G}$ a $\sigma$-field in $\mathscr{F}$.

(c)  **Conditional form of dominated convergence theorem.**  Let $\{X_n\}$ be a sequence of random variables and $Y$ an integrable random variable, all on $(\Omega, \mathscr{F}, P)$ such that $\mid X_n \mid \leqslant Y$ w.p.1 for all $n$. If $X_n \to X$ w.p.1 and if $\mathscr{G}$ is a $\sigma$-field in $\mathscr{F}$, then

$$E[X_n \mid \mathcal{G}] \to E[X \mid \mathcal{G}]$$

w.p.1 as $n \to \infty$.

Proofs of these three theorems may be found in Tucker (1967, pp. 215–216).

(11) Consider the sequence of discounted expected values of a stock, $v_{t+T}$, $T = 1, 2, \ldots$, described in application (3) of section 6 and assume that

$$r \geqslant E[x_{t+T} \mid \mathcal{F}_t]/v_t.$$

It is natural for a stockholder to want to know the probability that the maximum value of $v_{t+T}$ will exceed $\$\lambda$, where $\lambda$ is some positive number. Find conditions on $v_{t+T}$ which will allow the investor to compute $P[\omega: \max_T v_{t+T} \geqslant \lambda]$. Also, find conditions on $v_{t+T}$ that will allow $v_{t+T}$ to converge to, say, $v$ as $T \to \infty$, i.e. some random variable denoting a discounted expected value of the stock.

(12) Consider the *Wiener process with drift* $\{w_t, t \geqslant 0\}$ given by $w_t = z_t + \mu t$, $\mu \neq 0$, where $\{z_t, t \geqslant 0\}$ is a standard Wiener process. Compute that the increment $w_t - w_s$ for $0 \leqslant s < t$ has mean $\mu(t-s)$ and variance $t-s$.

(13) Suppose that $\{z_t, t \geqslant 0\}$ is a Wiener process. Show that $\{(z_t, \mathcal{F}_t), t = 0, 1, 2, \ldots\}$ is a martingale, where $\mathcal{F}_t = \sigma(z_0, z_1, z_2, \ldots, z_t)$.

(14) The job search application of section 8 can be extended to allow for discounting. Arguing as in the case with no discounting, when the job searcher seeks one more job offer, his expected return is $\beta[E(\max\{V, Y_1\} - c)]$, which implies a reservation wage $V$ which is the solution to

$$V = \beta[E(\max\{V, Y_1\} - c)],$$

where $\beta = 1/(1 + r)$, with $r$ being an interest rate. Observe that as the last equation illustrates, we assume that the search cost is incurred at the end of the period and that the wage offer is also received at the end of the period. Study this application to conclude that as the interest rate increases the reservation wage rate will decrease. For details see Lippman and McCall (1976a).

## 10. Further remarks and references

There are several good textbooks on probability. The material presented in this chapter, and particularly in sections 2, 3, 4 and 5, may be found in Ash (1972), Billingsley (1979), Chung (1974), Loève (1977), Neveu (1965), Papoulis (1965) and Tucker (1967) among other textbooks. Loève is a classic textbook now in its fourth edition. Tucker and Neveu are concise with Neveu having more advanced material than Tucker. Ash develops measure and integration theory, functional analysis, and topology in the first part of his book. He then applies these concepts in probability theory. On the other hand, Billingsley (1979) emphasizes the interplay between analysis and probability with the analytical issues being motivated by probabilistic questions. The standard reference on the various results on convergence is Billingsley (1968). For a brief introduction to some probabilistic concepts with applications in microeconomic theory, see McCall (1971).

It may be noted that probability theory began during the seventeenth century with Pascal and Fermat among others, when problems which arose in games of chance were formulated in mathematical terms. Modern probability theory, as it is essentially known today, was given a solid mathematical foundation by Kolmogorov in 1933. His path-breaking work was translated from Russian into English in 1950. See Kolmogorov (1950).

Section 6 on martingales reports definitions, examples and theorems from Ash (1972, section 7.3), Billingsley (1979, section 35), Doob (1953, ch. VI), Meyer (1966, part B) and Neveu (1975). The two classic sources for martingales used to be Doob (1953) and Meyer (1966) with the latter book making extensive use of the first book. To these two classics we must add Neveu (1975) and the updated Meyer and Dellacherie (1978). For a brief introduction to the subject see Doob's descriptive article in the *American Mathematical Monthly*, i.e. see Doob (1971) or Feller (1971). Karlin and Taylor (1975, ch. 6) also have a nice discussion on martingales. Note that Doob (1953) uses the terminology semimartingale and lower semimartingale in lieu of submartingale and supermartingale, respectively.

The economics and finance applications in section 6 are based on Samuelson's (1965, 1973) two papers and Hall's (1978) paper. See also MaCurdy (1978) and Malliaris (1981). The reader who is interested in a more detailed survey is referred to Fama (1970). Grossman and Stiglitz (1980) offer some constructive comments on the subject of efficient markets and proceed to redefine the notion of efficiency. A nontechnical presentation of the concepts of information, martingales, and prices is Alchian (1974) and a recent survey of macroeconomic martingales and econometric tests of the martingale property is O'Neill (1978).

Although in the application of futures pricing as a martingale we followed Samuelson (1965), another paper of significance is Mandelbrot (1966). Also, see the earlier paper by Houthakker (1961). It is safe to argue that the papers by Samuelson and Mandelbrot are among the primary sources which generated the recent interest in the martingale concept. Mandelbrot credits Bachelier's (1900) doctoral dissertation in mathematics on "Théorie de la Spéculation" as the first work in this area. Actually, Bachelier discovered the theory of Brownian motion five years before Einstein. For historical interest we mention that the name martingale is due to Ville (1939).

The martingale property has been tested not only for stock market prices but also elsewhere; for interest rates see Roll (1970), Sargent (1972, 1976), and Modigliani and Shiller (1973); for exchange rates see Cornell (1977); for price expectations see Mullineaux (1978) and McNees (1978). In the theory of continuous trading, Harrison and Pliska (1981) show that a security market is complete if and only if its vector price process has a certain martingale representation property.

The economics and finance literature related to the martingale property demonstrates an important interplay of several concepts: martingale property, random walk, market efficiency, rational expectations, arbitraged prices, stabilizing speculation, and statistical dependence among prices. Several theoretical papers attempt to clarify the precise relationship among these concepts. Among these papers we mention Fama (1970), Mandelbrot (1971), Danthine (1977, 1978), Shiller (1978) and Lucas (1978).

The basic reference on stochastic processes is Doob's (1953) classic book. Also, an exhaustive analysis of stochastic processes may be found in Gihman and Skorohod's (1974, 1975, 1979) three volumes. These are advanced books and for the reader who is interested in an introduction to this subject we suggest Karlin and Taylor (1975), Cox and Miller (1965), Prabhu (1965), Çinlar (1975), and Hoel, Port and Stone (1972) among others. The books of Billingsley (1979) and Tucker (1967) among several other textbooks have a chapter on stochastic processes.

Historically, the stochastic process which was first investigated in some detail is the Brownian motion. The English botanist R. Brown observed in 1827 that small particles immersed in a liquid exhibit ceaseless irregular motions. Much later, in 1905, Einstein described the process mathematically by postulating that the particles under observation are subject to perpetual collision with the molecules of the surrounding medium. The concise mathematical formulation of the theory of Brownian motion was given by Wiener in 1918. In this book and in many others the terms Brownian motion process and Wiener process are used interchangeably. In the past, a distinction was made between a Brownian motion process, which was one satisfying properties (1), (2) and (3) of the definition, and a Wiener process which was one satisfying conditions (1), (2), (3) and (4) of the definition.

The Markov chain is due to the Russian mathematician A.A. Markov who developed his ideas in 1907 in an effort to solve a problem originally posed by D. Bernoulli in 1769. Other mathematicians who contributed to the development of Markov chains and Markov processes are A. Kolmogorov and W. Feller. For some basic references the reader is referred to Karlin and Taylor (1975), Bharucha-Reid (1960), Cox and Miller (1965), Çinlar (1975), Hoel, Port and Stone (1972) and Feller (1968). For a more advanced and complete treatment see Dynkin (1965).

For an introductory exposition on optimal stopping the reader is encouraged to consult Robbins' (1970) descriptive paper in the *American Mathematical Monthly* and also Breiman (1964). As already mentioned in section 8, we used extensively the two books by DeGroot (1970) and by Chow, Robbins and Siegmund (1971). In these books several important results are presented which were first discovered by Chow and Robbins (1967), DeGroot (1968), Dvoretzy (1967), Yahav (1966) and several other mathematicians. In section 8 we did not present problems of optimal stopping related to Markov chains and processes. This important subject is covered in Çinlar (1975) and the classic book by Dynkin and Yushkevich (1969). At a more advanced level we recommend Shiryayev (1978) and Ruiz (1968). Actually, Shiryayev (1978, ch. 2) can be used to fully justify the optimal rule stated immediately after eq. (8.33). A useful book on optimal stopping, written essentially for students interested in business applications, is Leonardz (1974). For many applications of optimal stopping in job search models the standard reference is Lippman and McCall (1976a, 1976b, 1979).

We conclude by remarking that our original intentions were to present a much smaller subset of this chapter as an appendix. As the material converged to its current size we decided that a chapter is more appropriate organizationally than a long appendix. We make this remark to encourage the reader to use Chapter 1 as a reference or as a foundation chapter. Clearly, the main subject matter of this book is treated in Chapters 2, 3 and 4.

# CHAPTER 2

# STOCHASTIC CALCULUS

> The doctrine that knowledge of matters
> of fact is only probable is one of the cen-
> tral theses of contemporary analysis of
> scientific method.
>
> Nagel (1969, p. 5)

## 1. Introduction

This chapter presents various mathematical methods of stochastic calculus that are useful in modeling and analyzing the behavior of economic and financial phenomena under uncertainty. Such methods include Itô's lemma, stochastic differential equations, stochastic stability and stochastic control.

The study of such methods necessitates the presentation of various mathematical concepts and results, such as the notion of a stochastic integral, the properties of solutions of stochastic differential equations, various approaches to stochastic stability, and so on. These concepts and results are also presented in this chapter to accustom the reader to the theoretical underpinnings of stochastic calculus.

## 2. Modeling uncertainty

In modeling, analyzing and predicting aspects of economic reality, researchers are placing greater and greater emphasis upon stochastic methods. Such methods are expected to capture the various complexities, measurement errors and uncertainties that are associated with economic reality. The question that arises naturally is: How can combinations of complexity, uncertainty and ignorance, which are present in the process of economic theorizing, be incorporated into dynamic analysis? An answer to this question is described in this section, first for the discrete time case and next for the continuous time case. This analysis follows Åström (1970).

## 2.1.  Discrete time

Consider the discrete time index set $T$ consisting of the set of positive integers $\{0, 1, 2, 3, ...\}$ and let $x(t)$ denote a real state variable at time $t$, $t \in T$. A *dynamic deterministic system* may be described by the following difference equation:

$$x(t+1) = f(t, x(t)), \quad t \in T, \tag{2.1}$$

with initial condition $x(0) = x_0$. Needless to say, such deterministic difference equation models abound in economic theory. Consider, for example, the three-sector macroeconomic model,

$$Y_t = C_t + I_t + G_t,$$
$$C_t = a_0 + a_1 Y_{t-1}.$$

For given values of $I_t$ and $G_t$, denoted as $\bar{I}_t$ and $\bar{G}_t$, the difference equation obtained is

$$Y_t = (a_0 + \bar{I}_t + \bar{G}_t) + a_1 Y_{t-1}. \tag{2.2}$$

Eq. (2.2) describes a special case of eq. (2.1), namely a linear autonomous difference equation.

   In a way similar to the one used by econometricians in modeling uncertainty, we proceed to make the deterministic model of eq. (2.1) into a stochastic model. Recall that a standard way of introducing uncertainty into eq. (2.2) is to add a random variable $U_t$ to (2.2). It is often assumed that $U_t$ is a sequence of independent random variables, normally distributed with mean zero and finite variance. Thus (2.2) becomes

$$Y_t = (a_0 + \bar{I}_t + \bar{G}_t) + a_1 Y_{t-1} + U_t. \tag{2.3}$$

Eq. (2.3) is an example of a linear stochastic difference equation.

   Returning to eq. (2.1), one way it can be converted to incorporate uncertainty is to assume that $x(t+1)$ is a random variable having the expression

$$x(t+1) = f(t, x(t)) + v(t, x(t)), \quad t \in T, \tag{2.4}$$

where $f$ is the *conditional mean* of the random variable $x(t+1)$, conditioned upon $x(t)$, and $v$ is a random variable with mean zero and finite variance, denoted by $\sigma^2(t, x)$. We assume that the conditional distribution of the random variable $v$, given $x(t)$, is independent of $x(s)$ for $s < t$. Eq. (2.4) describes a stochastic difference equation and on the basis of the assumption of independence we conclude that the process $\{x(t), t \in T\}$ is a Markov process.

Next, suppose that the conditional distribution of $v$, given $x(t)$, is normal so that the random variable $u(t)$, where $u(t) = v(t)/\sigma(t, x)$, is normally distributed with mean zero and unit variance. Note that $u(t)$ is independent of $x$ and $\{u(t), t \in T\}$ can be taken to be a sequence of independent, identically distributed normal variables with mean 0 and unit variance. Eq. (2.4) thus becomes

$$x(t+1) = f(t, x(t)) + \sigma(t, x(t))u(t), \quad t \in T. \tag{2.5}$$

Eq. (2.5) is a *stochastic difference equation*. It is studied in several books such as Åström (1970) and Chow (1975).

## 2.2. Continuous time

Consider a *dynamic deterministic model* described by a differential equation

$$\frac{dx}{dt} = f(t, x(t)) \tag{2.6}$$

for $t \in T$, where $T = [0, T]$ or $T = [0, \infty)$, and with initial condition $x(0) = x_0$. Again, there are many examples of economic models described by differential equations such as (2.6); for example, consider

$$\frac{dk}{dt} = sf(k(t)) - nk(t), \quad k(0) = k_0, \tag{2.7}$$

i.e. Solow's differential equation of neoclassical growth analyzed in Solow (1956). Note that (2.7) is a special case of (2.6) because it is autonomous or, in other words, time-dependent.

To introduce uncertainty into (2.6) we proceed as follows. Consider the model at time $t$ and at time $t + \Delta t$; the deterministic equation (2.6) is obtained from

$$x(t + \Delta t) - x(t) = f(t, x(t)) \Delta t + o(\Delta t) \tag{2.8}$$

by dividing both sides by $\Delta t$ and taking limits as $\Delta t \to 0$. In (2.8) note that $o(\Delta t)$ means that $o(\Delta t)/\Delta t \to 0$ as $\Delta t \to 0$. Now observe that (2.8) is a difference equation and we may repeat the procedure followed in the discrete case. Let $\{v(t), t \in T\}$ be a stochastic process with independent increments, i.e. as in section 7 of Chapter 1 we assume that if $H_i \in \mathcal{R}$ and if $0 \leqslant t_0 \leqslant t_1 \leqslant ... \leqslant t_k$, then

$$P[v(t_i) - v(t_{i-1}) \in H_i, \ i \leqslant k] = \prod_{i \leqslant k} P[v(t_i) - v(t_{i-1}) \in H_i].$$

Eq. (2.8) now becomes

$$x(t + \Delta t) - x(t) = f(t, x(t)) \Delta t + v(t + \Delta t) - v(t) + o(\Delta t). \tag{2.9}$$

Assume that the conditional distribution of the increment of $v$, given $x(t)$, is normally distributed and write

$$v(t + \Delta t) - v(t) = \sigma(t, x(t)) [z(t + \Delta t) - z(t)], \tag{2.10}$$

where $\{z(t), t \in T\}$ is a Wiener process with mean zero and unit variance. Substituting (2.10) into (2.9) we obtain

$$x(t + \Delta t) - x(t) = f(t, x(t)) \Delta t + \sigma(t, x(t)) [z(t + \Delta t) - z(t)] + o(\Delta t). \tag{2.11}$$

In this last equation the term $o(\Delta t)$ is a random variable and usually denotes that $E |o(\Delta t)|^2/\Delta t \to 0$ as $\Delta t \to 0$. At this point observe that in (2.11) we are not able to divide both sides by $\Delta t$ and take limits as we did in (2.8). This is so

$$\lim \frac{z(t + \Delta t) - z(t)}{\Delta t}$$

does not exist. Recall theorem 7.2 of Chapter 1 which says that a Wiener process is nowhere differentiable outside some set of probability zero. Thus, instead of dividing (2.11) by $\Delta t$ and taking limits in the usual sense we only take limits as $\Delta t \to 0$ in a sense to be explained in the next section. In conclusion (2.11) can be written in the formal expression

$$dx = f(t, x) dt + \sigma(t, x) dz, \tag{2.12}$$

which is called *Itô's stochastic differential equation* with initial condition $x(0) = x_0$ and $t \in T$.

Equation (2.12) plays an important role in applied mathematics. See the books by Schuss (1980), Soong (1973) and Tsokos and Padgett (1974). For a brief historical evolution of eq. (2.12) since 1908, see Wonham (1970). It also plays an important role in economic theory and finance, as the next two chapters will illustrate. Because of the significance of eq. (2.12) in continuous stochastic state models in economics and finance, we proceed to give meaning to the formal expression in (2.12).


## 3. Stochastic integration

Consider a probability space $(\Omega, \mathscr{F}, P)$ on which both a stochastic process $x(t, \omega)$ and a Wiener process $z(t, \omega)$ are defined for $\omega \in \Omega$ and $t \in T$. Eq. (2.12) is a shorthand notation for

$$\mathrm{d}x(t, \omega) = f(t, x(t, \omega))\,\mathrm{d}t + \sigma(t, x(t, \omega))\,\mathrm{d}z(t, \omega), \tag{3.1}$$

which can also be written as

$$\mathrm{d}x(t) = f(t, x(t))\mathrm{d}t + \sigma(t, x(t))\,\mathrm{d}z(t) \tag{3.2}$$

or

$$\mathrm{d}x_t = f(t, x_t)\,\mathrm{d}t + \sigma(t, x_t)\,\mathrm{d}z_t, \tag{3.3}$$

where $\omega$ is being suppressed in (3.2) and (3.3).

Transforming eq. (3.2) into an *integral equation* we obtain

$$x(t) = x(0) + \int_0^t f(s, x(s))\mathrm{d}s + \int_0^t \sigma(s, x(s))\,\mathrm{d}z(s). \tag{3.4}$$

Note that $x(0)$ is a random variable since $x(0) = x(0, \omega)$ for $\omega \in \Omega$, which could degenerate into a constant. As a rule, the first integral in the right-hand side of (3.4) can be understood as a Riemann integral. The second integral is more of a problem because $\mathrm{d}z(t)$ does not exist. In other words, although $z(t)$ is continuous, it is a function of unbounded variation and the second integral cannot be interpreted as a Riemann–Stieljes integral. We need, therefore, to present a theory to make the second integral meaningful.

Stochastic integration was developed by Itô (1944) as he generalized a stochastic integral first introduced by Wiener in 1923. Parts of Itô's original work were presented, initially by Doob (1953), and later by the Russian mathematicians Gihman and Skorohod (1969). However, it is only recently that the ideas of stochastic integration, and stochastic calculus in general, have become accessible to the applied researcher by the publication of the books by Arnold (1974), Åström (1970) Gihman and Skorohod (1972), Bharucha–Reid (1972), and Ladde and Lakshmikantham (1980) among others.

In the analysis below we first give an intuitive analysis; we then proceed to a detailed exposition on stochastic integration.

## 3.1. Intuitive analysis

Consider the meaning of the ordinary differential equation (2.6) written in the form

$$\mathrm{d}x = f(t,x)\,\mathrm{d}t; \quad x(0) = x_0; \quad t \in T = [0, T],$$ (3.5)

we say that $y(s)$ is a solution of (3.5) if

$$\frac{\mathrm{d}y(s)}{\mathrm{d}s} = f(s, y(s))$$ (3.6)

for each $s \in T$. Or, what is the same thing, upon integrating both sides of (3.6) we say that $y(s)$ solves (3.5) if, for all times $t, s, s \leqslant t$,

$$y(t) - y(s) = \int_s^t f(r, y(r))\,\mathrm{d}r.$$ (3.7)

Now note that the right-hand side of (3.7) is an integral which can be approximated by dividing the time interval $[s, t]$, which is a subset of $[0, T]$, into small subintervals. Let $\epsilon > 0$ be given and consider the *partition*

$$s = t_0 < t_1 < t_2 < ... < t_n = t$$

$$\max_{0 \leqslant i \leqslant n-1} |t_{i+1} - t_i| \leqslant \epsilon, \quad [t_i, t_{i+1}) \subset [s, t].$$ (3.8)

We can then write:

$$y(t) - y(s) = \sum_{i=0}^{n-1} f(t_i, y(t_i))(t_{i+1} - t_i) + 0(\epsilon),$$

where $0(\epsilon) \to 0$ as $\epsilon \to 0$. This is a standard approximation procedure for calculating integrals in calculus. Now we follow a similar approach for stochastic calculus. We say that the stochastic process $y(t)$ solves (3.2) for all times $t$, $s$, for $s \leqslant t$, if

$$y(t) - y(s) = \int_s^t f(r, y(r)) \, dr + \int_s^t \sigma(r, y(r)) \, dz(r), \qquad (3.9)$$

where the right-hand side of (3.9) is defined by

$$\int_s^t f(r, y(r)) \, dr + \int_s^t \sigma(r, y(r)) \, dz(r) = \sum_0^{n-1} f(t_i, y(t_i)) \, (t_{i+1} - t_i) +$$

$$+ \sum_0^{n-1} \sigma(t_i, y(t_i)) \, [z(t_{i+1}) - z(t_i)] + 0(\epsilon). \qquad (3.10)$$

In (3.10), $0(\epsilon)$ is a random variable and $0(\epsilon) \to 0$ as $\epsilon \to 0$ means $E \, | \, 0(\epsilon) \, |^2 \to 0$ as $\epsilon \to 0$. Recall that E denotes unconditional expectation. There are some technical mathematical problems in justifying (3.10) and in proving the existence of stochastic processes $y(t)$ that satisfy

$$y(t) - y(s) = \sum_0^{n-1} f(t_i, y(t_i)) \, (t_{i+1} - t_i)$$

$$+ \sum_0^{n-1} \sigma(t_i, y(t_i)) \, [z(t_{i+1}) - z(t_i)] + 0(\epsilon)$$

for partitions of the form of (3.8). These technical aspects are treated in some detail in the next section. Let us proceed, however, to illustrate one possible difficulty. Given functions $g(t, z)$ and $z(t)$ when $\max | \, t_{i+1} - t_i | \to 0$, then the two limits

$$A = \lim_i \sum g(t_i, z(t_i)) \, [z(t_{i+1}) - z(t_i)]$$

and

$$B = \lim_i \sum g(t_{i+1}, z(t_{i+1})) \, [z(t_{i+1}) - z(t_i)]$$

are the same in the certainty case. This is not so in the stochastic case. By way of illustration let us consider the simple case where $g(t, z) = z(t)$, where $z(t)$ is a Wiener process with unit variance, and write as $\epsilon \to 0$:

$$A = \lim_i \sum_i z(t_i) [z(t_{i+1}) - z(t_i)]$$

and

$$B = \lim_i \sum_i z(t_{i+1}) [z(t_{i+1}) - z(t_i)].$$

In the certainty case, i.e. when $z(t)$ is deterministic rather than stochastic, $A$ and $B$ are ordinary Riemann–Stieljes integrals, and if we assume that $z(t)$ is integrable then $B-A \to 0$ as $\epsilon \to 0$. Thus, in the deterministic case there is no ambiguity concerning the concept of convergence. However, dealing with $z(t, \omega)$ as a random variable one must specify which limit concept is being used because there are several concepts of convergence. In the present case the appropriate interpretation of these limits is the following:

$$E \,|\, A - \sum_i z(t_i) [z(t_{i+1}) - z(t_i)] \,|^2 \to 0,$$

$$E \,|\, B - \sum_i z(t_{i+1}) [z(t_{i+1}) - z(t_i)] \,|^2 \to 0,$$

(3.11)

as $\epsilon \to 0$, where E denotes expectation. The limit defined in (3.11) is called the *limit in mean square* and describes one more convergence concept in addition to the other three convergence concepts defined in section 3 of Chapter 1. For $A$ and $B$ defined as in (3.11), it can be shown, using the fact that the Wiener process has independent stationary increments $z(t_{i+1}) - z(t_i)$ whose variance is $t_{i+1} - t_i$, that $B - A = t-s \neq 0$. Therefore we can get a multitude of stochastic integrals depending on whether we take limit expressions of type $A$ or $B$ or a convex combination of these two types. Itô's analysis uses limit expressions of type $A$ and this is the type of stochastic integral that we consider next. For another type of stochastic integral see section 15 at the end of this chapter.

## 3.2.  *Rigorous exposition*

This analysis follows Arnold (1974, pp. 57–75), Bharucha–Reid (1972, pp. 221–228), Gihman and Skorohod (1972, pp. 11–32) and Ladde and Lakshmikantham (1980, pp. 114–122). Our purpose now is to define carefully Itô's stochastic integral and to present its basic properties.

Let $z_t$ be a Wiener process for $t \geqslant 0$, defined on some probability space $(\Omega, \mathcal{F}, P)$. A family of σ-fields $\mathcal{F}_t$ in $\mathcal{F}$ defined for $t \geqslant 0$ is said to be *nonanticipating* with respect to $z_t$ if it satisfies:

(1)  $\mathcal{F}_{t_1} \subset \mathcal{F}_{t_2}$ for $0 \leqslant t_1 \leqslant t_2$;

(2) $\mathscr{F}_t$ contains the $\sigma$-field generated by $z_s$, $0 \leqslant s \leqslant t$, i.e. $\mathscr{F}_t \supset \sigma(z_s, 0 \leqslant s \leqslant t)$ $\equiv \mathscr{Z}_t$; and

(3) $\mathscr{F}_t$ is independent of the $\sigma$-field generated by the increment $z_u - z_t$, $t \leqslant u < \infty$, and denoted by $\sigma(z_u - z_t,\ t \leqslant u < \infty)$.

Note that condition (2) means that $z_t$ is measurable with respect to $\mathscr{F}_t$ for every $t \geqslant 0$. Condition (3) means that for $h = u - t, t \leqslant u < \infty$, the process $z(t+h)$ $- z(h)$ is independent of any of the events of the $\sigma$-field $\mathscr{F}_h$. In particular, condition (3) means that $\mathscr{F}_0$ can contain only events that are independent of the entire process $z_t, t \geqslant 0$.

As an example consider the family $\mathscr{F}_t = \mathscr{Z}_t$. It is the smallest possible non-anticipating family of $\sigma$-fields. It is often desirable to augment $\mathscr{Z}_t$ with other events that are independent of $\sigma(z_u - z_t,\ t \leqslant u < \infty)$, for example events describing initial conditions. In the case of stochastic differential equations we usually take $\mathscr{F}_t = \sigma(\mathscr{Z}_t, c)$, where $c$ is a random variable independent of $\sigma(z_u - z_t)$.

Consider now $\sigma(t, \omega): [0, T] \times \Omega \to R$ which is assumed to be measurable in $(t, \omega)$. It is said to be *nonanticipating* with respect to the family of $\sigma$-fields $\mathscr{F}_t$ if it satisfies the two conditions:

(1) the sample path $\sigma(t, \cdot)$ is $\mathscr{F}_t$-measurable for all $t \in [0, T]$, and

(2) the integral

$$\int_0^T |\sigma(t, \omega)|^2 \, \mathrm{d}t \tag{3.12}$$

is finite w.p.1.

Note that for a function $\sigma(t, \omega)$ with continuous sample paths w.p.1, the last integral is an ordinary Riemann integral. In more general cases (3.12) is taken to be a Lebesgue integral. This definition generalizes to matrix-valued functions, in which case the norm in (3.12) is defined as follows:

$$|\sigma(t, \omega)| = \left( \sum_i \sum_j \sigma_{ij}^2 \right)^{1/2} = (\mathrm{tr}\, \sigma \cdot \sigma')^{1/2},$$

where tr denotes trace and the prime denotes transpose.

A special class of nonanticipating functions is the class of nonanticipating step functions. The nonanticipating function $\sigma(t, \omega)$ is called a *step function* if there exists a partition of the interval $[0, T]$, say

$$0 = t_0 < t_1 < ... < t_n = T, \tag{3.13}$$

such that $\sigma(t, \omega) = \sigma(t_i, \omega)$ for $t \in [t_i, t_{i+1})$ for $i = 0, 1, 2, ..., n - 1$. For such step functions we now define Itô's stochastic integral.

Let $(\Omega, \mathscr{F}, P)$ be a probability space, $\sigma(t, \omega)$: $[0, T] \times \Omega \to R$ a nonanticipating step function for a partition of the form of (3.13) and $z(t, \omega)$: $[0, T] \times \Omega \to R$ a Wiener process. The *stochastic integral* of $\sigma$ with respect to $z$ over the interval $[0, T]$ is a real-valued random variable denoted by $I(\sigma)$ and defined as

$$
\begin{aligned}
I(\sigma) &= I(\sigma(\omega)) = \int_0^T \sigma(t, \omega)\,dz(t, \omega) \\
&= \int_0^T \sigma\,dz = \sum_{i=1}^n \sigma(t_{i-1}, \omega)\,[z(t_i, \omega) - z(t_{i-1}, \omega)] \\
&= \sum_1^n \sigma(t_{i-1})\,[z(t_i) - z(t_{i-1})] \\
&= \sum_0^{n-1} \sigma(t_i)\,[z(t_{i+1}) - z(t_i)].
\end{aligned}
\tag{3.14}
$$

The presence of $\omega \in \Omega$ emphasizes the fact that Itô's integral is a random variable and the omission of $\omega$ is sometimes done for notational convenience. It is important to remark that in the definition of the step function, and consequently in the definition of stochastic integral, the left-hand side endpoints of the subintervals are used for evaluation.

Next, consider an arbitrary nonanticipating function $\sigma(t, \omega)$. Suppose that $\{\sigma_n\}$ is a sequence of nonanticipating step functions. It is natural to ask if the function $\sigma$ could be approximated in some sense by $\{\sigma_n\}$. An affirmative answer is provided by the next lemma.

**Lemma 3.1.** Let $\sigma(t, \omega)$ be a nonanticipating function. Then there exists a sequence of nonanticipating step functions $\{\sigma_n\}$ such that as $n \to \infty$

$$
\int_0^T |\sigma_n(t) - \sigma(t)|^2\,dt \to 0 \quad \text{w.p.1.}
\tag{3.15}
$$

For a proof of this lemma see Arnold (1974, p. 67). The result of this lemma is strong and it implies the weaker result, i.e. as $n \to \infty$,

$$
P\left[ \omega: \int_0^T |\sigma_n(t, \omega) - \sigma(t, \omega)|^2\,dt > \epsilon \right] \to 0.
\tag{3.16}
$$

This follows from the fact that convergence w.p.1 implies convergence in probability, as is indicated in section 3 of Chapter 1. The general existence of Itô's stochastic integral will be established by showing that for a sequence of non-anticipating step functions $\{\sigma_n\}$ satisfying (3.16) the sequence of integrals $\int_0^T \sigma_n \, dz$ converges in probability to $\int_0^T \sigma \, dz$, i.e.

$$\int_0^T \sigma_n \, dz \xrightarrow{P} \int_0^T \sigma \, dz. \tag{3.17}$$

Suppose that $\{\sigma_n\}$ is a sequence of nonanticipating step functions approximating a nonanticipating function $\sigma$ in probability as in (3.16) and observe that

$$\int_0^T |\sigma_n - \sigma_m|^2 \, dt \leqslant 2 \int_0^T |\sigma - \sigma_n|^2 \, dt + 2 \int_0^T |\sigma - \sigma_m|^2 \, dt.$$

From our assumption about $\{\sigma_n\}$ satisfying (3.16), we conclude that

$$\int_0^T |\sigma_n - \sigma_m|^2 \, dt \xrightarrow{P} 0, \tag{3.18}$$

as $n, m \to \infty$. If we show that $\int_0^T \sigma_n \, dz$ is a Cauchy sequence converging in probability, then we will conclude that there exists a random variable to which it converges in probability. A technical lemma is needed in establishing that $\int_0^T \sigma_n \, dz$ is a stochastic Cauchy sequence.

**Lemma 3.2.** Let $\sigma(t, \omega)$ be a nonanticipating step function. Then for all $\epsilon > 0$ and $\delta > 0$,

$$P\left[\left| \int_0^T \sigma(t) \, dz(t) \right| > \delta \right] \leqslant \frac{\epsilon}{\delta^2} + P\left[ \int_0^T |\sigma(t)|^2 \, dt > \epsilon \right].$$

For a proof of this lemma see Arnold (1974, pp. 68–69).

Applying this lemma to $\sigma_n - \sigma_m$ we obtain

$$\limsup_{n,m \to \infty} P\left[ \left| \int_0^T \sigma_n(t) \, dz(t) - \int_0^T \sigma_m(t) \, dz(t) \right| > \delta \right]$$

$$\leqslant \frac{\epsilon}{\delta^2} + \limsup_{n,m \to \infty} P\left[ \int_0^T |\sigma_n(t) - \sigma_m(t)|^2 \, dt > \epsilon \right]. \tag{3.19}$$

From (3.18) and (3.19) we conclude that $\int_0^T \sigma_n\,dz$ is a stochastic Cauchy sequence, i.e.

$$P\left[\left|\int_0^T \sigma_n(t)\,dz(t) - \int_0^T \sigma_m(t)\,dz(t)\right| > \epsilon\right] \to 0,$$

as $n, m \to \infty$. Therefore (3.17) holds and the limit $\int_0^T \sigma\,dz$ is unique w.p.1 and independent of the choice of the sequence of nonanticipating step functions $\{\sigma_n\}$ satisfying (3.16). We summarize our results in

**Lemma 3.3.** Let $\sigma(t, \omega)$ be a nonanticipating function and suppose that the sequence of nonanticipating step functions $\{\sigma_n\}$ approximates $\sigma$ in the sense of (3.16). For each $n$, suppose that $I(\sigma_n)$ is defined by (3.14). Then, as $n \to \infty$ there exists a random variable $\int_0^T \sigma(t, \omega)\,dz(t, \omega)$ such that

$$P\left[\omega: \left|\int_0^T \sigma_n(t, \omega)\,dz(t, \omega) - \int_0^T \sigma(t, \omega)\,dz(t, \omega)\right| > \epsilon\right] \to 0. \quad (3.20)$$

The above analysis leads to a definition of Itô's stochastic integral. It is a definition based on the concept of convergence in probability as in (3.20). Other definitions can be given also based on a different concept of convergence. Shortly we will give such a definition based on convergence in the mean square and refer the reader to further sources on the subject of stochastic integration.

**Definition.** Let $(\Omega, \mathscr{F}, P)$ be a probability space and consider the Wiener process $z(t, \omega)$ and the nonanticipating function $\sigma(t, \omega)$, both defined on $[0, T] \times \Omega$. *Itô's stochastic integral* of $\sigma$ with respect to $z$ over the interval $[0, T]$, i.e.

$$\int_0^T \sigma(t, \omega)\,dz(t, \omega)$$

denoted by $I(\sigma)$, is a random variable, unique w.p.1 and defined as the *limit in probability*, of the stochastic Cauchy sequence

$$\int_0^T \sigma_n(t, \omega)\,dz(t, \omega) \xrightarrow{P} \int_0^T \sigma(t, \omega)\,dz(t, \omega) \equiv I(\sigma)$$

as lemma 3.3 indicates. Here $\{\sigma_n\}$ is a sequence of nonanticipating step functions that approximates $\sigma$ in the sense of convergence in probability, i.e.

$$\int_0^T | \sigma(t,\omega) - \sigma_n(t,\omega)|^2 \, dt \xrightarrow{P} 0.$$

We repeat that $I(\sigma)$ is unique w.p.1 and independent of the choice of the sequence $\{\sigma_n\}$. Uniqueness is proved in Friedman (1975, p. 66).

**An alternative definition.** As we remarked earlier in this section, an alternative definition of Itô's stochastic integral can be given using the concept of *convergence in the mean square* instead of *convergence in probability*. Since convergence in the mean square implies convergence in probability, the definition we are about to give is more general and it yields as a special case the definition already given. The alternative definition is based on the following

**Lemma 3.4.** Let $\sigma(t,\omega)$ be a nonanticipating function such that

$$\int_0^T E| \sigma(t,\omega)|^2 \, dt < \infty. \tag{3.21}$$

Then there exists a sequence of nonanticipating step functions $\{\sigma_n\}$ satisfying for each $n$

$$\int_0^T E| \sigma_n(t,\omega)|^2 \, dt < \infty,$$

and approximating $\sigma$ in the mean square limit, i.e. as $n \to \infty$

$$\int_0^T E | \sigma_n(t) - \sigma(t)|^2 \, dt \to 0.$$

Furthermore, as $n \to \infty$, there exists a random variable $\int_0^T \sigma(t) \, dz(t)$ such that

$$E \left| \int_0^T \sigma_n(t) \, dz(t) - \int_0^T \sigma(t) \, dz(t) \right|^2 \to 0. \tag{3.22}$$

For a proof see Arnold (1974, pp. 71–73) and Gihman and Skorohod (1972, pp. 11–15). Note that in order to obtain this convergence in the mean square we have to assume (3.21) which is more than is assumed in lemma 3.1. The alternative definition based on lemma 3.4 is now similar to the earlier definition with the only difference being the concept of convergence. Specifically, we define $\int_0^T \sigma(t, \omega)\,dz(t, \omega)$ as the mean square limit of the Cauchy sequence $\int_0^T \sigma_n$ $(t, \omega)\,dz(t, \omega)$ as in (3.22). Such a limit exists, is unique and independent of the choice of $\{\sigma_n\}$ because $\int_0^T \sigma_n\,dz$ is a Cauchy sequence satisfying

$$\mathrm{E} \left| \int\limits_0^T \sigma_n(t)\,dz(t) - \int\limits_0^T \sigma_m(t)\,dz(t) \right|^2 \to 0,$$

as $m, n \to \infty$.

This concludes our discussion on the concept of a stochastic integral. Before we state some of the basic properties of stochastic integrals we suggest some additional references for the reader with an interest in this area. They are Meyer (1967), McKean (1969), Metivier and Pellaumail (1980) and Friedman (1975).

Having given a meaning to $\int \sigma\,dz$ we state some of its properties.

### 3.3.  *Properties of Itô's stochastic integral*

Itô's stochastic integral has several useful properties. We summarize some of these properties in the following theorems.

**Theorem 3.1.**  Let $(\Omega, \mathscr{F}, P)$ be a probability space and $z(t, \omega)\colon [0, T] \times \Omega \to R$ a Wiener process.

(1)  If $\sigma_1\ (t, \omega)$ and $\sigma_2\ (t, \omega)$ are real-valued, nonanticipating functions defined on $[0, T] \times \Omega$ and $a_1 \in R$ and $a_2 \in R$, then

$$\int\limits_0^T (a_1\sigma_1 + a_2\sigma_2)\,dz = a_1 \int\limits_0^T \sigma_1\,dz + a_2 \int\limits_0^T \sigma_2\,dz.$$

(2)  Assume that $\sigma_1$ and $\sigma_2$ are as in (1) and let $a_1\ (\omega)$ and $a_2\ (\omega)$ be real-valued random variables such that $a_1\sigma_1 + a_2\sigma_2$ is a nonanticipating function on $[0, T]$. Then

$$\int\limits_0^T (a_1\sigma_1 + a_2\sigma_2)\,dz = a_1 \int\limits_0^T \sigma_1\,dz + a_2 \int\limits_0^T \sigma_2\,dz.$$

(3)  Let $[s, u]$ be a subset of $[0, T]$ and let $\chi_{[s,\ u]}$ denote the characteristic function of $[s, u]$. Then

$$\int_0^T \chi_{[s,u]} \, dz = z(u) - z(s).$$

(4) Suppose that $\sigma(t, \omega)$ is a nonanticipating function on $[0, T]$ such that $\int_0^T E |\sigma(s)|^2 \, ds < \infty$. Then $E \int_0^T \sigma(t) \, dz(t) = 0$ and

$$E \left| \int_0^T \sigma(t) \, dz(t) \right|^2 = \int_0^T E |\sigma(t)|^2 \, dt.$$

For proofs of these propositions see Gihman and Skorohod (1972, pp. 11–13). Note that properties (1)–(3) are similar to the properties of the ordinary Riemann–Stieljes integral.

Convergence in probability is established next for a sequence of nonanticipating *arbitrary* as opposed to step functions $\{\sigma_n\}$ approximating $\sigma$ in probability.

**Theorem 3.2.** Let $(\Omega, \mathscr{F}, P)$ be a probability space, $z(t, \omega): [0, T] \times \Omega \to R$ a Wiener process, $\{\sigma_n\}$ a sequence of arbitrary real-valued nonanticipating functions defined on $[0, T] \times \Omega$, and $\sigma(t, \omega): [0, T] \times \Omega \to R$ a nonanticipating function. Suppose that as $n \to \infty$

$$\int_0^T |\sigma_n(t) - \sigma(t)|^2 \, ds \overset{P}{\to} 0. \tag{3.23}$$

Then

$$\int_0^T \sigma_n \, dz \overset{P}{\to} \int_0^T \sigma \, dz, \tag{3.24}$$

where in both (3.23) and (3.24) we have convergence in probability.

For a proof see Arnold (1974, p. 74).

The last theorem we want to present is about Itô's stochastic integral as a function of its upper limit. Suppose that $\chi_{[0,t]}$ is the characteristic function of $[0, t]$ with $[0, t] \subset [0, T]$. Suppose that $\sigma$ is a nonanticipating function on $[0, t]$ for each $t$, such that $0 \leqslant t \leqslant T$, where $T$ is arbitrarily large. Define the process

$$x(t) = x(t, \omega) = \int_0^t \sigma(s, \omega) \, dz(s, \omega) = \int_0^T \sigma(s) \chi_{[0,t]} \, dz(s). \tag{3.25}$$

Note that $x(t)$ is a real-valued stochastic process defined uniquely up to stochastic equivalence for $t \in [0, T]$, with $x(0) = 0$ w.p.1. The following theorem states some of the properties of $x(t)$ under the assumption that we have selected a *separable version* of $x(t)$. (See section 7 of Chapter 1.)

**Theorem 3.3.** Let $(\Omega, \mathcal{F}, P)$ be a probability space, $z(t, \omega)$: $[0, T] \times \Omega \to R$ a Wiener process, $\sigma(t, \omega)$: $[0, T] \times \Omega \to R$ a nonanticipating function and define $x(t), t \in [0, T]$ as in (3.25). Then:

(1)  $x(t, \omega)$ is $\mathcal{F}_t$-measurable and thus nonanticipating.

(2)  If $\int_0^t E \mid \sigma(s, \omega) \mid^2 ds < \infty$ for all $t \in [0, T]$, then $(x_t, \mathcal{F}_t)$ is a real-valued martingale. Also,

$$Ex_t = 0 \quad \text{and} \quad E \mid x_t \mid^2 = \int_0^t E \mid \sigma(s, \omega) \mid^2 ds. \tag{3.26}$$

(3)  $x(t, \omega)$ has continuous sample paths w.p.1.

See Arnold (1974, pp. 81–84). Note that since $(x_t, \mathcal{F}_t)$ is a martingale, $(x_t - x_r, \mathcal{F}_t)$ is a martingale also for $0 \leqslant r \leqslant t \leqslant T$ and $(\mid x_t - x_r \mid, \mathcal{F}_t)$ is a submartingale. Therefore we may apply the various theorems on martingales and submartingales available to us to obtain useful results about $(x_t, \mathcal{F}_t)$ and $(x_t - x_r, \mathcal{F}_t)$.

The above analysis concludes our discussion on stochastic integration. For a more detailed exposition of this topic the reader is referred to McKean (1969) and the more recent book by Metivier and Pellaumail (1980).

The reader should also be informed that at a more advanced level a new theory of stochastic integration has been developed called the Strasbourg approach. The primary reference here is Meyer (1976). For a fascinating application of the Strasbourg approach to economic analysis see Harrison and Pliska (1981). They use the Strasbourg approach in developing a theory of a frictionless security market with continuous trading.

## 4.  Itô's lemma

We are now ready to state the definition of stochastic differential and then state and prove Itô's lemma which is the basic stochastic calculus rule for computing stochastic differentials of composite random functions.

Consider a probability space $(\Omega, \mathcal{F}, P)$, a stochastic process $x(t, \omega)$: $[0, T] \times \Omega \to R$ that is measurable for each $t \in [0, T]$ with respect to $\mathcal{F}_t$, and a Wiener process $z(t, \omega)$: $[0, T] \times \Omega \to R$. Assume that $\sigma(t, \omega)$: $[0, T] \times \Omega \to R$ is a nonanticipating function on $[0, T]$ and also that $f(t, \omega)$: $[0, T] \times \Omega \to R$ is measurable for each $t \in [0, T]$ with respect to $\mathscr{F}_t$ and also that $\int_0^T | f(t, \omega) | \, dt < \infty$ w.p.1. Let $0 \leqslant r \leqslant s \leqslant T$ and suppose

$$x(s) - x(r) = \int_r^s f(t, \omega) \, dt + \int_r^s \sigma(t, \omega) \, dz(t, \omega).$$

The *stochastic differential of the process* $x(t)$ is defined to be the quantity $f(t) \, dt + \sigma(t) \, dz(t)$ and is denoted as $dx(t)$, i.e.

$$dx(t) = f(t) \, dt + \sigma(t) \, dz(t).$$

Next we state and prove Itô's lemma which originally appeared in Itô (1951a) and later in Itô (1961). In our presentation below we follow Arnold (1974, pp. 96–99), Gihman and Skorohod (1969, pp. 387–389) and Ladde and Lakshmikantham (1980, pp. 122–126).

**Lemma 4.1** (Itô). Let $u(t, x)$: $[0, T] \times R \to R$ be a continuous nonrandom function with continuous partial derivatives $u_t$, $u_x$ and $u_{xx}$. Suppose that $x(t) = x(t, \omega)$: $[0, T] \times \Omega \to R$ is a process with stochastic differential

$$dx(t) = f(t) \, dt + \sigma(t) \, dz(t).$$

Let $y(t) = u(t, x(t))$. Then the process $y(t)$ has also a differential on $[0, T]$ given by

$$
\begin{aligned}
dy(t) &= [u_t(t, x(t)) + u_x(t, x(t)) f(t) + \tfrac{1}{2} u_{xx}(t, x(t)) \sigma^2(t)] \, dt \\
&\quad + u_x(t, x(t)) \sigma(t) \, dz(t).
\end{aligned}
$$

*Proof.* It is sufficient to prove the lemma for step functions $f$ and $\sigma$ because the general case may be obtained by taking limits. Also, it is sufficient to prove the lemma for a subinterval where $f$ and $\sigma$ are constant as functions of $t$, i.e. $f(t, \omega) = f(\omega)$ and $\sigma(t, \omega) = \sigma(\omega)$.

Consider the partition

$$0 < t_1 < t_2 < \ldots < t_n = t \leqslant T.$$

Since $y(t) = u(t, x(t))$ with $y(0) = u(0, x(0))$, then

$$y(t) - y(0) = \sum_{k=1}^n [u(t_k, x(t_k)) - u(t_{k-1}, x(t_{k-1}))]. \tag{4.1}$$

Using Taylor's theorem we can write

$$
\begin{aligned}
u(t_k, x(t_k)) - u(t_{k-1}, x(t_{k-1})) =\ & u_t(t_{k-1} + \theta_k(t_k - t_{k-1}), x(t_{k-1}))(t_k - t_{k-1}) \\
& + u_x(t_{k-1}, x(t_{k-1}))(x(t_k) - x(t_{k-1})) \\
& + \tfrac{1}{2} u_{xx}(t_{k-1}, x(t_{k-1}) + \overline{\theta}_k[x(t_k) - \\
& - x(t_{k-1})])(x(t_k) - x(t_{k-1}))^2, \qquad (4.2)
\end{aligned}
$$

where $0 < \theta_k < 1$ and also $0 < \overline{\theta}_k < 1$. The first observation we want to establish is that the limit of $y(t) - y(0)$ does not change in (4.1) as $\delta_n \to 0$. From the continuity assumption of $x(t)$, $u_t$ and $u_{xx}$, and the boundedness of $x(t)$, we conclude that there are random variables $\alpha_n$ and $\beta_n$ that approach 0 w.p.1 as $\delta_n = \max(t_k - t_{k-1}) \to 0$, satisfying the two inequalities

$$
|u_t(t_{k-1} + \theta_k(t_k - t_{k-1}), x(t_{k-1})) - u_t(t_{k-1}, x(t_{k-1}))| \leqslant \alpha_n
$$

and

$$
|u_{xx}(t_{k-1}, x(t_{k-1}) + \overline{\theta}_k[x(t_k) - x(t_{k-1})]) - u_{xx}(t_{k-1}, x(t_{k-1}))| \leqslant \beta_n.
$$

Taking summations in the first inequality above we obtain

$$
\begin{aligned}
|\Sigma u_t(t_{k-1} & + \theta_k(t_k - t_{k-1}), x(t_{k-1}))(t_k - t_{k-1}) - \\
& - \Sigma u_t(t_{k-1}, x(t_{k-1}))(t_k - t_{k-1})| \leqslant \Sigma \alpha_n(t_k - t_{k-1}).
\end{aligned}
$$

For the second inequality we remark that

$$
\Sigma [x(t_k) - x(t_{k-1})]^2 \leqslant 2f^2 \, \Sigma(t_k - t_{k-1})^2 + 2\sigma^2 \Sigma [z(t_k) - z(t_{k-1})]^2
$$

and also that

$$
E \Sigma [z(t_k) - z(t_{k-1})]^2 = t - 0 = t.
$$

Therefore, taking summations and using the results from the last two equations we conclude

$$
\begin{aligned}
|\Sigma u_{xx}&(t_{k-1}, x(t_{k-1}) + \overline{\theta}_k[x(t_k) - x(t_{k-1})]) \, [x(t_k) - x(t_{k-1})]^2 \\
& - \Sigma u_{xx}(t_{k-1}, x(t_{k-1})) \, [x(t_k) - x(t_{k-1})]^2 | \\
& \leqslant \beta_n \, \Sigma [x(t_k) - x(t_{k-1})]^2 \xrightarrow{P} 0,
\end{aligned}
$$

as $\delta_n \to 0$. Thus, the limit of $y(t) - y(0)$ in (4.1) does not change as $\delta_n \to 0$.

Putting (4.2) into (4.1) our proof will be completed if we show that

$$\Sigma\{u_t(t_{k-1}, x(t_{k-1}))(t_k - t_{k-1}) + u_x(t_{k-1}, x(t_{k-1}))[x(t_k) - x(t_{k-1})]$$

$$+ \tfrac{1}{2} u_{xx}(t_{k-1}, x(t_{k-1}))[x(t_k) - x(t_{k-1})]^2\} \overset{P}{\to}$$

$$\int_0^t [u_t(s, x(s)) + u_x(s, x(s))f + \tfrac{1}{2} u_{xx}(s, x(s))\sigma^2]\,ds$$

$$+ \int_0^t u_x(s, x(s))\sigma dz(s). \tag{4.3}$$

Observe that as $\delta_n \to 0$,

$$\Sigma u_t(t_{k-1}, x(t_{k-1}))(t_k - t_{k-1}) \to \int_0^t u_t(s, x(s))\,ds \quad \text{w.p.1} \tag{4.4}$$

and also

$$\Sigma u_x(t_{k-1}, x(t_{k-1}))[x(t_k) - x(t_{k-1})] \overset{P}{\to}$$

$$\int_0^t u_x(s, x(s))f\,ds + \int_0^t u_x(s, x(s))\sigma dz(s). \tag{4.5}$$

Both (4.4) and (4.5) hold because of the continuity assumption. This means that in (4.3) it remains to study

$$\Sigma u_{xx}(t_{k-1}, x(t_{k-1}))[x(t_k) - x(t_{k-1})]^2$$

$$= f^2 \Sigma u_{xx}(t_{k-1}, x(t_{k-1}))(t_k - t_{k-1})^2$$

$$+ 2f\sigma \Sigma u_{xx}(t_{k-1}, x(t_{k-1}))(t_k - t_{k-1})[z(t_k) - z(t_{k-1})]$$

$$+ \sigma^2 \Sigma u_{xx}(t_{k-1}, x(t_{k-1}))[z(t_k) - z(t_{k-1})]^2. \tag{4.6}$$

By the assumption of the continuity of $u_{xx}$ and the continuity of $z(t)$, the first two terms in (4.6) converge to 0 w.p.1. That leaves only to show that as $\delta_n \to 0$,

$$\Sigma u_{xx}(t_{k-1}, x(t_{k-1}))[z(t_k) - z(t_{k-1})]^2 \overset{P}{\to} \int_0^t u_{xx}(s, x(s))\,ds. \tag{4.7}$$

To prove (4.7) first note that as $\delta_n \to 0$

$$\Sigma u_{xx}(t_{k-1}, x(t_{k-1}))(t_k - t_{k-1}) \to \int_0^t u_{xx}(s, x(s))\,ds$$

w.p.l. Therefore (4.7) will be established if it is shown that as $\delta_n \to 0$

$$\Sigma u_{xx}(t_{k-1}, x(t_{k-1}))([z(t_k) - z(t_{k-1})]^2 - (t_k - t_{k-1})) \xrightarrow{P} 0. \qquad (4.8)$$

For notational convenience denote the left-hand side of (4.8) by $S_n$. Thus (4.8) can be written as

$$S_n \xrightarrow{P} 0 \qquad (4.9)$$

for $\delta_n \to 0$. For (4.9) we use a truncation technique. Define

$$I_k^N(\omega) = \begin{cases} 1 & \text{if } |x(t_i)| \leqslant N \quad \text{for all } i \leqslant k, \\ 0 & \text{otherwise,} \end{cases}$$

and let

$$\epsilon_k = [z(t_k) - z(t_{k-1})]^2 - (t_k - t_{k-1}),$$

and finally,

$$S_n^N = \sum_1^n u_{xx}(t_{k-1}, x(t_{k-1})) I_{k-1}^N \epsilon_k.$$

Observe that $ES_n^N = 0$ and $E(S_n^N)^2 \to 0$ as $\delta_n \to 0$ because the $\epsilon_k$ are independent of each other and of $u_{xx}(t_{k-1}, x(t_{k-1})) I_{k-1}^N$. Furthermore, this implies that $S_n^N \xrightarrow{P} 0$ as $\delta_n \to 0$ for each $N$. Also, the error of the truncation is given by

$$P[S_n \neq S_n^N] = P\left[ \max_s |x(s)| > N \right]$$

and such error can be made arbitrarily small by choosing $N$ large because $x(s)$ is finite w.p.l. In conclusion,

$$P[|S_n| > \epsilon] \leqslant P[|S_n^N| > \epsilon] + P[S_n \neq S_n^N] \qquad (4.10)$$

with both terms of the right-hand side of (4.10) being arbitrarily small. Thus (4.9) holds. This proves the theorem in the special case. Turning next to the general case we observe that we may choose sequences $\{f_n\}$ and $\{\sigma_n\}$ of step functions such that w.p.l

$$\int_0^t |f_n(s) - f(s)| \, ds \to 0$$

$$\int_0^t |\sigma_n(s) - \sigma(s)|^2 \, ds \to 0$$

and the sequence of processes

$$x_n(t) = x_n(0) + \int_0^t f_n(s) \, ds + \int_0^t \sigma_n(t) \, dz(t)$$

converges uniformly w.p.1 to $x(t)$. Then the sequence of processes $y_n(t) = u(t, x_n(t))$ also converges uniformly w.p.1 to $y(t)$. Taking the limit as $n \to \infty$ below we obtain Itô's lemma

$$
\begin{aligned}
y_n(t) - y_n(0) = \ & \int_0^t [u_t(s, x_n(s)) + u_x(s, x_n(s)) f_n(s) \\
& + \tfrac{1}{2} u_{xx}(s, x_n(s)) \sigma_n^2(s)] \, ds \\
& + \int_0^t u_x(s, x_n(s)) \sigma_n(s) \, dz(s).
\end{aligned}
$$

Before we move on to some examples using Itô's lemma we state a straightforward generalization of this result.

**Lemma 4.2** (Generalized Itô).  Let $u(t, x): [0, T] \times R^d \to R^k$ denote a continuous nonrandom function such that its partial derivatives $u_t, u_{x_i}, u_{x_i x_j}$ are continuous, where

$$u_t = \frac{\partial}{\partial t} u(t, x),$$

$$u_{x_i} = \frac{\partial}{\partial x_i} u(t, x), \qquad i = 1, 2, \ldots, d,$$

$$u_{x_i x_j} = \frac{\partial}{\partial x_i \partial x_j} u(t, x), \qquad i, j \leq d,$$

Suppose that $x(t) = x(t, \omega): [0, T] \times \Omega \to R^d$ is a process with stochastic differential

$$dx(t) = f(t) \, dt + \sigma(t) \, dz(t),$$

where $f(t) = f(t, \omega)$: $[0, T] \times \Omega \to R^d$ is measurable in $(t, \omega)$ i.e. measurable in both arguments, and

$$\sigma(t) = \sigma(t, \omega)\colon [0, T] \times \Omega \to R^d \times R^m.$$

Here $\sigma$ is a $(d \times m)$-matrix-valued function, nonanticipating in $[0, T]$ and finally $z(t) = z(t, \omega)$: $[0, T] \times \Omega \to R^m$ is a $m$-dimensional Wiener process. Let $y(t) = u(t, x(t))$. Then the process $y(t)$ also has a differential on $[0, T]$ given by

$$\begin{aligned}
\mathrm{d}y(t) &= [u_t(t, x(t)) + u_x(t, x(t)) f(t) \\
&\quad + \tfrac{1}{2} \sum_i \sum_j u_{x_i x_j}(t, x(t)) [\sigma(t) \sigma'(t)]_{ij}] \, \mathrm{d}t \\
&\quad + u_x(t, x(t)) \sigma(t) \, \mathrm{d}z(t).
\end{aligned} \tag{4.11}$$

This completes the statement of the generalized Itô's lemma. Note that the double summation can also be written as

$$\sum_i \sum_j u_{x_i x_j}(\sigma\sigma')_{ij} = \mathrm{tr}\,(u_{xx}\,\sigma\sigma') = \mathrm{tr}\,(\sigma\sigma' u_{xx})$$

where $u_{xx} = u_{x_i x_j}$ is a $(d \times d)$-matrix whose elements are $k$-vectors. Thus, an alternative expression for $\mathrm{d}y(t)$ is

$$\mathrm{d}y(t) = u_t \mathrm{d}t + u_x \mathrm{d}x(t) + \tfrac{1}{2} \, \mathrm{tr}\,(\sigma\sigma' u_{xx}) \, \mathrm{d}t.$$

We recapitulate the analysis of this section by emphasizing three points.

First, Itô's lemma is a useful result because it allows us to compute stochastic differentials of arbitrary functions having as an argument a stochastic process which itself is assumed to possess a stochastic differential. In this respect Itô's formula is as useful as the chain rule of ordinary calculus.

Secondly, given an Itô stochastic process $x(t)$ with respect to a given Wiener process $z(t)$ and letting $y(t) = u(t, x(t))$ to be a new process, Itô's formula gives us the stochastic differential of $y(t)$, where $\mathrm{d}y(t)$ is also given with respect to the same Wiener process.

Thirdly, an inspection of the proof of Itô's lemma reveals that it consists of an application of Taylor's theorem of advanced calculus and several probabilistic arguments to establish the convergence of certain quantities to appropriate integrals. Therefore, the reader may obtain Itô's formula by applying Taylor's theorem instead of remembering the specific result. This is illustrated below in three cases.

*Case 1:* This corresponds to the generalized Itô's lemma. Given

$$f(t, \omega): [0, T] \times \Omega \to R^d,$$

$$\sigma(t, \omega): [0, T] \times \Omega \to R^d \times R^m,$$

$$z(t, \omega): [0, T] \times \Omega \to R^m,$$

that satisfy the assumptions of Itô's lemma and the process $x(t)$ having stochastic differential

$$dx_i(t) = f_i(t) \, dt + \sigma_{ir}(t) \, dz_r(t),$$

with $i = 1, 2, ..., d$, and $r = 1, 2, ..., m$, we define

$$y(t) = u(t, x(t)): [0, T] \times R^d \to R^k.$$

Note that consistency among the ranges of the three processes holds

$$dx(t) \quad = \quad f(t) \, dt \quad + \quad \sigma(t) \qquad dz(t).$$
$$(R^d \times R^1) \quad (R^d \times R^1) \quad (R^d \times R^m) \quad (R^m \times R^1)$$

Now apply Taylor's theorem, make the necessary substitutions, and rearrange terms. Note that $t$ is dropped and summations are from 1 to $d$:

$$
\begin{aligned}
dy(t) &= u_t \, dt + \Sigma u_{x_i} \, dx_i + \tfrac{1}{2} \Sigma\Sigma u_{x_i x_j} \, dx_i \, dx_j \\
&= u_t \, dt + \Sigma u_{x_i} (f_i \, dt + \sigma_{ir} \, dz_r) \\
&\quad + \tfrac{1}{2} \Sigma\Sigma u_{x_i x_j} (f_i \, dt + \sigma_{ir} \, dz_r)(f_j \, dt + \sigma_{jr} \, dz_r) \\
&= u_t \, dt + \Sigma u_{x_i} f_i \, dt + \Sigma u_{x_i} \sigma_{ir} \, dz_r + \tfrac{1}{2} \Sigma\Sigma u_{x_i x_j} [\sigma_{ir} \sigma'_{jr}] \, dz_r \, dz'_r \\
&= u_t \, dt + u_x f \, dt + \tfrac{1}{2} \Sigma\Sigma u_{x_i x_j} [\sigma \sigma']_{ij} \, dt + u_x \sigma \, dz \\
&= (u_t + u_x f + \tfrac{1}{2} \Sigma\Sigma u_{x_i x_j} [\sigma \sigma']_{ij}) \, dt + u_x \sigma \, dz.
\end{aligned}
$$

Note that the following multiplication rules have been used

$$dt \times dt = (dt)^2 = 0,$$

$$dt \times dz_r = 0,$$

$$dz_r \times dz_q = 0 \quad \text{for } r \neq q,$$

$$dz_r \times dz_q = dt \quad \text{for } r = q.$$

*Case 2:*   In this case we partially specialize the results of case 1. Suppose that

$$f(t, \omega): [0, T] \times \Omega \to R^d,$$
$$\sigma(t, \omega): [0, T] \times \Omega \to R^d,$$
$$z(t, \omega): [0, T] \times \Omega \to R^1.$$

Also, let the process $x(t)$ have differential

$$dx_i(t) = f_i(t)\, dt + \sigma_i(t)\, dz(t),$$

with $i = 1, 2, ..., d$. Define

$$y(t) = u(t, x(t)) = u(t, x_1(t), ..., x_d(t)): [0, T] \times R^d \to R.$$

Note symbolically that consistency of dimensionalities holds:

$$dx(t) \quad = \quad f(t)\, dt \quad + \quad \sigma(t) \qquad dz(t).$$
$$(R^d \times R^1) \quad (R^d \times R^1) \quad (R^d \times R^1) \quad (R^1 \times R^1)$$

As in case 1, we apply Taylor's theorem, make the necessary substitutions, re-arrange terms, and use the multiplication rules

$$(dt)^2 = 0; \quad (dz)^2 = dt; \quad dt \times dz = 0 \tag{4.12}$$

to obtain (all summations are from 1 to $d$; $t$ is dropped):

$$
\begin{aligned}
dy(t) &= u_t\, dt + \Sigma u_{x_i}\, dx_i + \tfrac{1}{2} \Sigma\Sigma u_{x_i x_j}\, dx_i dx_j \\
&= u_t\, dt + \Sigma u_{x_i}(f_i\, dt + \sigma_i\, dz) + \tfrac{1}{2} \Sigma\Sigma u_{x_i x_j}(f_i\, dt + \sigma_i dz)(f_j\, dt + \sigma_j dz) \\
&= u_t\, dt + \Sigma u_{x_i} f_i\, dt + \Sigma u_{x_i} \sigma_i dz + \tfrac{1}{2} \Sigma\Sigma u_{x_i x_j} \sigma_i \sigma_j\, dt \\
&= (u_t + \Sigma u_{x_i} f_i + \tfrac{1}{2} \Sigma\Sigma u_{x_i x_j} \sigma_i \sigma_j)\, dt + (\Sigma u_{x_i} \sigma_i)\, dz.
\end{aligned}
$$

*Case 3:*   In this case we further specialize case 2 by assuming that $f$ and $\sigma$ are real-valued functions. This case is the one stated in Itô's lemma. We proceed immediately with the calculations:

$$dy(t) = u_t dt + u_x dx + \tfrac{1}{2} u_{xx} (dx)^2$$
$$= u_t dt + u_x (f dt + \sigma dz) + \tfrac{1}{2} u_{xx} (f dt + \sigma dz)^2$$
$$= u_t dt + u_x f dt + u_x \sigma dz + \tfrac{1}{2} u_{xx} \sigma^2 dt$$
$$= (u_t + u_x f + \tfrac{1}{2} u_{xx} \sigma^2) dt + u_x \sigma dz.$$

This is the result of Itô's lemma. Note that the multiplication rules of (4.12) are applied to yield

$$(f dt + \sigma dz)^2 = \sigma^2 dt.$$

## 5. Examples

Several examples from economics and finance which illustrate the use of Itô's lemma are presented in Chapters 3 and 4. Here we give some examples of Itô's lemma which are of a mathematical nature.

**Example 1.** Let $y(t) = u(x(0), t) = x(0) e^{at}$ with $dx(t) = ax(t) dt$, where $a$ is a nonzero real number. Then,

$$dy(t) = u_t dt + u_x dx + \tfrac{1}{2} u_{xx} (dx)^2$$
$$= ax(0) e^{at} dt.$$

Thus, $y(t)$ solves $dx(t)$.

**Example 2.** Suppose that $y(t) = u(x(t)) = e^{x(t)}$ with $y(0) = e^{x(0)} > 0$. Let

$$dx(t) = -\tfrac{1}{2} \sigma^2 (t) dt + \sigma(t) dz(t).$$

Itô's lemma yields

$$dy(t) = u_t dt + u_x dx(t) + \tfrac{1}{2} u_{xx} (dx(t))^2$$
$$= e^{x(t)} [-\tfrac{1}{2} \sigma^2 (t) dt + \sigma(t) dz(t)]$$
$$\qquad + \tfrac{1}{2} e^{x(t)} [-\tfrac{1}{2} \sigma^2 (t) dt + \sigma(t) dz(t)]^2$$
$$= -\tfrac{1}{2} e^{x(t)} \sigma^2 (t) dt + e^{x(t)} \sigma(t) dz(t) + \tfrac{1}{2} e^{x(t)} \sigma^2 (t) dt$$
$$= e^{x(t)} \sigma(t) dz(t).$$

Note that the process

$$y(t) \;=\; y(0)\,e^{x(t)}$$

$$=\; y(0)\exp\left\{-\tfrac{1}{2}\int_0^t \sigma^2(s)\,ds + \int_0^t \sigma(s)\,dz(s)\right\}$$

satisfies the equation of $dy(t)$ for $t \in [0, T]$.

**Example 3.** Suppose that $y(t) = u(x(t)) = e^{x(t)}$, where

$$dx(t) = -\tfrac{1}{2}\,dt + dz(t)$$

and $x(0) = 0$. Applying Itô's lemma we obtain:

$$\begin{aligned}
dy(t) &= u_t dt + u_x\,dx(t) + \tfrac{1}{2}\,u_{xx}\,[dx(t)]^2 \\
&= e^{x(t)}[-\tfrac{1}{2}\,dt + dz(t)] + \tfrac{1}{2}\,e^{x(t)}[-\tfrac{1}{2}\,dt + dz(t)]^2 \\
&= e^{x(t)}\,dz(t).
\end{aligned}$$

This is a special case of example 2 and it illustrates that, given the stochastic differential equation

$$dy(t) = e^{x(t)}\,dz(t) = y(t)\,dz(t) \tag{5.1}$$

with initial condition $y(0) = e^{x(0)} = 1$, its solution is

$$y(t) = 1\,\exp\{-\tfrac{1}{2}\,t + z(t)\}. \tag{5.2}$$

This result is useful because it illustrates the difference between ordinary differential equations and stochastic differential equations. Observe that if (5.1) were an ordinary differential equation its solution would be $y(t) = \exp[z(t)]$, which is different from (5.2).

**Example 4.** Suppose that $y(t) = u(x_1, x_2) = tz(t)$ with

$$dx_1(t) = dt \quad \text{and} \quad dx_2(t) = dz(t).$$

From Itô's lemma we compute for $i = 1, 2$

$$\begin{aligned}
\mathrm{d}y(t) &= u_t \mathrm{d}t + \Sigma u_{x_i} \mathrm{d}x_i(t) + \tfrac{1}{2} \Sigma\Sigma u_{x_i x_j} \mathrm{d}x_i(t) \mathrm{d}x_j(t) \\
&= z(t)\mathrm{d}t + t\mathrm{d}z(t) + \tfrac{1}{2}\left[\mathrm{d}t\mathrm{d}z(t) + \mathrm{d}z(t)\mathrm{d}t\right] \\
&= z(t)\mathrm{d}t + t\mathrm{d}z(t).
\end{aligned}$$

**Example 5.** Suppose that $y(t) = u(t, x_1, x_2) = t x_1 x_2$ with

$$\mathrm{d}x_1(t) = f_1(t)\mathrm{d}t + \sigma_1(t)\mathrm{d}z(t),$$
$$\mathrm{d}x_2(t) = \sigma_2(t)\mathrm{d}z(t).$$

Applying Itô's lemma we compute,

$$\begin{aligned}
\mathrm{d}y(t) &= u_t \mathrm{d}t + \Sigma u_{x_i} \mathrm{d}x_i(t) + \tfrac{1}{2} \Sigma\Sigma u_{x_i x_j} \mathrm{d}x_i(t) \mathrm{d}x_j(t) \\
&= x_1 x_2 \mathrm{d}t + t x_2 \mathrm{d}x_1(t) + t x_1 \mathrm{d}x_2(t) + \\
&\quad + \tfrac{1}{2}\left[t\mathrm{d}x_1(t)\mathrm{d}x_2(t) + t\mathrm{d}x_2(t)\mathrm{d}x_1(t)\right] \\
&= x_1 x_2 \mathrm{d}t + t x_2 \left[f_1 \mathrm{d}t + \sigma_1 \mathrm{d}z\right] + t x_1 \sigma_2 \mathrm{d}z + \\
&\quad + \tfrac{1}{2}\left[t\sigma_1 \sigma_2 \mathrm{d}t + t\sigma_1 \sigma_2 \mathrm{d}t\right] \\
&= (x_1 x_2 + t x_2 f_1 + t\sigma_1 \sigma_2)\mathrm{d}t + (t x_2 \sigma_1 + t x_1 \sigma_2)\mathrm{d}z.
\end{aligned}$$

**Example 6.** Suppose $y(t) = u(x(t)) = u(z(t))$, where $u$ is assumed to be twice continuously differentiable with respect to $x$, $x(0) = z(0) = 0$ and $\mathrm{d}x(t) = \mathrm{d}z(t)$. Itô's lemma gives

$$\begin{aligned}
\mathrm{d}y(t) &= u_t \mathrm{d}t + u_x \mathrm{d}x(t) + \tfrac{1}{2} u_{xx} [\mathrm{d}x(t)]^2 \\
&= u'(z(t))\mathrm{d}z(t) + \tfrac{1}{2} u''(z(t))\mathrm{d}t. \tag{5.3}
\end{aligned}$$

Here a prime denotes ordinary derivative. If we write (5.3) in integral form it becomes

$$y(t) = u(z(t)) = u(0) + \int_0^t u'(z(s))\mathrm{d}z(s) + \tfrac{1}{2} \int_0^t u''(z(s))\mathrm{d}s. \tag{5.4}$$

From (5.4), solving for the second term in the right-hand side, we obtain:

$$\int_0^t u'(z(s))\mathrm{d}z(s) = u(z(t)) - u(z(0)) - \tfrac{1}{2} \int_0^t u''(z(s))\mathrm{d}s. \tag{5.5}$$

Eq. (5.5) is usually called the fundamental theorem of calculus for Itô's stochastic integral and it is useful because it expresses the integral in the left-hand side of (5.5) only in terms of an ordinary integral.

The above examples illustrate the use of Itô's lemma in a mathematical context which may seem far removed from applications in economic analysis. However, we hasten to add that Itô's lemma has found numerous applications in finance and economics and here we mention three broad areas of application: first, the modern theory of contingent claim valuation, with its primary model being the celebrated option pricing formula of Black and Scholes; second, the formulation of appropriate budget constraints in a stochastic context with Merton's work on optimum consumption and portfolio rules being a primary model, and third, the formulation and analysis of optimal stochastic control problems with section 10 in this chapter providing a useful illustration.

## 6. Stochastic differential equations

In this section we study conditions that guarantee the existence and uniqueness of solutions of stochastic differential equations of the Itô type. Such equations were first introduced in section 2 above in (2.12). Recall that such equations can be transformed into an integral equation as in (3.4) and that it was (3.4) that motivated the analysis on stochastic integration.

Consider a probability space $(\Omega, \mathcal{F}, P)$ and the interval $[0, T]$ which can also be $[0, \infty)$, and let us write an Itô stochastic differential equation, such as

$$dx(t) = f(t, x(t)) dt + \sigma(t, x(t)) dz(t), \tag{6.1}$$

with initial condition

$$x(0, \omega) = c(\omega) = c. \tag{6.2}$$

Eqs. (6.1) and (6.2) define a *stochastic initial value problem* and can be expressed as a stochastic integral equation

$$x(t) = c + \int_0^t f(s, x(s)) ds + \int_0^t \sigma(s, x(s)) dz(s). \tag{6.3}$$

As before, $x(t)$ is a real-valued stochastic process, $f$ and $\sigma$ are real-valued functions measurable on $[0, T] \times R$ and $z(t)$ is a Wiener process. We assume that $x(t)$ is $\mathcal{F}_t$-measurable and note that $\mathcal{F}_t$ must be independent of the $\sigma$-field generated by $z_u - z_t, u \geqslant t$, for all $t \geqslant 0$. For our analysis it is enough to choose for $\mathscr{F}_t$ the smallest $\sigma$-field with respect to which the initial random variable condition $c(\omega)$, and the random variables $z_s, s \leqslant t$, are measurable. We are now ready to give a definition.

A stochastic process $x(t)$ is called a *solution* of (6.1) and (6.2), i.e. *of an Itô stochastic differential equation* on $[0, T]$, if it satisfies the following three properties:

(1)  $x(t)$ is $\mathscr{F}_t$-measurable, i.e. nonanticipating for $t \in [0, T]$;
(2)  the functions $f$ and $\sigma$ are such that w.p.1

$$\int_0^T |f(t, x(t))| \, dt < \infty \quad \text{and} \quad \int_0^T |\sigma(t, x(t))|^2 \, dt < \infty; \text{ and}$$

(3)  eq. (6.3) holds for all $t \in [0, T]$ w.p.1.

In the theory and applications of stochastic differential equations questions of existence and uniqueness are important for analytical reasons. Usually, researchers are interested in establishing the existence and uniqueness of their model described by a stochastic differential equation before they proceed to study the various probabilistic properties of the solution. Below we state the existence and uniqueness theorem for stochastic differential equations. In the next section we state various fundamental properties of such solutions.

**Theorem 6.1.**  Consider eq. (6.1) with initial condition given by (6.2) and suppose that the following assumptions are satisfied.

(1)  The functions $f(t, x)$ and $\sigma(t, x)$ are defined for $t \in [0, T]$ and $x \in R$ and are measurable with respect to all their arguments.
(2)  There exists a constant $K > 0$ such that for $t \in [0, T]$ and $x \in R, y \in R$,

$$|f(t, x) - f(t, y)| + |\sigma(t, x) - \sigma(t, y)| \leqslant K |x - y|, \tag{6.4}$$

$$|f(t, x)|^2 + |\sigma(t, x)|^2 \leqslant K^2 (1 + |x|^2). \tag{6.5}$$

(3)  The initial condition $x(0, \omega)$ in (6.2) does not depend on $z(t)$ and $Ex(0, \omega)^2 < \infty$.

Then there exists a solution of (6.1) satisfying the initial condition in (6.2) which is unique w.p.1, has continuous sample paths $x(t)$ w.p.1, and $\sup_t Ex(t)^2 < \infty$.

For a proof of this theorem see Gihman and Skorohod (1972, pp. 40–43). The theorem is established by using a Picard–Lindelöf iteration argument similar to the one used in ordinary differential equations and the Borel–Cantelli lemma.

Obviously, the theorem holds for vector-valued functions $x(t), z(t)$ and $f(t)$, and the matrix-valued function $\sigma(t)$ with appropriate dimensionalities.

Equation (6.4) is called the *Lipschitz condition* and (6.5) is called the *Restriction on growth condition*. *Uniqueness of solutions* means that if $x_1(t)$ and $x_2(t)$ are two solutions then

$$P\left[\sup_t |x_1(t) - x_2(t)| = 0\right] = 1. \tag{6.6}$$

Note that if the functions $f(t, x)$ and $\sigma(t, x)$ are defined on $[0, T] \times R$ and the assumptions of theorem 3.1 hold on every finite subinterval $[0, T]$ of $[0, \infty)$, then we say that the stochastic differential equation has a *global solution* defined for $t \in [0, \infty)$. This is the case for autonomous stochastic differential equations, where $f(t, x) = f(x)$ and $\sigma(t, x) = \sigma(x)$.

For eq. (6.4) to hold it is sufficient that the functions $f(t, x)$ and $\sigma(t, x)$ have both continuous partial derivatives of first order with respect to $x$ for every $t \in [0, T]$ which are bounded on $[0, T] \times R$.

As a special case of (6.1) consider the autonomous stochastic differential equation, treated in Gihman and Skorohod (1972, p. 106),

$$dx(t) = f(x(t)) \, dt + \sigma(x(t)) \, dz(t), \tag{6.7}$$

where $f(x)$ and $\sigma(x)$ are functions defined for $x \in R$. Suppose that $f(x)$ and $\sigma(x)$ satisfy the following conditions:

$$|f(x)| + |\sigma(x)| \leqslant K(1 + |x|) \tag{6.8}$$

holds for some constant $K > 0$. Also, for each $C > 0$ there is a positive real number $L_C$ such that for $|x| \leqslant C$ and $|y| \leqslant C$ we have

$$|f(x) - f(y)| + |\sigma(x) - \sigma(y)| \leqslant L_C |x - y|. \tag{6.9}$$

Then for each random initial condition $c(\omega)$ not dependent on $z(t)$ there exists a unique solution of (6.7) satisfying the initial condition. This result is useful in many applications which yield autonomous stochastic differential equations. In such cases, eqs. (6.8) and (6.9) are used to establish the existence and uniqueness of solutions.

Finally, let us consider an important special case of the class of stochastic differential equations, namely the linear equations. Suppose that $x(t)$ is a $d$-dimensional process, $a(t)$ is a $(d \times d)$-matrix-valued function, $\sigma(t)$ is a $(d \times d)$-matrix-valued function, and $z(t)$ is a $d$-dimensional Wiener process. Consider

$$dx(t) = a(t)x(t)dt + \sigma(t)dz(t) \tag{6.10}$$

with initial condition $x(0) = c$. Assume that the conditions for existence and uniqueness of solutions are satisfied and let us ask the question: What does the solution look like in this special case? We claim that the solution of (6.10) on $[0, T]$ has the form

$$x(t) = \phi(t)\left[c + \int_0^t \phi^{-1}(s)\sigma(s)dz(s)\right], \tag{6.11}$$

where $\phi(t)$ is the fundamental matrix of

$$\dot{x}(t) = a(t)x(t). \tag{6.12}$$

Let us verify the claim. Start with equation

$$y(t) = c + \int_0^t \phi^{-1}(s)\sigma(s)dz(s) \tag{6.13}$$

which can be written in differential form as

$$dy(t) = \phi^{-1}(t)\sigma(t)dz(t). \tag{6.14}$$

Observe from (6.11) and (6.13) that

$$x(t) = \phi(t)y(t). \tag{6.15}$$

Now we are in a position to apply Itô's lemma to show that $dx(t)$ has the appropriate form as in (6.10). To do so we use the two equations (6.15) and (6.14). We have

$$
\begin{aligned}
dx(t) &= \frac{\partial}{\partial t}[\phi(t)y(t)]dt + \frac{\partial}{\partial y}[\phi(t)y(t)]dy(t)) \\
&\quad + \frac{1}{2}2\frac{\partial^2}{\partial y^2}[\phi(t)y(t)][dy(t)]^2 \\
&= \dot{\phi}(t)y(t)dt + \phi(t)dy(t) + 0 + 0 \\
&= a(t)\phi(t)y(t)dt + \sigma(t)dz(t) \\
&= a(t)\phi(t)\phi^{-1}(t)x(t)dt + \sigma(t)dz(t) \\
&= a(t)x(t)dt + \sigma(t)dz(t).
\end{aligned}
$$

To illustrate (6.11) consider the linear case given by

$$dx(t) = ax(t)\,dt + \sigma(t)\,dz(t),$$                                                                    (6.16)

with $x(0) = c$. In this case since the matrix $a$ is autonomous we know from ordinary differential equations that the system $\dot{x}(t) = ax(t)$ has as fundamental matrix

$$\phi(t) = e^{at}, \quad t \in [0, T],$$

where $\phi^{-1}(t) = e^{-at}$. Applying (6.11) we get

$$
\begin{aligned}
x(t) &= \phi(t)\left[c + \int_0^t \phi^{-1}(s)\,\sigma(s)\,dz(s)\right] \\
&= e^{at}\left[c + \int_0^t e^{-as}\sigma(s)\,dz(s)\right] \\
&= ce^{at} + \int_0^t e^{a(t-s)}\sigma(s)\,dz(s).
\end{aligned}
$$

For a more detailed discussion of the issues discussed above and for additional related topics see Arnold (1974) and Gihman and Skorohod (1972).

## 7. Properties of solutions

The solutions of ordinary differential equations satisfy, under certain conditions, two well-known properties called the *property of dependence of solutions on parameters and initial data* and the *property of differentiability of solutions*. It is natural to ask whether stochastic differential equations satisfy similar properties.

In this section we show that the property of dependence of solutions on parameters and initial data holds, but the property of differentiability of solutions does not hold. The latter is due to the fact that if $x(t)$ is the unique solution of an Itô stochastic differential equation then $x(t)$ depends on the Wiener process $z(t)$ whose nondifferentiability anywhere implies the nondifferentiability of $x(t)$. However, if we define differentiability in the mean square we can show that under certain assumptions $x(t)$ is mean square-differentiable. Beyond these two properties, solutions of stochastic differential equations satisfy some additional properties that are probabilistic in nature, i.e. the solutions are Markov processes and under certain assumptions they are diffusion processes. We now proceed to establish precisely the various properties.

Consider the stochastic differential equation which depends on a parameter $p$, written as

$$dx(p, t) = f(p, t, x)dt + \sigma(p, t, x)dz(t), \tag{7.1}$$

with initial condition $x(p, 0) = c(p)$ for $t \in [0, T]$ and $p \in \mathscr{P}$, where $\mathscr{P}$ is the parameter set. Denote by $x(p, t)$ the solution of (7.1) and let $p_0 \in \mathscr{P}$. The *property of dependence of solutions on parameters* is established by the next theorem.

**Theorem 7.1.** Let $x(p, t)$ be the solution of (7.1) and suppose that $f(p, t, x)$ and $\sigma(p, t, x)$ satisfy for all $p$ the conditions of existence and uniqueness, i.e. (6.4) and (6.5). Also, assume the following:

(1) as $p \to p_0$ then $c(p) \overset{P}{\to} c(p_0)$; $\tag{7.2}$

(2) for every $N > 0$, $|x| \leqslant N$, $t \in [0, T]$, as $p \to p_0$ then

$$\lim \sup_{x} (|f(p, t, x) - f(p_0, t, x)| + |\sigma(p, t, x) - \sigma(p_0, t, x)|) = 0;$$

(3) there is a constant $K$, independent of $p$ such that for $t \epsilon [0, T]$

$$|f(p, t, x)|^2 + |\sigma(p, t, x)|^2 \leqslant K^2 (1 + |x|^2).$$

Then, as $p \to p_0$,

$$\sup_{t} |x(t, p) - x(t, p_0)| \overset{P}{\to} 0. \tag{7.3}$$

For a proof of this theorem see Gihman and Skorohod (1972, pp. 54–55).

As a special case of theorem 7.1 suppose that the functions $f$ and $\sigma$ are independent of the parameter $p$, i.e. let

$$dx(p, t) = f(t, x)dt + \sigma(t, x)dz(t), \tag{7.4}$$

with initial condition $x(p, 0) = c(p)$ for $t \in [0, T]$. Here we maintain the dependence of initial condition on the parameter $p$. As a corollary of theorem 7.1 we now obtain the property of the stochastically continuous dependence of solutions on initial data, i.e. for eq. (7.4) assuming existence and uniqueness of its solution, (7.2) implies (7.3).

We now define mean-square-differentiability for a stochastic process. Let $x(t)$ be a stochastic process with $t \in [0, T]$ and let $u \in [0, T]$ be a specific point. We say that $x(t)$ is *mean-square-differentiable* at $t = u$, with random variable $y(u)$ as its derivative, if the second moments of $x(t)$ and $y(u)$ exist and satisfy for $h > 0$

$$E \mid [x(u+h) - x(u)]/h - y(u) \mid^2 \to 0$$

as $h \to 0$.

We apply the above definition to a stochastic process depending upon a parameter and given by

$$x(s, t, c) = c + \int_s^t f(v, x(s, v, c)) \, dv$$
$$+ \int_s^t \sigma(v, x(s, v, c)) \, dz(v), \tag{7.5}$$

where $v \in [s, T]$, $0 \leqslant s \leqslant t \leqslant T$ and $x(s, s, c) = c \in R$. A question naturally arises: Under what assumptions is $x(s, t, c)$ of (7.5) mean-square-differentiable with respect to initial data $c$? The answer is given by the next theorem.

**Theorem 7.2.** Suppose that $f$ and $\sigma$ of (7.5) are continuous with respect to $(t, x)$ and they have bounded first and second partial derivatives with respect to the argument $x$. Then for given $t = u$, $u \in [s, T]$, the solution $x(s, u, c)$ is twice mean-square-differentiable with respect to $c$ and

$$\frac{\partial}{\partial c} x(s, u, c) = 1 + \int_s^u f_x(v, x(s, v, c)) \frac{\partial}{\partial c} x(s, v, c) \, dv$$
$$+ \int_s^u \sigma_x(v, x(s, v, c)) \frac{\partial}{\partial c} x(s, v, c) \, dz(v).$$

For a proof of this theorem and its generalization see Gihman and Skorohod (1972, pp. 59–62). Next we recall the definition of a Markov process and establish the fact that solutions of stochastic differential equations are Markov processes.

A real-valued process $x(t)$ for $t \in [0, T]$ defined on the probability space $(\Omega, \mathscr{F}, P)$ is called a *Markov process* if for $0 \leqslant s \leqslant t \leqslant T$ and for every set $B \in \mathscr{R}$, with $\mathscr{R}$ denoting the Borel $\sigma$-field of $R$, the following equation holds w.p.1:

$$P[x(t) \in B \mid \mathscr{F}_s] = P[x(t) \in B \mid \sigma(x(s))].$$

Here $\mathscr{F}_s$ is the $\sigma$-field generated by $x(s)$ for $s \in [0, T]$ and $\sigma(x(s))$ denotes the $\sigma$-field generated by the single random variable $x(s)$. In words, the definition says that for a Markov process the past and the future are statistically independent when the present is known. For a Markov process $x(t)$ we obtain from the theory of conditional probability the existence of a transition probability function, denoted by $P(s, x, t, B)$, which yields the probability that $x(t) \in B$, given that $x(s) = s$, for $0 \leqslant s \leqslant t \leqslant T$.

**Theorem 7.3.** Consider the stochastic differential equation (6.1) with initial condition as in (6.2) and suppose that a unique solution $x(t)$, $t \in [0, T]$ exists. Then $x(t)$ is a Markov process whose initial probability distribution at $t = 0$ is $c$ and whose transition probability is given by

$$P(s, x, t, B) = P[x(t) \in B \mid x(s) = x].$$

For a proof see Arnold (1974, p. 147). A special class of Markov processes is the class of diffusion processes. A real-valued Markov process $x(t)$, $t \in [0, T]$, with almost certainly continuous sample paths, is called a *diffusion process* if its transition probability $P(s, x, t, B)$ satisfies the following three conditions for every $s \in [0, T]$, $x \in R$ and $\epsilon > 0$:

(1) $\displaystyle \lim_{t \downarrow s} \frac{1}{t-s} \int_{|y-x| > \epsilon} P(s, x, t, \mathrm{d}y) = 0;$  (7.6)

(2) there exists a real-valued function $f(s, x)$ such that

$$\lim_{t \downarrow s} \frac{1}{t-s} \int_{|y-x| \leqslant \epsilon} (y-x) P(s, x, t, \mathrm{d}y) = f(s, x); \tag{7.7}$$

(3) there exists a real-valued function $h(s, x)$ such that

$$\lim_{t \downarrow s} \frac{1}{t-s} \int_{|y-x| \leqslant \epsilon} (y-x)^2 P(s, x, t, \mathrm{d}y) = h(s, x). \tag{7.8}$$

Note that $f$ is called the drift coefficient and $h$ is called the *diffusion coefficient* and they are obtained from conditions (2) and (3). We remark that condition (1) says that large changes in $x(t)$ over a short period of time are improbable.

**Theorem 7.4.** Consider the stochastic differential equation (6.1) with initial condition as in (6.2) and suppose that a unique solution $x(t)$, $t \in [0, T]$ exists. Suppose also that the functions $f$ and $\sigma$ are continuous with respect to $t$. Then the solution $x(t)$ is a diffusion process with drift coefficient $f(t, x)$ and diffusion coefficient $h(t,x) = \sigma^2 \ (t,x)$.

For a proof of this result see Arnold (1974, p. 153).

The usefulness of this last theorem can be explained as follows. Note from the definition of a diffusion process that the transition probability function $P(s, x, t, B)$ is crucial in defining the drift coefficient $f(t, x)$ and the diffusion coefficient $h(t, x)$. A question arises. Suppose that a diffusion process is given with coefficients $f$ and $h$. Can we obtain $P(s, x, t, B)$ from $f$ and $h$? The answer is yes. Actually the decisive property of a diffusion process is that the transition probability $P(s, x, t, B)$ is uniquely determined under certain assumptions by the coefficients $f$ and $h$. This fact is surprising because $f$ and $h$ involve only the first and second moments which, in general, are not sufficient to define a distribution. Therefore given a stochastic differential equation that satisfies the hypotheses of theorem 7.4 we know that its solution is a diffusion process which under certain assumptions could yield a transition probability without actually having to find an explicit solution. How is this done? We answer the question by making certain definitions and by stating relevant theorems.

To each diffusion process with coefficients $f = (f_i)$, $i = 1, 2, ..., d$ and $h = (h_{ij})$, $i, j = 1, 2, ..., d$ we assign the second-order *differential operator*

$$\mathscr{D} = \sum_i f_i(s, x) \frac{\partial}{\partial x_i} + \tfrac{1}{2} \sum_i \sum_j h_{ij}(s, x) \frac{\partial^2}{\partial x_i \partial x_j} \ . \tag{7.9}$$

$\mathscr{D}g$ can formally be written for every twice partially differentiable function $g(x)$. For real-valued $f$ and $h$ (7.9) becomes

$$\mathscr{D} = f(s, x) \frac{\mathrm{d}}{\mathrm{d}x} + \tfrac{1}{2} h(s, x) \frac{\mathrm{d}^2}{\mathrm{d}x^2} \ . \tag{7.10}$$

Denote by $\mathrm{E}_{s,x}$ expectation conditioned upon $x$ at time $s$. The following theorem is important in our analysis.

**Theorem 7.5.** Suppose that $x(t)$ for $t \in [0, T]$ is a $d$-dimensional diffusion process with continuous coefficients $f(s, x)$ and $h(s, x)$ and that conditions (1), (2) and (3) in (7.6), (7.7) and (7.8) hold uniformly in $s \in [0, T]$. Let $g(x)$ denote a continuous, bounded, real-valued function such that for $s < t$, $t$ fixed, $x \in R^d$,

$$u(s, x) = \mathrm{E}_{s,x} g(x(t)) = \int_{R^d} g(y) P(s, x, t, \mathrm{d}y). \tag{7.11}$$

Suppose $u$ has continuous bounded partials

$$\frac{\partial u}{\partial x_i} \; ; \quad \frac{\partial^2 u}{\partial x_i \partial x_j} \; ; \quad 1 \leqslant i, j \leqslant d.$$

Then $u(s, x)$ is differentiable with respect to $s$ and satisfies the partial differential equation

$$\frac{\partial u}{\partial s} + \mathcal{D} u = 0, \tag{7.12}$$

with boundary condition $u(s, x) \rightarrow g(x)$ as $s \rightarrow t$.

For a proof of this theorem see Gihman and Skorohod (1969, p. 373). Eq. (7.12) is called *Kolmogorov's backward equation* because differentiation is with respect to the backward arguments $s$ and $x$, where $s < t$ and $x < y$. Kolmogorov's backward equation (7.12) enables us to determine $P(s, x, t, y)$ which is uniquely defined if we know (7.11). If (7.12) has a unique solution for $g$ a real-valued continuous and bounded function, we can for given $f$ and $h$ calculate $u(s, x)$ and then from it calculate $P(s, x, t, y)$. More specifically, for stochastic differential equations a corollary of theorem 7.5 is the following result.

**Theorem 7.6.** Suppose that eq. (6.1) with initial condition as in (6.2) satisfies the hypotheses of theorem 7.4 and therefore has as a unique solution $x(t)$ which is a diffusion process. Also, assume that $f$ and $\sigma$ have continuous and bounded first- and second-order partial derivatives with respect to $x$. Let $g(x)$ denote a continuous bounded real-valued function with continuous and bounded first- and second-order derivatives and let for $0 \leqslant s \leqslant t \leqslant T$ and $x \in R$,

$$u(s, x) = \mathrm{E}_{s,x} \, g(x(s, t, x)).$$

Then $u$ and its first- and second-order partial derivatives with respect to $x$ and its partial derivative with respect to $s$ are continuous and bounded and satisfy the equation

$$\frac{\partial}{\partial s} u(s, x) + \mathcal{D} u(s, x) = 0 \quad \text{as } s \rightarrow t, u(s, x) \rightarrow g(x).$$

The proof follows from theorem 7.5. This result shows that the study of stochastic differential equations and their solutions is closely related to the study of second-order partial differential equations of the form of (7.12).

For the special class of autonomous stochastic differential equations such as (6.7) we immediately obtain as a corollary of theorems 7.3 and 7.4 that their solutions are Markov and diffusion processes.

The last result of this section refers to the moments of solutions.

**Theorem 7.7.** Consider the stochastic differential equation (6.1) with initial condition as in (6.2) and suppose that it has a unique solution. Also, assume that

$$E \mid c \mid^{2n} < \infty,$$

where $n$ is a positive integer. Then the solution $x(t)$, $t \in [0, T]$, of (6.1) satisfies

$$E \mid x(t) \mid^{2n} \leqslant (1 + E \mid c \mid^{2n}) e^{Ct},$$

$$E \mid x(t) - c \mid^{2n} \leqslant D(1 + E \mid c \mid^{2n}) t e^{Ct},$$

where $C = 2n(2n + 1)K^2$ and $D$ are constant depending only on $n$, $T$ and $K$ of (6.4).

For a proof see Arnold (1974, pp. 116–118).


## 8. Point equilibrium and stability

A particular solution of a stochastic differential equation that plays an important role in many applications is the equilibrium solution.

Consider eq. (6.1) defined for $t \in [0, T]$. If there exists a nonrandom constant $c$ such that for all $t \in [0, T]$,

$$f(t, c) = \sigma(t, c) = 0, \tag{8.1}$$

with $x(t, \omega) = c$ for all $t \in [0, T]$ and w.p.1, then we say that $c$ is a *point equilibrium solution*, or simply an *equilibrium solution*. Another terminology for the same concept is *stationary point* or *steady state point*. Observe that the definition specifies $c$, if it exists, as a nonrandom constant, or equivalently, as a point distribution.

Various other definitions of the concept "equilibrium solution" exist. For a survey of such definitions the reader may consult Majumdar (1975). A different definition than the one presented above will be given in the next section.

Suppose that a point equilibrium solution exists for an autonomous stochastic differential equation of the Itô type such as

$$\mathrm{d}x(t) = f(x(t))\,\mathrm{d}t + \sigma(x(t))\,\mathrm{d}z(t) \tag{8.2}$$

for $t \in [0, T]$. In this section we study the stability of the equilibrium solution of an autonomous stochastic differential equation such as (8.2). For convenience and without loss of generality we assume that 0 is an equilibrium solution. We use two approaches to stochastic stability that yield similar results to acquaint the reader with both methods. We call these two approaches the Gihman—Skorohod approach and the Liapunov—Kushner approach.

## 8.1. The Gihman—Skorohod approach

This method is presented in Gihman—Skorohod (1972, pp. 145—151).

Consider eq. (8.2) and suppose that for $t \in [0, \infty)$ it has a unique equilibrium solution which is 0 such that

$$f(0) = \sigma(0) = 0. \tag{8.3}$$

We say that *the 0-equilibrium is stable* if for any $\epsilon > 0$ there exists an $\eta > 0$ such that if $|x| < \eta$ then as $t \to \infty$

$$P[\lim x(t) = 0] \geqslant 1 - \epsilon. \tag{8.4}$$

Here $x = x(0, \omega)$, i.e. $x$ is the initial condition and in (8.4) the probability is conditioned on $x$. If for any $\epsilon > 0$ there exists an $\eta > 0$ such that (8.4) holds for $x \in (0, \eta)$ or $x \in (-\eta, 0)$, then we say that the 0-equilibrium is *right-stable* and *left-stable*, respectively. Conditions for stability are given in the next theorem.

**Theorem 8.1.** Assume that for eq. (8.2) the conditions in (8.3) hold and that there exists an $\eta > 0$ such that $\sigma(x) > 0$ for $0 < |x| < \eta$. For the stability of the 0-equilibrium (1) from the right; (2) from the left; and (3) both, it is necessary and sufficient that for some $\eta > 0$, (1) $I_1 < \infty$; (2) $I_2 < \infty$; and (3) $I_1 + I_2 < \infty$, respectively, where

$$I_1 = \int_0^\eta \exp\left\{ \int_u^\eta \frac{2f(y)}{\sigma^2(y)}\,\mathrm{d}y \right\} \mathrm{d}u \tag{8.5}$$

and

$$I = \int_{-\eta}^0 \exp\left\{ - \int_{-\eta}^u \frac{2f(y)}{\sigma^2(y)}\,\mathrm{d}y \right\} \mathrm{d}u. \tag{8.6}$$

For a proof see Gihman and Skorohod (1972, pp. 146–148). This theorem is useful because it reduces the study of stochastic stability to a computation of (8.5) and (8.6).

Sometimes it is difficult to establish the stochastic stability property of an equilibrium solution, possibly because such a property may not hold in a given model. In such a case the researcher may ask whether the solutions are bounded. In many applications the *property of boundedness of solutions* is the next most desirable property to that of stochastic stability. It comes as no surprise that equations similar to (8.5) and (8.6) are used to establish boundedness from above, as the next theorem states.

**Theorem 8.2.** Assume that eq. (8.2) with initial condition $x(0) = c$ has a unique solution denoted by $x(t)$ with $\sigma(x) > 0$ for $x \in R$. Let

$$I_1(x) = \int_{-\infty}^{x} \exp\left\{-\int_{0}^{u} \frac{2f(y)}{\sigma^2(y)}\,dy\right\} du \tag{8.7}$$

and

$$I_2(x) = \int_{x}^{\infty} \exp\left\{-\int_{0}^{u} \frac{2f(y)}{\sigma^2(y)}\,dy\right\} du. \tag{8.8}$$

If $I_1(x) < \infty$ and $I_2(x) = \infty$, then for $t \in [0, T]$

$$P\left[\omega: \sup_t x(t, \omega) < \infty\right] = 1.$$

For a proof of this theorem see Gihman and Skorohod (1972, p. 119).

We now present the elements of the Liapunov–Kushner approach.

## 8.2 The Liapunov–Kushner approach

This method is developed in Kushner (1967, pp. 27–76) and is an extension of the deterministic indirect stability method of Liapunov. Before we state the main theorem we discuss some preliminary notions.

First, we introduce the function $V(x)$ which plays the role of a *Liapunov function*. Its properties are described in theorem 8.3.

Secondly, we define the set $Q_m \equiv \{x: V(x) < m < \infty\}$. It is assumed that $Q_m$ is a bounded set.

Thirdly, $\mathcal{D}_m V(x)$ is the differential operator given by (7.10) and it plays a role equivalent to the role of the trajectory derivative in deterministic stability. It denotes the average time rate of change of the process $V(x(t))$ at time $t$ given $x(t) = x$. For an equation such as (8.2) we have

$$\mathcal{D}_m V(x) = f(x) \frac{dV(x)}{dx} + \tfrac{1}{2} \sigma^2(x) \frac{d^2 V(s)}{dx^2} . \tag{8.9}$$

**Theorem 8.3.** Let $x(t)$ be a unique solution of (8.2) and suppose that $V(x)$ is a continuous non-negative function which is bounded, and has bounded, continuous first- and second-order derivatives in $Q_m$. If $\mathcal{D}_m V(x) \leqslant 0$ in $Q_m$ the $x(t) \to \bar{x}$, where $\bar{x} \in \{x: \mathcal{D}_m V(x) = 0\} \cap Q_m$ with probability at least $1 - V(x)/m$.

For a proof of this theorem see Kushner (1967, p. 42). To illustrate theorem 8.3 consider the simple example given by Kushner (1967, pp. 55–56). Let $x(t)$ be the solution of an Itô equation of the form

$$dx = ax\,dt + \sigma x\,dz. \tag{8.10}$$

Choose as $V(x) = x^2$ and note that for each set $Q_m = \{x: x^2 < m < \infty\}$, $V(x)$ is non-negative, bounded and has bounded first- and second-order derivatives in $Q_m$. Applying (8.9) we get

$$\mathcal{D}_m V(x) = ax\,2x + \tfrac{1}{2}\,2(\sigma x)^2 = x^2\,(2a + \sigma^2).$$

If $2a + \sigma^2 < 0$, then $\mathcal{D}_m V(x) \leqslant 0$, and by theorem 8.3, $x(t) \to 0$ w.p.1 because $m$ can be made arbitrarily large.

Two remarks are in order. First, both approaches to stability yield a result that shows the solution of an Itô equation not just staying within a small neighborhood of the 0-equilibrium solution but actually the solution approaching the 0-equilibrium. Technically we describe this stability as a *local asymptotic stochastic stability*. If the 0-equilibrium is stable and $x(t) \to 0$ w.p.1 for any nonrandom initial condition $x \in R$, then we say that the 0-equilibrium is *globally asymptotically stable* or that *stochastic stability in the large* holds.

Secondly, as in the deterministic case, finding an appropriate Liapunov function to use in establishing stability is usually difficult. Kushner (1967, pp. 60–61) describes a method presented below to guide the researcher in his efforts for Itô equations such as (8.2). A similar method is also presented by Feller (1954). As this method is described note its similarity to the Gihman–Skorohod method, and particularly eq. (8.5).

Assume that a Liapunov function exists with $V(0) = 0$ and $V(x) > 0$ for $x \neq 0$. Suppose that $V(x)$ is a continuous and bounded function and has bounded continuous first- and second-order derivatives so that $\mathscr{D}_m V(x)$ is defined in some open bounded set $Q_m$ with (8.9) holding. Suppose that

$$\mathscr{D}_m V(x) = f(x) V_x(x) + \tfrac{1}{2} \sigma^2(x) V_{xx}(x) \leqslant 0. \tag{8.11}$$

From (8.11) we obtain:

$$\frac{V_{xx}(x)}{V_x(x)} \leqslant - \frac{2f(x)}{\sigma^2(x)} . \tag{8.12}$$

The function that satisfies (8.12) when strict equality holds is

$$V(x) = \int\limits_0^x \exp\left\{ - \int^u \frac{2f(y)}{\sigma^2(y)} \, dy \right\} du, \tag{8.13}$$

provided that the integral in (8.13) exists. Thus, we see that eqs. (8.5) and (8.6) used in the Gihman–Skorohod method play the role of Liapunov functions. For further study of stability see Ladde and Lakshmikanthan (1980, pp. 56–91).

## 9. Existence of stationary distribution

A broader concept than that of the equilibrium solution discussed in the previous section is the concept of *stationary distribution*, or *steady state distribution*, or *equilibrium distribution*. Such a stationary distribution is obtained from the solution of a stochastic differential equation which is a time independent random variable, i.e. $x(t, \omega) = x(\omega)$.

The problem of the existence of a stationary distribution for a stochastic differential equation has been solved in the mathematical literature only for certain special cases. Some of the pioneering work in this area includes Feller (1954), Tanaka (1957) and Khas'minskii (1962) among others. Their results are summarized in Mandl (1968) and have been applied in economics in the work of Bourguignon (1974) and Merton (1975a). Below we present conditions for the existence of a stationary distribution of an autonomous Itô stochastic equation of the form of (8.2). In the next chapter we will apply the result to obtain the stationary distribution for the neoclassical economic growth model under uncertainty.

Consider eq. (8.2) and suppose that $f(x)$ and $\sigma(x)$ are continuously differentiable on $[0, \infty)$ with $\sigma(x) > 0$ on $(0, \infty]$ and (8.3) holding. From theorem 7.4 we know that the solution of (8.2), $x(t)$, is a diffusion process taking on values in the interval $[0, \infty]$. Furthermore, the endpoints of the interval $[0, \infty]$ are *absorbing states*, i.e. if $x(t) = 0$, then $x(u) = 0$ for $u > t$ and similarly if $x(t) = \infty$, then $x(u) = \infty$ for $u > t$.

In general, a stationary distribution will always exist in the sense that $x$ will either (1) be absorbed at one of the boundaries or (2) it will have a finite density function on the interval $(0, \infty)$ or (3) it will have a discrete probability mix of (1) and (2). Possibility (2) is the nontrivial case which is of interest for (8.2) under the stated assumptions. Note, however, that in case (2) the boundaries are *inaccessible* in the sense that, as $\epsilon \to 0$,

$$P[x(t) \leq \epsilon] \to 0, \tag{9.1}$$

and also

$$P[x(t) \geq 1/\epsilon] \to 0. \tag{9.2}$$

A necessary and sufficient condition for (9.1) and (9.2) to hold is that eqs. (9.3) $-$(9.5) are satisfied, where

$$\int_0^x I_2(u)\,\mathrm{d}u = \infty, \tag{9.3}$$

$$\int_x^\infty I_2(u)\,\mathrm{d}u = \infty, \tag{9.4}$$

$$\int_0^\infty I_1(u)\,\mathrm{d}u < \infty, \tag{9.5}$$

where $I_1(u)$ and $I_2(u)$ are defined as follows:

$$I_1(u) = \frac{1}{\sigma^2(u)} \exp\left\{ \int^u \frac{2f(y)}{\sigma^2(y)}\,\mathrm{d}y \right\}$$

and

$$I_2(u) = \frac{1}{\sigma^2(u)} \int^u \exp\left\{ \int_y^u \frac{2f(v)}{\sigma^2(v)}\,\mathrm{d}v \right\}\,\mathrm{d}u.$$

If conditions (9.3)–(9.5) are satisfied a stationary distribution, denoted by $\pi(x)$, exists and is given by

$$\pi(x) = \frac{m}{\sigma^2(x)} \exp\left\{\int^x \frac{2f(y)}{\sigma^2(y)}\,\mathrm{d}y\right\},\qquad (9.6)$$

with $m$ chosen so that $\int_0^\infty \pi(x)\,\mathrm{d}x = 1$.

## 10.  Stochastic control

In this section we establish various propositions from stochastic control which have been found useful in economic applications. The analysis is intuitive. For a rigorous analysis see Fleming and Rishel (1975).

Consider the problem:

$$J(k(t), t, \infty) = \max_{v(\cdot)} \mathrm{E}_t \int_t^\infty \mathrm{e}^{-\rho s}\, u(k(s), v(s))\,\mathrm{d}s \qquad (10.1)$$

subject to the conditions

$$\mathrm{d}k(t) = T(k(t), v(t))\,\mathrm{d}t + \sigma(k(t), v(t))\,\mathrm{d}z(t), \quad k(t) \text{ given}. \qquad (10.2)$$

Here $v = v(t) = v(t, \omega)$ is the *control variable*, $k = k(t) = k(t, \omega)$ is the state variable, $\rho \geqslant 0$ is the *discount on future utility*, $u$ denotes a *utility function*, $T$ is the *drift component* of technology, and $\sigma$ is the *diffusion component*. Note that $\mathrm{E}_t$ denotes expectation conditioned on $k(t)$ and $v(t)$. The problem described in (10.1) and (10.2) is a stochastic analogue stated and studied by Arrow and Kurz (1970, pp. 27–51). A standard technique for our problem, as in the case of Arrow and Kurz, is *Bellman's Principle of Optimality* according to which "an optimal policy has the property that, whatever the initial state and control are, the remaining decisions must constitute an optimal policy with regard to the state resulting from the first decision". See Bellman (1957, p. 83). Problem (10.1) and (10.2) is studied here for the undiscounted, finite horizon case, i.e. for $\rho = 0$ and $N < \infty$. This case is further subdivided in certain subcases as the analysis below indicates.

## 10.1. Maximum principle in one dimension

Consider the special case of (10.1) and (10.2) where the problem becomes

$$J(k(t), t, N) = \max_{v} E_t \int_t^N u(k, v)\, ds \tag{10.3}$$

subject to conditions

$$dk = T(k, v)\, dt + \sigma(k, v)\, dz, \quad k(t) \text{ given.} \tag{10.4}$$

Using Bellman's technique of dynamic programming, problem (10.3) and (10.4) can be analyzed as follows:

$$
\begin{aligned}
J(k(t), t, N) &= \max_{v} E_t \int_t^N u(k, v)\, ds \\
&= \max_{v} E_t \int_t^{t+\Delta t} u(k, v)\, ds + \max_{v} E_{t+\Delta t} \int_{t+\Delta t}^N u(k, v)\, ds \\
&= \max_{v} E_t \int_t^{t+\Delta t} u(k, v)\, ds + J(k(t+\Delta t), t+\Delta t, N) \\
&= \max_{v} E_t \left[ \int_t^{t+\Delta t} u(k, v)\, ds + J(k(t+\Delta t), t+\Delta t, N) \right] \\
&= \max_{v} E_t [ u(k(t), v(t))\, \Delta t + J(k(t), t, N) \\
&\quad + J_k \Delta k + J_t \Delta t + \tfrac{1}{2} J_{kk} (\Delta k)^2 \\
&\quad + J_{kt} (\Delta k)(\Delta t) + \tfrac{1}{2} J_{tt} (\Delta t)^2 + o(\Delta t) ].
\end{aligned}
\tag{10.5}
$$

Observe that Taylor's theorem is used to obtain (10.5) and therefore it is assumed that $J$ has continuous partial derivatives of all orders less than 3 in some open set containing the line segment connecting the two points $(k(t), t)$ and $(k(t+\Delta t), t+\Delta t)$. Let (10.4) be approximated and write

$$\Delta k = T(k, v)\, \Delta t + \sigma(k, v)\, \Delta z + o(\Delta t). \tag{10.6}$$

Insert (10.6) into (10.5) and recall the multiplication rules of eq. (4.12) of Chapter 2. The result is

$$0 = \max_{v} \mathrm{E}_t \left[ u(k(t), v(t)) \, \Delta t + (J_k T + J_t + \tfrac{1}{2} J_{kk} \sigma^2) \, \Delta t \right.$$

$$\left. + J_k \sigma \Delta z + o(\Delta t) \right]. \tag{10.7}$$

For notational convenience let

$$\Delta J = [J_t + J_k T + \tfrac{1}{2} J_{kk} \sigma^2] \, \Delta t + J_k \sigma \Delta z. \tag{10.8}$$

Using (10.8), eq. (10.7) becomes

$$0 = \max_{v} \mathrm{E}_t \left[ u(k(t), v(t)) \, \Delta t + \Delta J + o(\Delta t) \right]. \tag{10.9}$$

Eq. (10.9) is a partial differential equation with boundary condition

$$\frac{\partial J}{\partial k} (k(N), N, N) = 0. \tag{10.10}$$

Pass $\mathrm{E}_t$ through the parenthesis of (10.9) and after dividing both sides by $\Delta t$, let $\Delta t \to 0$ to conclude

$$0 = \max_{v} \left[ u(k(t), v(t)) + J_t + J_k T(k(t), v(t)) + \tfrac{1}{2} J_{kk} \sigma^2 (k(t), v(t)) \right]. \tag{10.11}$$

Eq. (10.11) is usually written as

$$-J_t = \max_{v} \left[ u(k(t), v(t)) + J_k T(k(t), v(t)) + \right.$$

$$\left. + \tfrac{1}{2} J_{kk} \sigma^2 (k(t), v(t)) \right] \tag{10.12}$$

and is known as the *Hamilton–Jacobi–Bellman equation of stochastic control theory*.

Let us proceed further with our analysis. We define the *costate variable $p(t)$* in the next equation as

$$p(t) = J_k(k(t), t, N). \tag{10.13}$$

From (10.13) it follows immediately that

$$p_k = \frac{\partial p}{\partial k} = J_{kk}. \tag{10.14}$$

Using (10.13) and (10.14) we may rewrite (10.12) as

$$-J_t = \max_v H\left(k, v, p, \frac{\partial p}{\partial k}\right), \tag{10.15}$$

where $H$ is the functional notation of the expression inside the brackets of (10.12). Assume next that a function $v$ exists that solves the maximization problem of (10.15) and denote such a function by

$$v^0 = v^0\left(k, p, \frac{\partial p}{\partial k}\right). \tag{10.16}$$

Note that $v^0$ is a function of $k(t)$ and $t$ alone along the optimum path, because $J_k$ is a function of $k(t)$ and $t$ alone. In the applied control literature, and more specifically in economic applications, $v^0$ is called a *policy function*. Assuming then that a policy function $v^0$ exists, (10.15) may be rewritten as

$$
\begin{aligned}
-J_t &= \max_v H\left(k, v, p, \frac{\partial p}{\partial k}\right) \\
&= H\left(k, v^0\left(k, p, \frac{\partial p}{\partial k}\right), \frac{\partial p}{\partial k}\right) \\
&= H^0\left(k, p, \frac{\partial p}{\partial k}\right).
\end{aligned} \tag{10.17}
$$

This last equation, (10.17), is again a functional notation of the right-hand side expression of (10.12) under the assumption of the existence of an optimum control $v^0$, i.e.

$$H^0\left(k, p, \frac{\partial p}{\partial k}\right) = u(k, v^0) + pT(k, v^0) + \frac{1}{2}\frac{\partial p}{\partial k}\sigma^2(k, v^0). \tag{10.18}$$

Equipped with the above analysis, our final goal in this subsection is to derive a system of stochastic differential equations describing the behavior of the state and costate variables, i.e. find expressions for $dk$ and $dp$. An expression for $dk$ is almost readily available to us from (10.4), (10.16) and (10.18); in particular

$$
\begin{aligned}
dk &= T(k, v^0)dt + \sigma(k, v^0)dz \\
&= H_p^0\left(k, p, \frac{\partial p}{\partial k}\right)dt + \sigma\left(k, p, \frac{\partial p}{\partial k}\right)dz \\
&= H_p^0\,dt + \sigma dz.
\end{aligned} \tag{10.19}
$$

Note that $\partial H^0/\partial p = H_p^0 = T$, used in the derivation of (10.19), is obtained from (10.18). Next we derive an expression for $dp$. Use the definition of $p(t)$ given in (10.13), eq. (10.4), and Itô's lemma to get

$$
\begin{aligned}
dp &= J_{kt}\,dt + J_{kk}\,dk + \tfrac{1}{2}J_{kkk}(dk)^2 \\
&= J_{kt}\,dt + J_{kk}(T\,dt + \sigma\,dz) + \tfrac{1}{2}J_{kkk}(T\,dt + \sigma\,dz)^2 \\
&= [J_{kt} + J_{kk}T + \tfrac{1}{2}J_{kkk}\sigma^2]\,dt + J_{kk}\sigma\,dz.
\end{aligned}
\tag{10.20}
$$

To simplify eq. (10.20) we compute $J_{kt}$ from (10.17) assuming that equality holds for mixed partial derivatives

$$
\begin{aligned}
-J_{tk} &= H_k^0 + H_p^0\,\frac{\partial p}{\partial k} + H_{pk}^0\,\frac{\partial^2 p}{\partial k^2} \\
&= H_k^0 + TJ_{kk} + \tfrac{1}{2}\sigma^2 J_{kkk}.
\end{aligned}
\tag{10.21}
$$

Note that to obtain (10.21), first we use the facts that $H_p^0 = T$ and $H_{pk}^0 = \tfrac{1}{2}\sigma^2$, both resulting from partial differentiation of (10.18), and secondly, the definition in (10.14). Substituting (10.21) into (10.20) we reach our desired result

$$
\begin{aligned}
dp &= [-H_k^0 - J_{kk}T - \tfrac{1}{2}\sigma^2 J_{kkk} + J_{kk}T + \tfrac{1}{2}\sigma^2 J_{kkk}]\,dt + J_{kk}\sigma\,dz \\
&= -H_k^0\,dt + \sigma J_{kk}\,dz.
\end{aligned}
\tag{10.22}
$$

We summarize the above analysis in a proposition.

**Proposition 10.1** (Pontryagin Stochastic Maximum Principle). Suppose that $k(t)$ and $v^0(t)$ solve for $t \in [0, N]$

$$
\max_{v} E_0 \int_0^N u(k(t), v(t))\,dt
$$

subject to the conditions

$$
dk = T(k(t), v(t))\,dt + \sigma(k(t), v(t))\,dz, \quad k(t)\text{ given.}
$$

Then, there exists a costate variable $p(t)$ such that for each $t, t \in [0, N]$:
(1) $v^0$ maximizes $H(k, v, p, \partial p/\partial k)$ where

$$H\left(k, v, p, \frac{\partial p}{\partial k}\right) = u(k, v) + pT(k, v) + \tfrac{1}{2}\,\sigma^2 \cdot \frac{\partial p}{\partial k}\;;$$

(2) the costate function $p(t)$ satisfies the stochastic differential equation $dp = -H_k^0\,dt + \sigma(k, v^0)\,J_{kk}\,dz$; and

(3) the transversality condition holds

$$p(k(N), N) = \frac{\partial J}{\partial k}\,(k(N), N, N) \geqslant 0,$$

$$p(N)k(N) = 0.$$

Next we proceed to make some generalizations.

## 10.2. Generalizations

First, mathematically speaking the optimal path obtained from Pontryagin's Stochastic Maximum Principle is the solution of two stochastic differential equations subject to certain conditions. More specifically, we rewrite these equations together

$$dk = H_p^0\,dt + \sigma dz, \tag{10.19}$$

$$dp = -H_k^0\,dt + \sigma J_{kk}\,dz, \tag{10.22}$$

$$k(0) \text{ given}, \tag{10.23}$$

$$p(k(N), N) = 0. \tag{10.24}$$

Secondly, it is easy to generalize proposition 10.1 by introducing a bequest function $B(k(t), t)$. If this were to be the case the maximization problem would become

$$\max_v \mathrm{E}_0\!\left[\int_0^N u(k(t), v(t))\,dt + B(k(N), N))\right]$$

subject to the same conditions (10.4) as before. Proposition 10.1 holds with a new transversality condition, i.e.

$$p(k(N), N) = \frac{\partial B}{\partial k}\,(k(N), N).$$

Thirdly, generalizing (10.3) and (10.4) in another direction by allowing discounting, we need not repeat the previous analysis. Let

$$J(k(t), t, N) = \max_v E_t \int_t^N e^{-\rho s} u(k, v) \, ds$$

subject to (10.4) and $k(t)$ given. Write

$$W(k(t), t, N) = e^{\rho t} J(k(t), t, N)$$

and also

$$J_t = \frac{d}{dt} \{e^{-\rho t} W\} = -\rho e^{-\rho t} W.$$

Thus, the Hamilton–Jacobi–Bellman equation is now transformed into

$$\rho W = \max_v [u(k(t), v(t)) + W_k T(k(t), v(t)) + \tfrac{1}{2} \sigma^2 (k(t), v(t)) W_{kk}].$$

The remaining analysis can then be patterned after what is done above.

Finally, consider the general multidimensional time dependent case described by

$$J(k(t), t, N)) = \max_v E_t \left[ \int_t^N u(k(s), v(s), s) \, ds + B(k(N), N) \right]$$

subject to

$$dk_i(t) = T_i(k(t), v(t), t) \, dt + \sigma_i(k(t), v(t), t) \, dz_i(t), \quad i = 1, 2, ..., n,$$

$k(t)$ given.

For this problem the Hamilton–Jacobi–Bellman equation is

$$\begin{aligned} -J_t(k(t), t, N) &= \max_v [u(k(t), v(t), t) + J_k' T(k(t), v(t), t) \\ &\quad + \tfrac{1}{2} \operatorname{tr}(J_{kk} \sigma(k(t), v(t), t) \sigma'(k(t), v(t), t))] \\ &= \max_v H(k(t), v(t), p(t), p_k(t), t) \\ &= H^0(k(t), p(t), p_k(t), t). \end{aligned}$$

As before, primes denote transpose. The multidimensional analogue of the costate stochastic differential equation becomes

$$dp_i = H^0_{k_i} \, dt + \sum_{j=1}^{n} J_{k_i k_j} \, \sigma_j \, dz_j$$

provided that $dz_i$ is not correlated with $dz_j$ for $i \neq j$. A detailed analysis may be found in Bismut (1973).

## 10.3.  Constraints on controls

We now study a more general problem in stochastic control theory which is formulated as follows:

$$J(k(t), t, N) = \max_v \mathrm{E}_t \left[ \int_t^N u(k(s), v(s), s) \, ds + B(k(N), N) \right] \tag{10.25}$$

subject to the conditions

$$dk_i(t) = T_i(k(t), v(t), t) \, dt + \sum_{j_i=1}^{n_i} \sigma_{ij_i}(k(t), v(t), t) \, dz_{ij_i},$$

$$i = 1, 2, ..., n, \tag{10.26}$$

$k(t)$ given;

$$g_\ell(k(t), v(t), t) \geqslant 0, \quad \ell = 1, 2, ..., L. \tag{10.27}$$

In other words, the control stochastic process $v(t)$ satisfies a set of $L$ inequality constraints. Note that it is assumed that $dz_{ij_i}$ are Wiener processes that satisfy

$$\text{covariance } (dz_{rj_r}, dz_{sj_s}) = \rho_{rj_r sj_s} \, dt,$$

where $\rho_{rj_r sj_s}$ is the correlation coefficient which is independent of $k(t)$ and $v(t)$.

We proceed with the analysis of this problem and at the end we summarize the results in a proposition, as we did earlier.

We write the recursive equation by Bellman's Principle of Optimality:

$$J(k(t), t, N) = \max_v \mathrm{E}_t \int_t^{t+\Delta t} u(k(s), v(s), s) \, ds + J(k(t + \Delta t), t + \Delta t, N).$$

Now, if the approximation

$$\mathrm{E}_t \int_t^{t+\Delta t} u(k(s), v(s), s) \, ds = u(k(t), v(t), t) \Delta t + o(\Delta t)$$

is valid, then we may write

$$J(k(t), t, N) = \max_v [u(k(t), v(t), t) \Delta t$$
$$+ E_t J(k(t + \Delta t), t + \Delta t, N) + o(\Delta t)].$$

Put,

$$\Delta J(t) = J(k(t + \Delta t), t + \Delta t, N) - J(k(t), t, N).$$

Thus we obtain

$$0 = \max_v [u(k(t), v(t), t) \Delta t + E_t \Delta J(t) + o(\Delta t)]. \tag{10.28}$$

Use Taylor's Theorem to expand $\Delta J(t)$ around $(k(t), t)$, assuming $J$ is twice differentiable. Then

$$\Delta J(t) = J_t \Delta t + J'_k \Delta k + \tfrac{1}{2} \Delta k' J_{kk} \Delta k + o(\Delta t). \tag{10.29}$$

By taking conditional expectation of $\Delta J(t)$ in (10.29) we have

$$E_t \Delta J(t) = J_t \Delta t + E_t [J'_k \Delta k] + E_t [\tfrac{1}{2} \Delta k' J_{kk} \Delta k] + o(\Delta t). \tag{10.30}$$

So, we have to compute $E_t [J'_k \Delta k]$ and $E_t [\tfrac{1}{2} \Delta k' J_{kk} \Delta k]$. We do so next. From (10.26) write

$$\Delta k_i = T_i \Delta t + \sum_{j_i=1}^{n_i} \sigma_{ij_i} \Delta z_{ij_i} + o(\Delta t), \tag{10.31}$$

and by taking conditional expectation of (10.31) we have

$$E_t \Delta k_i = E_t \left[ T_i \Delta t + \sum_{j_i=1}^{n_i} \sigma_{ij_i} \Delta z_{ij_i} + o(\Delta t) \right] = T_i \Delta t + o(\Delta t),$$

because $E_t \Delta z_{ij_i} = 0$ for all $i$ and $j_i$. With this information available we compute:

$$E_t \{J'_k \Delta k\} = E_t \left\{ \sum_{i=1}^{n} J'_{k_i} \Delta k_i \right\} = \sum_{i=1}^{n} J'_{k_i} E_t (\Delta k_i)$$

$$= \sum_{i=1}^{n} J'_{k_i} T_i \Delta t + o(\Delta t).$$

Furthermore,

$$
\begin{aligned}
E_t\{(\Delta k)'J_{kk}(\Delta k)\} &= E_t\left\{\sum_{r=1}^{n}\sum_{s=1}^{n}(\Delta k_r)J_{k_r k_s}(\Delta k_s)\right\} \\
&= E_t\left\{\sum_{r=1}^{n}\sum_{s=1}^{n}\sum_{j_r=1}^{n_r}\sum_{j_s=1}^{n_s}J_{k_r k_s}\sigma_{rj_r}\sigma_{sj_s}\Delta z_{rj_r}\Delta z_{sj_s}+o(\Delta t)\right\} \\
&= \sum_{r=1}^{n}\cdot\sum_{s=1}^{n}\sum_{j_r=1}^{n_r}\sum_{j_s=1}^{n_s}J_{k_r k_s}\sigma_{rj_r}\sigma_{sj_s}\rho_{rj_r sj_s}\Delta t+o(\Delta t),
\end{aligned}
$$

because $E_t\{\Delta z_{rj_r}\Delta z_{sj_s}\}=\rho_{rj_r sj_s}\Delta t+o(\Delta t)$.

Collecting the results above (10.30) becomes

$$
\begin{aligned}
E_t\Delta J(t) &= J_t\Delta t+\sum_{i=1}^{n}J_{k_i}T_i\Delta t \\
&\quad+\tfrac{1}{2}\sum_{r=1}^{n}\sum_{s=1}^{n}\sum_{j_r=1}^{n_r}\sum_{j_s=1}^{n_s}J_{k_r k_s}\sigma_{rj_r}\sigma_{sj_s}\rho_{rj_r sj_s}\Delta t+o(\Delta t),
\end{aligned}
$$

and also

$$
\begin{aligned}
-J_t(k(t),t,N)\Delta t = \max_{v}\{&u(k(t),v(t),t)\Delta t \\
&+\phi(k(t),v(t),t,N)\Delta t+o(\Delta t)\},
\end{aligned}
$$

where $\phi$ is defined by

$$
\phi(k(t),v(t),t,N)=\sum_{i=1}^{n}J_{k_i}T_i+\tfrac{1}{2}\sum_{r,s,j_r,j_s}J_{k_r k_s}\sigma_{rj_r}\sigma_{sj_s}\rho_{rj_r sj_s}. \tag{10.32}
$$

But notice now that $\phi$ is a function of $(k,v,t,N)$ since each $T_i$ and $\sigma_{ij_i}$ are functions of $(k,v,t,N)$. Therefore we can replace the maximization problem by the simpler one, i.e.

$$
-J(k(t),t,N)\Delta t=\max_{v\in C_g(k(t),t)}[u(k(t),v(t),t)\Delta t+\phi(k(t),v(t),t,N)\Delta t+
$$

$$
+o(\Delta t)],
$$

where the constraint set is defined by

$$
C_g(k(t),t)=\{v\in R^m:g(k(t),v,t)\geqslant 0\}.
$$

We summarize the results in

**Proposition 10.2** (The stochastic maximum principle with constraints).  Suppose that $J(k, t, N)$ is twice continuously differentiable in $(k, t)$ and that the optimal $k(t)$ and $v(t)$ are such that for all $r$

$$E_r \int_r^{r+\Delta r} u(k(s), v(s), s)\, ds = u(k(r), v(r), r)\, \Delta r + o(\Delta r).$$

Then, at each time $t$, the optimal control $v(t)$ solves

$$\max_{v \in C_g(k(t), t)} [u(k(t), v, t) + \phi(k(t), v, t, N)],$$

where

$$C_g(k(t), t) = \{v \in R^m : g_\varrho(k(t), v, t) \geq 0, \quad \varrho = 1, 2, ..., L\}$$

and $J(k, t, N)$ must solve the partial differential equation

$$-J_t(k, t, N) = \max_{v \in C_g} [u(k, v, t) + \phi(k, v, t, N)],$$

with $\phi$ defined as in (10.32) with boundary condition

$$J(k, N, N) = B(k, N, N)$$

for all $k$.


## 11.  Bismut's approach

Bismut (1973) has applied the general methods of convex analysis developed by Rockafellar (1970) to problems of optimal stochastic control. In this section we present an intuitive exposition of Bismut's ideas as found in Bismut (1975). Although a specific application of Bismut's method is postponed until the next chapter, the analysis below of optimal stochastic control attempts to interpret in terms of economic concepts the methods and results of mathematical theory. In particular, there are two important concepts in optimal stochastic control that are useful in economic applications. These are the concept of risk-taking and information processing.

As before the problem can be formulated as:

$$\max E_0 \int_0^T u(k, v, t, \omega)\, dt \tag{11.1}$$

subject to the conditions

$$dk = f(k, v, t, \omega)\, dt + \sigma(k, v, t, \omega)\, dz,$$

$$k(0) = k_0 \text{ given.} \tag{11.2}$$

Here $u$ can represent an *instantaneous utility* or *profit function*, $k$ denotes *capital stock*, $v$ is the *investment* decision, and $\omega$ is the *environmental factor*. For simplicity we study the one-dimensional case. We assume that an increasing system of information is available from the family of $\sigma$-fields $\{\mathcal{F}_t : t \in [0, T]\}$ and that this information includes the past values of $k$ and $z$ and also $f$ and $\sigma$. Here, as before, $z$ is a Wiener process. Under this assumption the expected mean and variance of the capital increment consecutive to any decision $v$ are thus known. Let $p_t$ denote the *marginal value of capital* at time $t$ which is given by

$$p_t = \frac{\partial E_t}{\partial k} \int_t^T u(k, v, s, \omega)\, ds, \tag{11.3}$$

i.e. $p_t$ is the partial derivative with respect to $k$ of the conditional expectation of the utility function from time $t$ with $v$ being an optimal policy. Bismut (1973, p. 387) and Bismut (1975, p. 242) assume that $p_t$ may be written as

$$p_t = p_0 + \int_0^t \dot{p}_s\, ds + \int_0^t H_s\, dz_s + M_t, \tag{11.4}$$

where $\dot{p}_s$ is the infinitesimal expected rate of growth of $p$, $H_s$ is the infinitesimal conditional covariance of $p$ with $z$, $M$ is a predictive term, $M_0 = 0$, which is the best estimate at time $t$ of a given random variable and which is independent of $z$. Its infinitesimal increments have then a null conditional expected value at each time.

This decomposition corresponds to the idea that $p_t$ can be decomposed in the sum of $p_0$, of the second term that gives the expected infinitesimal increment of $p_t$ at each time, of the third term which integrates uncertainties in the accumulation process, and finally of the last term which integrates the information on environmental factors $M_t$.

Define now $\mathcal{H}$ by

$$\mathcal{H} = u(k, v, t, \omega) + pf(k, v, s, \omega) + H\sigma(k, v, s, \omega). \tag{11.5}$$

Bismut (1973, p. 401) proves that the following relations hold for optimal $v$:

$$\frac{\partial \mathscr{H}}{\partial v} = 0, \tag{11.6}$$

$$dp = -\frac{\partial \mathscr{H}}{\partial k} \, dt + H \, dz + dM, \tag{11.7}$$

$$p_T = 0. \tag{11.8}$$

As in the preceding section $dk$ is given by (11.2) with $v$ being optimal. Note that eqs. (11.6)–(11.8) and (11.2) are very similar to the equations stated in proposition 10.1. More specifically, eq. (11.6) follows from the maximality of $v$, while eq. (10.19) and (11.2) with $v$ maximal are identical. Eqs. (10.24) and (11.8) are the same and they denote the transversality condition. Finally, (10.22) and (11.7) are the only equations that differ. Note that Bismut's random variable $H$ in (11.7) corresponds to the random variable $J_{kk}\sigma$ in (10.22), while the term $dM$ in (11.7) has no analogous term in (10.22). Roughly speaking, Bismut's correspondence of primal and dual variable is as follows:

$$f \rightarrow p, \tag{11.9}$$

$$\sigma \rightarrow H, \tag{11.10}$$

$$\mathscr{F}_t \rightarrow M_t. \tag{11.11}$$

Let us now interpret the various variables above to uncover their economic meaning. First note that $\mathscr{H}$ in (11.5) is the sum of instantaneous utility or profit, plus the expected infinitesimal increment of capital valued at its marginal expected value, minus the risk associated with a given investment policy valued at its cost. The instantaneous attitude towards risk is given by $H$ and it is positive if the individual is risk-taking and negative if he is risk-averting. Next we interpret $dp$ or $-dp$, i.e. the conditional expected rate of depreciation in the marginal value of capital. From (11.7) we see that $-dp$ is the sum of capital's contribution to utility or profits plus capital's contribution to enhancing the expected value of the increment of the capital stock, minus its contribution to increasing the conditional standard deviation of the increment of the capital stock valued at the cost of risk, minus two other terms: $H \, dz$ and $dM$. To interpret $H \, dz$ note from (11.2) that

$$dz = (1/\sigma)(dk - f \, dt) \tag{11.12}$$

which, upon multiplication with $H$, yields

$$Hdz = (H/\sigma)(dk - f dt). \qquad (11.13)$$

The term $Hdz$ in (11.7) is then a correction term in the evolution of the marginal value of capital which evaluates in terms of $p$ the difference between $dk$ and $E(dk)$, where $E(dk) = f dt$. The last term, $dM$, denotes changes in the predic-tion of long-term uncertainties which can either increase or decrease the value of capital. Intuitively speaking, $M$ incorporates information which is not contained in past values of $z$ and while $z$ contains all short-term uncertainties that appear in the accumulation process, $M$ is a prediction of the long-term uncertainties.

This concludes the discussion of Bismut's approach.

## 12. Jump processes

In this section we develop the generalized Itô formula for *jump processes*. Jump processes will be modeled by a Poisson process that describes the arrival of random events. When a Poisson event arrives a jump in the state variable takes place and this jump will be distributed according to a preassigned density function. After the generalized Itô formula is developed, we will then develop the maximum principle by following much the same approach that we did for the case of diffusion processes. Our approach remains intuitive.

### 12.1. Generalized Itô formula

We will develop the generalized formula for the case of mixed Poisson and Brownian processes. Consider the following:

$$dx(t) = f(t,x) dt + \sigma(t,x) dz(t) + g(t,x) dq(t). \qquad (12.1)$$

Here, for $R = (-\infty, \infty)$,

$$f(t,x): [0,\infty) \times R \to R, \sigma(t,x): [0,\infty) \times R \to R,$$

$$g(t,x): [0,\infty) \times R \to R.$$

Also, $\{z(t)\}_{t=0}^{\infty}$ is a standardized Wiener process and $\{q(t)\}_{t=0}^{\infty}$ is a Poisson process assumed to be distributed independently of $\{z(t)\}_{t=0}^{\infty}$ in order to keep things simple.

Let $\lambda\Delta t + o(\Delta t)$ be the probability that $q(t)$ jumps once in $(t, t + \Delta t)$. Let the amplitude $A$ of the jump be random with density function $p(a)$, i.e. $p(a)da$ is the probability of a jump amplitude contained in $(a, a + da)$ up to higher order terms in $da$. Assume that the probability that $q(t)$ jumps more than once in $(t, t + \Delta t)$ is $o(\Delta t)$. Thus, the probability that $q(t)$ is constant on $(t, t + \Delta t)$ is $1 - \lambda\Delta t + o(\Delta t)$. The generalized Itô formula for the one-dimensional case may be stated. See Kushner (1967, p. 18) for a rigorous statement and proof. Gihman and Skorohod (1972, p. 263) have a more complete treatment of generalized Itô formulae than Kushner.

**Proposition 12.1** (Generalized Itô formula). Let $F(t, x)$ be twice continuously differentiable in $(t, x)$. Let there exist a closed and bounded interval such that $\{a \mid p(a) > 0\} \subset I$. Let $\Delta F = F(t + \Delta t, x + \Delta x)$ and let $E_t \Delta F$ denote conditional expectation conditioned on $x(t) = x$. Then

$$E_t \Delta F = \left\{ F_t(t,x) + F_x(t,x)f(t,x) + \tfrac{1}{2} F_{xx}(t,x)\sigma^2(t,x) \right.$$
$$\left. + \lambda \left( \int_{a \in I} [F(t, x + g(t,x)a) - F(t,x)]p(a)\,da \right) \right\}\Delta t + o(\Delta t).$$

To establish the theorem note that

$$E_t \Delta F = (E_t^* \Delta F)\lambda\Delta t + (1 - \lambda\Delta t) E_t^{**} \Delta F + o(\Delta t), \tag{12.2}$$

where $E_t^*$ means expectation conditioned on the occurrence of the Poisson event and $E_t^{**}$ means expectation conditioned on the Poisson event not occurring. Eq. (12.2) may be written

$$E_t \Delta F = E_t^{**} \Delta F + \lambda\Delta t (E_t^* \Delta F - E_t^{**} \Delta F) + o(\Delta t). \tag{12.3}$$

The first term in eq. (12.3) is the conditional expectation of the change in the function $F$ given that the Poisson event does not occur. The second term of (12.3) is where the action is. Notice first that the term $E_t^{**}\Delta F \to 0$, as $\Delta t \to 0$. Thus, we only have to look at the term $E_t^* \Delta F$ to calculate the second term on the right-hand side of eq. (12.3) up to $o(\Delta t)$. Performing these calculations we get

$$E_t^{**}\Delta F = F_t\Delta t + F_x E_t^{**}\Delta x + \tfrac{1}{2} F_{xx}E_t^{**}\Delta x^2$$
$$+ \lambda\Delta t \left( \int_{a \in I} [F(t, x + g(t,x)a) - F(t,x)]p(a)\,da \right) + o(\Delta t).$$

$$\tag{12.4}$$

We are able to write $\lambda \Delta t E_t^{**} \Delta F = o(\Delta t)$ because $E_t^{**} \Delta F \to 0$ as $\Delta t \to 0$. Also,

$$\lambda \Delta t E_t^* \Delta F = \lambda \Delta t E_t^* \int_{a \in I} [F(t + \Delta t, x + f\Delta t + \sigma \Delta z + ga + o(\Delta t))$$

$$- F(t, x)] p(a) \, da = \lambda \Delta t \int_{a \in I} [F(t, x + ga) - F(t, x)] p(a) \, da + o(\Delta t),$$

because $f\Delta t + \sigma \Delta z \to 0$ as $\Delta t \to 0$, since $\Delta z$ is normally distributed with mean 0 and variance $\Delta t$.

The formula (12.4) reduces to

$$E_t \Delta F = F_t \Delta t + F_x f \Delta t + \tfrac{1}{2} F_{xx} \sigma^2 \Delta t$$

$$+ \lambda \left( \int_{a \in I} (F(t, x + ga) - F(t, x)) p(a) \, da \right) \Delta t + o(\Delta t), \quad (12.5)$$

which is the generalized Itô formula.

## 12.1. The maximum principle for jump processes

Consider the problem

$$J(k(t), t, N) = \max_{v(\cdot)} E_t \left[ \int_t^N u(k(s), v(s), s) \, ds + B(k(N), N) \right] \quad (12.6)$$

subject to the conditions,

$$dk(s) = T(k(s), v(s), s) \, ds + \sigma(k(s), v(s), s) \, dz(s)$$

$$+ g(k(s), v(s), s) \, dq(s), \quad (12.7)$$

where all notation is as in section 10 except that the jump component

$$g(k(s), v(s), s) \, dq(s)$$

has been added. Here the jump amplitude $A$ is distributed with density $p(a)$, i.e. independent of $\{z(t)\}_{t=0}^{\infty}$ as in the statement of the generalized Itô formula. This assumption is made to keep matters simple.

Let us follow the procedure in section 10 together with the generalized Itô formula to discover the stochastic maximum principle.

Calculate for $(k, v, t)$ given at $t$

$$\mathrm{E}_t \left[ \int_t^{t+\Delta t} u \, \mathrm{d}s + J(k(t+\Delta t), t+\Delta t, N) - J(k(t), t, N) \right]$$

$$= u(k, v, t) \Delta t + J_t(k, t, N) \Delta t + J_k(k, t, N) T(k, v, t) \Delta t$$

$$+ \tfrac{1}{2} J_{kk}(k, t, N) \sigma^2(k, v, t) \Delta t$$

$$+ \lambda \Delta t \int_{a \in I} [J(k + g(k, v, t)a, t, N) - J(k, t, N)] p(a) \, \mathrm{d}a + o(\Delta t). \quad (12.8)$$

Eq. (12.8) is obtained from the generalized Itô formula applied to $\mathrm{E}_t \Delta J$. Let $\phi(k, v, t, N)$ denote the right-hand side of (12.8) divided by $\Delta t$. Then we write

**Proposition 12.2** (Maximum principle). Assume that $J$ satisfies the hypotheses of the generalized Itô formula. Then the optimal control function $v^0(k, t, N)$ is found by choosing $v$ to solve

$$\max_v \phi(k, v, t, N)$$

and $J$ is determined by the partial differential equation

$$0 = \max_v \phi(k, v, t, N) = \phi(k, v^0(k, t, N), t, N)$$

with the boundary condition

$$J(k, N, N) = B(k, N).$$

Obviously the maximum principle can be generalized in a straightforward way to multidimensions, constraint sets on the controls $v$, correlated Wiener and Poisson process, and so on, at this level of heuristic argument. Of course, to do these generalizations rigorously would take a lot of detailed mathematics.

## 13. Optimal stopping and free boundary problems

In Chapter 1, section 8, we introduced the reader to some basic notions on optimal stopping. In that chapter we considered a sequence of random variables $Y_1$, $Y_2$, ... and their corresponding rewards $X_1, X_2, \dots$ . By imposing certain assumptions on these sequences of random variables we were able to obtain the existence of an optimal stopping rule. In this section we rely on van Moerbeke (1974) to treat the continuous time case, where, instead of a sequence of random variables $Y_1$, $Y_2$, ..., we consider a Wiener process $z_t$ starting at $z_0(\omega) = 0$. Note that we write the *reward* function as $g(z, t)$ and by this we mean that at time $t$ when the state of affairs is $z = z_t(\omega)$, the value of the reward is given by $g(z, t)$. The *average reward* for a random period of time $\tau = \tau(\omega)$ is written as

$$Eg(z + z_\tau, t + \tau), \tag{13.1}$$

where the arguments of $g$, i.e. $(z + z_\tau, t + \tau)$, denote the space − time of the Wiener process starting at $(z, t)$. In other words, suppose that we start playing a game at time $t$ with a state of affairs $z$ and play for a random period $\tau$, where $\tau \leqslant T - t$. The state of affairs corresponding to this random period is determined by the underlying Wiener process and we denote it by $z_\tau$. Having started at $t$ with $z$ and played for $\tau$ periods giving us $z_\tau$, eq. (13.1) estimates the expected reward of such a game. Our primary interest is with a finite time interval $[0, T]$, where $T < \infty$ and $t \in [0, T]$. We assume that $g$ and all its partials are continuous for $t < T < \infty$ and have limits as $t \to T$. A discontinuity is permitted at $t = T$, but in this case we require that $h(z) \equiv g(z, T) - g(z, T-)$ is infinitely differentiable except for a few isolated jumps. Here $g(z, T-)$ denotes the left limit of $g(z, t)$ at $t = T$. The function $h$ is called the *final gain* and with no loss of generality we assume that it is non-negative. The assumption that $h(z) \geqslant 0$ is not restrictive because if $h(z) < 0$ in a given interval it will make sense to stop earlier, before hitting the final $t = T$.

Having defined the functions $g$ and $h$, and stated the assumptions made about them, we now impose upon $g$ a *growth* condition called the *Tychonov condition*. If the functions

$$g, \frac{\partial g}{\partial t}, \frac{\partial g}{\partial z}, \frac{\partial^2 g}{\partial z^2}, \frac{\partial^2 g}{\partial z \partial t}, \frac{\partial^3 g}{\partial z^3},$$

$$h, \frac{\partial h}{\partial z}, \frac{\partial^2 h}{\partial z^2}, \frac{\partial^3 h}{\partial z^3},$$

are bounded by $e^{o(z^2)}$ when $|z|$ tends to $\infty$, uniformly in any finite strip $[t, T]$, then $g$ is said to satisfy the *Tychonov condition*.

Suppose we start at time $t$ with a state of affairs $z$ and we play a game up to time $T < \infty$. Under these circumstances the *optimal reward* $\hat{g}(z, t)$ is obtained by maximizing (13.1) over all stopping times $\tau$, with $\tau \leqslant T - t$. The stopping time $\tau$ achieving this maximum is called the *optimal strategy*. Obviously, the optimal reward function $\hat{g}$ is important for our analysis and we would like to characterize it in some way. This is done by means of the concept of an excessive function. A function $f$, bounded below, is called *excessive* in an open domain $D$ of $R^2$ if

   (1)  $Ef(z + z_\tau, t + \tau) \leqslant f(z, t)$ for every stopping time $\tau$ not exceeding the first exit time $\tau_D$ from $D$, and

   (2)  $Ef(z + z_{\tau_n}, t + \tau_n) \to f(z, t)$ for every sequence of stopping times $\tau_n < \tau_D$ such that $P[\tau_n \to 0] = 1$.

Note that if $f$ is sufficiently differentiable, excessivity in the domain $D$ is the same thing as

$$\frac{\partial f}{\partial t} + \frac{1}{2} \frac{\partial^2 f}{\partial z^2} \leqslant 0$$

for all $(z, t)$ in $D$. Equipped with the definition of excessivity we characterize $\hat{g}$ as the smallest excessive function exceeding $g$. The Tychonov condition implies that $\hat{g}$ is finite and continuous.

Next we distinguish between two regions: a *continuation region* $C$ where $\hat{g} > g$, which means that it pays to play the game, and a *stopping region* $S$ where $\hat{g} = g$, which means that quitting is best. Since $\hat{g}$ is continuous the continuation region $C$ is open. We assume that the continuation region $C$ has a continuously differentiable boundary $z = s(t)$, except possibly for a few isolated points where $| ds/dt |$ may blow up. The boundary separating these two regions is the *optimal stopping boundary*. The Tychonov condition imposed on $g$ helps us to conclude that the optimal strategy is to play as long as you remain in $C$ and stop when you hit the optimal boundary. Let $\tau_0$ denote this hitting time. Then

$$\hat{g}(z, t) = Eg(z + z_{\tau_0}, t + \tau_0). \tag{13.2}$$

Therefore, our purpose is to find the optimal stopping boundary.

  At this point it is appropriate to remark that there is a beautiful interplay between our problem and the theory of partial differential equations. Although we proceed rapidly to reach our present goal of finding the optimal stopping boundary, we would like to encourage the interested reader to consult van Moerbeke (1974) and some of the many references he cites. Having said this, note that the problem of finding $\hat{g}$ and the optimal strategy can be solved by converting it into a *free boundary* problem for the *heat equation*. From (13.2) we conclude that $\hat{g}$ is *parabolic* in the continuation region $C$ which means that first $\hat{g}$ is excessive, and secondly $E\hat{g}(z + z_{\tau_U}, t + \tau_U) = \hat{g}(z, t)$, where $\tau_U$ is the first exit time from any open set $U$ with compact closure. Parabolic functions satisfy the backwards heat equation and conversely solutions of the backwards heat equation which are bounded below are all parabolic functions. Therefore,

$$\frac{\partial \hat{g}}{\partial t} + \frac{1}{2} \frac{\partial^2 \hat{g}}{\partial z^2} = 0 \quad \text{in } C, \tag{13.3}$$

$$\hat{g} = g \quad \text{at the boundary of } C \tag{13.4}$$

and

$$\hat{g}(z, T) = g(z, T). \tag{13.5}$$

Furthermore, because of our earlier assumption about the continuously differentiable boundary of the continuation region $C$, the optimality of $\hat{g}$ implies that

$$\frac{\partial \hat{g}}{\partial z}(s(t), t) = \frac{\partial g}{\partial z}(s(t), t), \tag{13.6}$$

and also for $(y, u) \in C$, as $(y, u) \to (s(t), t)$ we have that

$$\frac{\partial \hat{g}}{\partial z}(y, u) \to \frac{\partial \hat{g}}{\partial z}(s(t), t), \tag{13.7}$$

with both (13.6) and (13.7) holding at points $(s(t), t)$, where $|\,ds/dt\,| < \infty$. Eqs. (13.5), (13.6) and (13.7) are called the *smooth fit* equations.

Let us reflect for a moment. What eqs. (13.3)–(13.7) describe is an initial and boundary value problem with two boundary conditions, which means that our problem is overdetermined unless we choose to keep the boundary free. Suppose that we keep the boundary free. Then it seems plausible that eqs. (13.3)–(13.7) can determine both the boundary $s(t)$ and the optimal reward $\hat{g}$.

We close this section by stating

**Theorem 13.1.** Let $C$ be an open set in $t \leqslant T < \infty$ with a continuously differentiable boundary curve $z = s(t)$, except possibly for a finite number of isolated points where $ds/dt$ blows up. Let a Tychonov-type function $u$ satisfy

$$\frac{\partial u}{\partial t} + \frac{1}{2} \frac{\partial^2 u}{\partial z^2} = 0 \quad \text{in } C,$$

$$u = g \quad \text{at } (z, t) = (s(t), t),$$

$$u(z, T) = g(z, T),$$

$$\frac{\partial u}{\partial z} = \frac{\partial g}{\partial z} \quad \text{at } (z, t) = (s(t), t), \quad \text{if } |\,ds/dt\,| < \infty,$$

$u > g$  in  $C$   and   $u = g$  elsewhere,

$$H \equiv \frac{\partial g}{\partial t} + \frac{1}{2} \frac{\partial^2 g}{\partial z^2} \leqslant 0 \quad \text{in the complement of } C.$$

Then $u$ is actually $\hat{g}$ and $s(t)$ is the optimal stopping boundary.

For a proof of this theorem see van Moerbeke (1974). The above brief analysis can serve as an introduction to the topic of optimal stopping in continuous time.

## 14.  Miscellaneous applications and excercises

(1)  Suppose that $\{X(t), t \in T\}$ and $\{Y(t), t \in T\}$ are real-valued stochastic processes defined on the same probability space and such that $E(X(t)) < \infty$ and $E(Y(t)) < \infty$ for all $t \in T$. The *covariance function* of these two processes denoted by $r_{XY}(s, t)$, is defined for $s < t$ by

$$r_{XY}(s, t) = \text{cov}(X(s), Y(t))$$

$$= E([X(s) - E(X(s))] [Y(t) - E(Y(t))]).$$

This definition extends (4.12) of Chapter 1. When the two processes are the same then the covariance function is called the *autocovariance function* and is denoted by $r_X(s, t)$. Show the following two simple facts:

$$\text{var } X(t) = r_X(t, t) \quad \text{for } t \in T,$$

$$r_X(s, t) = r_X(t, s) \quad \text{for } s, t \in T.$$

(2)  A stochastic process $\{X(t), t \in T\}$ such that $E(X(t)) < \infty$ for all $t \in T$, is *continuous in the mean square* at time $t$ if as $h \to 0$.

$$\lim E(X(t + h) - X(t))^2 \to 0.$$

Let $\mu_X(t) = E(X(t))$ and suppose that $\mu_X(t)$ is continuous in $t$, $t \in T$, and that $\mu_X(s, t)$, $s \in T$ and $t \in T$, is jointly continuous in $s$ and $t$. Show that under these hypotheses $\{X(t), t \in T\}$ is continuous in the mean square.

(3)  Consider eq. (2.11) of this chapter which we rewrite for convenience below

$$x(t + \Delta t) - x(t) = f(t, x(t)) \Delta t + \sigma(t, x(t)) [z(t + \Delta t) - z(t)] + o(\Delta t).$$

Show that:

$$E(x(t+\Delta t) - x(t)) \;=\; f(t, x(t)) \Delta t + o(\Delta t),$$

$$\text{var}\,(x(t+\Delta t) - x(t)) = \sigma^2\,(t, x(t))\,E\,(z(t+\Delta t) - z(t))^2 + o(\Delta t)$$

$$= \sigma^2\,(t, x(t)) \Delta t + o(\Delta t).$$

(4)  Consider eq. (2.11) with *backward differences* which we write as follows:

$$x(t) - x(t-\Delta t) = f(t, x(t)) \Delta t + \sigma(t, x(t))\,(z(t) - z(t-\Delta t)) + o(\Delta t).$$

Assume that the functions $f$ and $\sigma$ are continuous. Show that:

$$E(x(t) - x(t-\Delta t)) \;=\; f(t-\Delta t, x(t-\Delta t)) \Delta t$$

$$+ \sigma_x\,(t-\Delta t, x(t-\Delta t))\sigma\,(t-\Delta t, x(t-$$

$$-\,\Delta t)) \Delta t + o(\Delta t),$$

$$\text{var}\,(x(t) - x(t-\Delta t)) \;=\; \sigma^2\,(t, x(t)) \Delta t + o(\Delta t).$$

Compare the results of this exercise with the results of the preceding exercise to conclude that the mean of the increment of the process depends on the type of the difference that is used while the variance of the increment of the process is not so affected.

(5)  Consider the partition of $[s, t] \subset [0, T]$

$$s_0 = t_0 < t_1 < \ldots < t_n = t,$$

$$\max_{0 < i < n-1} |t_{i+1} - t_i| \leqslant \epsilon, \quad [t_i, t_{i+1}) \subset [s, t),$$

where $\epsilon > 0$ and arbitrarily small. Suppose that $z(t, \omega)$ is a Wiener process with unit variance and suppose that as $\epsilon \to 0$, $A$ and $B$ satisfy:

$$E \left| A - \sum_i z(t_i)\,[z(t_{i+1}) - z(t_i)] \right|^2 \to 0,$$

$$E \left| B - \sum_i z(t_{i+1})\,[z(t_{i+1}) - z(t_i)] \right|^2 \to 0.$$

Show that $B - A = t - s$.

(6)  Suppose that $z(t)$ is a Wiener process with unit variance and consider the

stochastic integral $\int_s^t z(u)\,dz(u)$. Use the definitions of the Itô integral and the Stratonovich integral in sections 3 and 15 of this chapter to obtain: first,

$$\int_s^t z(u)\,dz(u) = \tfrac{1}{2}\left[z^2(t) - z^2(s)\right] - \tfrac{1}{2}(t - s),$$

when the integral is interpreted as an Itô integral; secondly,

$$\int_s^t z(u)\,dz(u) = \tfrac{1}{2}\left[z^2(t) - z^2(s)\right],$$

when the integral is interpreted as a Stratonovich integral. In particular, note that the Stratonovich integral satisfies the integration by parts formula of ordinary calculus, while the Itô integral does not. Note that integration by parts yields

$$\int_s^t z(u)\,dz(u) = z(t)z(t) - z(s)z(s) - \int_s^t z(u)\,dz(u),$$

or equivalently

$$\int_s^t z(u)\,dz(u) = \tfrac{1}{2}\left[z^2(t) - z^2(s)\right].$$

See Stratonovich (1966, p. 365).

(7)  Suppose that

$$dx_1(t) = x_1^2(t)\,dt + dz(t),$$

$$dx_2(t) = x_2\,dz(t).$$

Use Itô's lemma to compute $dy(t)$ for each of the following:

(a)  $y(t) = u(t, x_1, x_1) = x_1(t)x_2(t)$,

(b)  $y(t) = u(t, x_1, x_2) = t[x_1(t)x_2(t)]$,

(c)  $y(t) = u(t, x_1, x_2) = z(t)[x_1(t)x_2(t)]$.

(8)  Consider the system of stochastic differential equations with initial conditions

$$x_1(0) = x_2(0) = z(0) = 0,$$

$$dx_1(t) = dz(t),$$

$$dx_2(t) = x_1 dz(t),$$

where $z(t)$ is a Wiener process with unit variance. Use Itô's lemma to verify that the solution of this system of equations is given by

$$x_1(t) = z(t),$$

$$x_2(t) = \int_0^t z(t) \, dz(t).$$

(9)   Solve the following two stochastic differential equations of first order:

(a)   $dx(t) = a_0 x(t) \, dt + \sigma_0 \, dz(t),$

where $a_0$ and $\sigma_0$ are nonzero constants.

(b)   $dx(t) = a(t) x(t) \, dt + \sigma(t) \, dz(t),$

where $a(t)$ and $\sigma(t)$ are arbitrary nonrandom functions of time.

(10)   Consider the stochastic equation

$$dx(t) = -x(t) \, dt + \sigma(t) x(t) \, dz(t),$$

where $\sigma(t)$ is an arbitrary nonrandom function of time. Discover sufficient conditions for the stability of the 0-equilibrium.

(11)   Consider the second-order stochastic system

$$dx_1(t) = x_2(t) \, dt,$$

$$dx_2(t) = -x_1(t) \, dt - \sigma x_1(t) \, dz(t).$$

Kozin and Prodromou (1971) study the sample stability of this system and the interested reader is encouraged to consult their paper. Unlike the straightforward stability analysis of the nonstochastic system where $\sigma = 0$, the stochastic stability analysis of the above system is quite technical.

(12)   Consider a stochastic control problem having a linear quadratic objective function:

$$- W(x(t)) = \min_{v} E_t \int_{s=t}^{\infty} e^{-\rho(s-t)} \{a(x(s))^2 + b(v(s))^2\} ds$$

subject to

$$dx(t) = v(t)dt + \sigma x(t)dz(t),$$

where, $\sigma > 0$, $a > 0$, $b > 0$ and $\rho > 0$. Discover an optimum solution. Note that this problem and one of its generalizations are discussed in the next chapter.

## 15.  Further remarks and references

Most economists are familiar with the analysis of discrete time stochastic models from studying econometrics. In this chapter we use briefly the discrete time case in order to motivate the analysis of the continuous time case. Problems related to discrete time stochastic models, beyond the introductory econometrics level such as in Intriligator (1978), are presented in some detail in Chow (1975), Aoki (1976), and Bertsekas and Shreve (1978). Also, note that Sargent (1979) has a chapter on linear stochastic difference equations. The modeling of uncertainty in continuous time is presented in Åström (1970), Balakrishnan (1973), Friedman (1975), Soong (1973), Tsokos and Padgett (1974), and Gihman and Skorohod (1969, 1972).

Itô's stochastic differential equation appeared in Itô (1946, 1950) and was later studied in some detail in Itô (1951b). In his 1946 paper Itô discovered that certain mathematical questions, raised by Kolmogorov and later by Feller about partial differential equations related to diffusion processes, could be studied by solving stochastic differential equations. Later in his 1951 Memoir, Itô expanded the Picard iteration method of ordinary differential equations to establish theorems of existence and uniqueness of Itô stochastic differential equations.

It is worth remarking that Itô's stochastic differential equation is one among many stochastic equations. Syski (1967) considers a basic system of random differential equations of the form

$$\frac{dx}{dt} = f(x(t), y(t), t)$$

for $t \in T$, with initial condition $x(t_0) = x_0$, where $f$, $x$ and $y$ can be vectors of appropriate dimensions, and he classifies such equations into three basic types: random differential equations with (1) random initial conditions, (2) random inhomogeneous parts, and (3) random coefficients. The Itô stochastic differential equation is a special class of random differential equations which, however, is important for several reasons. First, the conditional mean and the conditional variance as functions are sufficient statistics for Itô equations. Thus, for Itô equations the calculations of the conditional mean and the conditional variance functions completely determine the whole process. This is analogous to the statistical fact that the mean and variance are sufficient statistics in normal distribution theory.

Secondly, the Itô equation exhibits a nonanticipating property which is useful in modeling uncertainty. In other words, if the true source of uncertainty in a system is d$z$ and we want the differential equations not to be clairvoyant, then the evolution of the state variable $x$ in the next instant should depend only on uncertainty evolving in that instant. An Itô equation is consistent with this lack of clairvoyance.

Thirdly, the Itô equation has solutions which when they exist have nice properties. Section 7 describes these properties and it suffices to say here that the Markov property and the diffusion property are useful properties with a well developed theory available about them. Finally, although economists are just discovering the usefulness of the Itô equation, it is worth noting that this equation has found important applications in the engineering literature, and particularly in the areas of control, filtering and communication theory.

The discussion of section 3 concentrates on Itô's integral. Basic references are Itô (1944, 1951b). Doob (1953) presents in some detail Itô's original ideas with some extensions. See also Doob (1966) for an explanation of the connection between Wiener's integral and Itô's generalization. Recently, Åström (1970) and Arnold (1974) have presented simplified discussions of Itô's integral. A detailed account about stochastic integrals is given in McKean (1969). A different approach to stochastic integration and in general to stochastic calculus may be found in McShane (1974). Wong and Zakai (1965) discuss the convergence of ordinary integrals to stochastic integrals. We note that section 3 is a brief introduction to stochastic integration having as its purpose to motivate and supply a definition for the classical Itô stochastic integral. For the reader who is interested in pursuing his study on stochastic integrals we suggest Metivier and Pellaumail (1980), McKean (1969) and Kussmaul (1977). These books develop the theory according to the Itô prototype. The idea of defining stochastic integrals with respect to square integrable martingales, suggested in Doob (1953), was extended in Kunita and Watanabe (1967). Further extensions of the stochastic integral with respect to a special class of Banach-valued processes is presented in Meyer (1976) and with respect to Hilbert-valued processes in Kunita (1970). Naturally, the various extensions of the concept of a stochastic integral have implications on Itô's lemma and the study of stochastic differential equations. Metivier and

Pellaumail (1980) provide detailed references of such extensions.

In this book we use Itô's integral because we chose to analyze Itô's stochastic differential equations. The reader, however, should be informed that an alternative approach to stochastic integration has been proposed by Stratonovich (1966). To illustrate the difference between the Itô and the Stratonovich integral we consider the simple special case

$$\int_s^t z(u)\,\mathrm{d}z(u),$$

$(15.1)$

where $z(t)$ is a Wiener process with unit variance. In this special case, for a partition of the form of (3.8) the analysis of section 3 in particular lemma 3.4 in this chapter showed that Itô's integral denoted by $A$ is defined such that as $\epsilon \to 0$, then

$$\mathrm{E}\left|A - \sum_i z(t_i)\left[z(t_{i+1}) - z(t_i)\right]\right|^2 \to 0.$$

$(15.2)$

Recall that (15.2) comes from the two equations in (3.11) of this chapter. Stratonovich (1966, p. 363) defines his stochastic integral, in our special case, as

$$\mathrm{E}\left|S - \sum_i\left[\frac{z(t_i) + z(t_{i+1})}{2}\right]\left[z(t_{i+1}) - z(t_i)\right]\right|^2 \to 0,$$

$(15.3)$

where $S$ denotes the Stratonovich stochastic integral. Note that the Stratonovich stochastic integral is a particular linear combination of the integrals $A$ and $B$ defined in (3.11). More specifically,

$$S = \tfrac{1}{2}\,A + \tfrac{1}{2}\,B,$$

where $A$ is Itô's stochastic integral, and $A$ and $B$ are as in (3.11). The Itô and Stratonovich integrals are not greatly different. Actually, Stratonovich has developed formulae representing one integral in items of the other, and these formulae are not complicated. See Stratonovich (1966, p. 365). However, because of the nature of stochastic calculus these two integrals have different properties. The Itô integral and the Itô differential equation maintain the intuitive idea of a state model. Also, the Itô integral has the useful properties that it is a martingale and it preserves the interpretations that the expectation of $\mathrm{d}x$ in (2.12) is $f\mathrm{d}t$ and the conditional variance of $\mathrm{d}x$ is $\sigma^2\,\mathrm{d}t$. The main disadvantage of the Itô integral is that it does not preserve the differentiation rules of ordinary calculus as Itô's lemma demonstrates. The Stratonovich integral preserves many computational rules of ordinary calculus but it does not have the just stated advantages of the Itô integral. For further details about the relation between these two integrals see Meyer (1976).

So far, all economics and finance applications of stochastic calculus have used the Itô integral because the associated Itô differential equations provide a meaningful modeling of uncertainty, as explained earlier. Consequently, Itô's lemma is important for computing stochastic differentials of composite random functions. Put differently, Itô's lemma is a formula of a change of a variable for processes that are stochastic integrals with respect to a Wiener process. Note that Itô's lemma first appeared in Itô (1951a) and later in Itô (1961). We repeat here what we said earlier, namely that Itô's lemma and Itô's integral were discovered by Itô as he was working on partial differential equations related to diffusion processes. This original research led also to the development of stochastic differential equations. It was Itô, and later Doob, Gihman and Skorohod among others, who established the field of stochastic differential equations.

At this point we will discuss briefly one method for solving stochastic differential equations of first order as presented in Gihman and Skorohod (1972, pp. 33–39). Consider the equation on $[0, T]$,

$$dx(t) = f(t, x(t)) dt + \sigma(t, x(t)) dz(t) \tag{15.4}$$

and note that it can be written in integral form as

$$x(t) = x(0) + \int_0^t f(s, x(s)) ds + \int_0^t \sigma(s, x(s)) dz(s), \tag{15.5}$$

where $x(0)$ is the initial condition. The idea of the technique is to discover appropriate transformations of (15.4) so that the right-hand side of (15.5) has a more convenient form, i.e. the unknown function does not appear on the right-hand side of (15.5). We illustrate this technique with an example and refer the reader to Gihman and Skorohod (1972, pp. 33–39) for the detailed mathematics and for more examples.

Consider the equation

$$dx(t) = a_0 x(t) dt + \sigma_0 x(t) dz, \tag{15.6}$$

with $a_0$ and $\sigma_0$ constants. To solve this equation consider the substitution $y = \log x$ and use Itô's lemma to obtain:

$$dy = \frac{1}{x} \, dx + \frac{1}{2} \left( -\frac{1}{x^2} \right) (dx)^2 = \frac{1}{x} \, (a_0 x \, dt + \sigma_0 x \, dz) - \frac{1}{2} \frac{1}{x^2} \, \sigma_0^2 x^2 \, dt$$

$$= (a_0 - \tfrac{1}{2} \sigma_0^2) \, dt + \sigma_0 \, dz.$$

Integrating this last equation we get:

$$y(t) - y(0) = \int_0^t (a_0 - \tfrac{1}{2} \sigma_0^2) \, dt + \int_0^t \sigma_0 \, dz,$$

which can be written as:

$$y(t) = y(0) + (a_0 - \tfrac{1}{2} \sigma_0^2) \, t + \sigma_0 z(t). \tag{15.7}$$

Recalling the substitution $y = \log x$, i.e. $x = e^y$, for $x(0) = e^{y(0)}$, we conclude that the solution of (15.6) is:

$$x(t) = e^{y(t)} = x(0) \exp \left\{ \left( a_0 - \frac{\sigma_0^2}{2} \right) t + \sigma_0 z(t) \right\}.$$

Thus, finding the appropriate substitution and using integration can lead to finding the solution of stochastic differential equations.

In sections 8 and 9 we discussed the problem of stochastic stability and we distinguished between stability of a point equilibrium and stability in the sense of convergence in distribution to a steady state distribution. Mathematicians have studied stability for a point equilibrium where noise has disappeared, while the area of stability in the sense of convergence in distribution to a steady state distribution independent of initial conditions has not received much attention. Even so, both areas remain quite open for future research. One of the first papers in the stochastic stability of point equilibrium is the work of the Russian mathematicians Kats and Krasovskii (1960) where they extend the deterministic Liapunov (1949) method. Note that Antosiewicz (1958), Borg (1949), Hahn (1963, 1967), Cesari (1963), Hartman (1964), Krasovskii (1965), La Salle and Lefschetz (1961), La Salle (1964), Massera (1949, 1956), and Yoshizawa (1966) are among the basic references for Liapunov deterministic method to stability. In the United States, Bucy (1965) and Wonham (1966a, 1966b), among others, contributed to the solution of problems of stochastic stability in the spirit of the Liapunov method. Kushner (1967a) gives a complete account of the major stochastic stability results following the Liapunov method. The reader may find Kushner's (1972) stochastic stability survey useful for an introduction to the subject. Although general results about stochastic stability are not abundant, some progress has been achieved in the stability of linear stochastic systems. Some of these results, with many references, appear in Kozin (1972).

Our distinction between stochastic stability of a point equilibrium and stability in the sense of convergence in distribution independent of initial conditions should not imply that these are the only two concepts of stochastic stability. Within the area of stochastic stability of point equilibrium several definitions and theorems exist. Our section 8 gives only one notion of stochastic stability of point equilibrium. Now we mention two more definitions of stochastic stability of a point equilibrium to illustrate the scope of this important area of mathematical research. For example, if we let $x(t)$ denote the solution of a stochastic differential equation such as (8.1) on $[0, \infty)$, with $x = 0$ being the equilibrium solution, we then say that *the 0-equilibrium is stable in the mean* if the expectation exists, and given $\epsilon > 0$ there exists an $\eta > 0$ such that $|x| \equiv |x(0, \omega)| < \eta$ implies

$$E \left( \sup_t |x(t)| \right) < \epsilon.$$

Another definition is this: *the 0-equilibrium is exponentially stable in the mean* if the expectation exists, and if there exist constants $\alpha$, $\beta$ and $\eta$, all greater than zero, such that $|x| \equiv |x(0, \omega)| < \eta$ implies

$$E(|x(t)|) < \beta |x(0, \omega)| e^{-\alpha t}$$

for all $t > 0$. Several other definitions are available in the mathematical literature. See Kozin (1972) and Kushner (1971).

A simple illustration reported in Kozin (1972, pp. 142–192) may be appropriate at this point to give the reader an indication of the mathematical curiosities that arise in stochastic stability.

Consider the Itô equation

$$dx(t) = ax(t)\,dt + \sigma x(t)\,dz,$$

where $a$ and $\sigma$ are constants and $z$ is a Wiener process with unit variance; suppose that the initial condition is $x(0) = x_0$. The solution of this equation, w.p.1, is

$$x(t) = x(0) \exp \left[ \left( a - \frac{\sigma^2}{2} \right) t + \sigma z(t) \right],$$

and the $n$th moments of this solution are given by

$$\mathrm{E}(x^n(t)) = x^n(0) \exp\left[\left(a - \frac{\sigma^2}{2}\right) nt + \frac{\sigma^2 n^2}{2} t\right].$$

From this last equation Kozin (1972, p. 192) concludes that there is exponential stability of the $n$th moment provided

$$a < \frac{\sigma^2}{2}(n - n^2).$$

Thus, for $n = 1$, $a < 0$ implies that the first moment is exponentially stable, but note that higher moments are unstable. For $n = 2$, $a < -\sigma^2$ guarantees the exponential stability of the first and second moments, but higher moments are unstable. Recall that the Liapunov–Kushner method of section 8 applied to the same equation showed stability of $x(t)$, i.e. of the sample paths w.p.1 provided $a < -\sigma^2/2$, but nothing was said there about the stability of the moments. It is hard to give an economic interpretation to an economic model described by an Itô equation whose point equilibrium is stable but whose higher moments are unstable. Therefore we need to distinguish between the sample path behavior and the moment behavior and actually choose to give priority regarding stability characteristics to one or the other behavior.

The area of stochastic stability in the sense of convergence in distribution to a steady state distribution remains an open field for research. It is difficult to establish the existence of an equilibrium distribution in a general setting, let alone prove its stability. For special cases we have existence theorems for an equilibrium distribution. The basic results in this area are reported in the book by Mandl (1968) and some of the original work first appeared in Feller (1954), Tanaka (1957) and Khasminskii (1962). Merton (1975a) has a stochastic stability result in a special case of a continuous growth model which he obtains by assuming constant savings functions being maximized over the set of constant savings functions. This is a very special stability result. The problem of stochastic stability in a continuous stochastic optimal growth, i.e. the problem of proving that optimal stochastic processes converge in distribution to a steady state distribution, remains open. Brock and Mirman (1972) proved this result for the discrete stochastic growth model.

Concerning the topic of stochastic control, the problem stated in eqs. (10.1) and (10.2) is a stochastic version of the deterministic control problem studied by Arrow and Kurz (1970, pp. 27–51). We chose to study (10.1) and (10.2) because most economists are familiar with the deterministic control problem analyzed by Arrow and Kurz (1970) and the present stochastic version will enable the reader to compare the familiar with the new mathematical issues and results of the stochastic extension.

The analysis of section 10 uses Bellman's principle of optimality as it first appeared in Bellman (1957), and subsequently popularized in books such as Dreyfus (1965), Hadley (1964) and Mangasarian (1969), along with stochastic analysis to derive stochastic conditions of optimality.

Aoki (1967), Åström (1970), Kushner (1971) and Bertsekas (1976) have presentations of stochastic control at the introductory level. However, unlike the area of deterministic optimal control where many books are available, such as Anderson and Moore (1971), Athans and Falb (1966), Berkovitz (1974), Bryson and Ho (1979), Kwakernaak and Sivan (1972), Strauss (1968), Pontryagin *et al.* (1962), Hestenes (1966), and Lee and Markus (1967), the literature on stochastic optimal control is not yet very large. At an advanced level the reader is encouraged to consult the book by Fleming and Rishel (1975) and the papers of Benes (1971), Bismut (1973, 1976), Davis (1973), Fleming (1969, 1971), Kushner (1965, 1967b, 1975), Rishel (1970) and Wonham (1970). We note that Fleming and Rishel (1975, ch. 5) provide a rigorous analysis for the stochastic optimal control problem which supplements our heuristic approach. In what they call the *verification theorem*, Fleming and Rishell (1975, p. 159) give sufficient conditions for an optimum, supposing that a well-behaved solution exists for the Hamilton–Jacobi–Bellman nonlinear partial differential equation with the appropriate boundary conditions.

In section 12 we present a generalized Itô formula and a maximum principle for jump processes. Our approach is intuitive and its aim is to familiarize the reader with some of the mathematics used in Merton (1971). A rigorous analysis would require an analysis of the exact mathematical properties of the term $g(t, x)\,dq(t)$ of eq. (12.1) and the meaning of the jump process integrals. Kushner (1967, p. 18) and Doob (1953, p. 284) discuss some of these issues. For a more detailed treatment see Dellacherie (1974).

It is appropriate to note that as economists apply the techniques of stochastic control theory to economic models a need will soon develop for a stability analysis of such stochastic models. Recent economic research in the deterministic case by Araujo and Scheinkman (1977), Benveniste and Scheinkman (1977, 1979), Scheinkman (1976, 1978), Brock and Scheinkman (1977, 1976), Brock (1976, 1977), Cass and Shell (1976), McKenzie (1976), Magill (1977a), Samuelson (1972), and Levhari and Liviatan (1972), among others, has demonstrated how a wide class of economic problems arise from deterministic optimal control whose stability properties are crucial for correctly specifying such models. The availability of mathematical results concerning the stability of deterministic control systems such as Gal'perin and Krasovskii (1963), Hale (1969), Hartman (1961), Hartman and Olech (1962), Lefschetz (1965), Mangasarian (1963, 1966), Markus and Yamabe (1960), Rockafellar (1973, 1976), and Roxin (1965, 1966), have helped economic researchers. On the other hand, the nonavailability of enough mathematical results on the stability of systems of stochastic differential equations arising from a stochastic control context may delay economic research in this area. For a sample of such economic problems see Brock and Magill (1979) and Magill (1977b).

Many applications of stochastic methods in economics and finance deal with stochastic control in discrete time as opposed to the continuous time case discussed in this chapter. We chose to include in the next two chapters discrete time stochastic applications to enrich the reader's education, although we have not discussed explicitly discrete stochastic methods in this chapter. There are many similarities, however, and the various applications will illustrate them in what follows. It suffices to mention here that Kushner (1971) establishes various similarities and relationships between discrete and continuous time stochastic models. Discrete time stochastic problems are usually less complicated than the corresponding continuous time problems. The latter may be approximated by the discrete time procedure using time intervals of length $h$. Since $h$ may be made arbitrarily small, relationships between discrete and continuous time models can be established. In finance, Merton (1978) approximates the continuous time model by using only elementary probability methods to derive the continuous time theorems. In so doing he uncovers the economic assumptions imbedded in the continuous time mathematical theory.

# CHAPTER 3

# APPLICATIONS IN ECONOMICS

> For the person who thinks in mathematics, and does not simply translate his verbal thoughts or his images into mathematics, mathematics is a language of discovery as well as a language of verification.
>
> H.A. Simon (1977, p. xv)

## 1. Introduction

In this chapter we present several examples to illustrate the use of stochastic methods in economic analysis. Some applications use specific results from the previous chapters, while some other applications introduce new techniques.

## 2. Neoclassical economic growth under uncertainty

The research of Bourguignon (1974) and Merton (1975a) has extended the neoclassical model of growth developed by Solow (1956) to incorporate uncertainty. Such an extension uses Itô's lemma as a tool for introducing uncertainty into the deterministic model.

Consider a homogeneous production function $F(K, L)$ of degree 1, where $K$ denotes units of capital input and $L$ denotes units of labor input. From homogeneity we obtain that $F(K/L, 1) = f(K/L) = f(k)$, where $k = K/L$. For equilibrium to obtain, investment must equal saving, i.e.

$$\dot{K} = \frac{dK}{dt} = sF(K, L), \quad 0 < s < 1,$$

where $s$ is the marginal propensity to save. *The Solow neoclassical differential equation of growth* for the certainty case is obtained as follows (a dot above a variable denotes its time derivative):

$$\dot{k} = \frac{\mathrm{d}k}{\mathrm{d}t} = \frac{\mathrm{d}}{\mathrm{d}t}\left(\frac{K}{L}\right) = \frac{\dot{K}L - \dot{L}K}{L^2} = \frac{\dot{K}}{L} - \frac{\dot{L}}{L} \cdot \frac{K}{L} = sf(k) - nk, \tag{2.1}$$

where $\dot{L}/L = n$, i.e. it is assumed that

$$L(t) = L(0)e^{nt}, \quad L(0) > 0, 0 < n < 1. \tag{2.2}$$

The existence, uniqueness and global asymptotic stability properties of the steady state solution of the neoclassical differential equation in (2.1) are presented in Burmeister and Dobell (1970, pp. 23–30).

Suppose now that instead of $\dot{L}/L = n$, labor growth is described by the stochastic differential equation,

$$\mathrm{d}L = nL\,\mathrm{d}t + \sigma L\,\mathrm{d}z. \tag{2.3}$$

The stochastic part is $\mathrm{d}z$, where $z = z(t, \omega) = z(t)$ is a Wiener process defined on some probability space $(\Omega, \mathscr{F}, P)$. In the engineering literature $\mathrm{d}z$ is usually called white noise. The drift of the process, $n$, is the expected rate of labor growth per unit of time and the variance of the process per unit of time is $\sigma^2$. Note that $\mathrm{d}L/L = n\,\mathrm{d}t + \sigma\,\mathrm{d}z$ says that over a short period of time the proportionate rate of change of the labor force is normally distributed with mean $n\,\mathrm{d}t$ and variance $\sigma^2\,\mathrm{d}t$.

The new specification of the growth of labor in (2.3) alters the neoclassical differential equation of growth. To compute *the stochastic neoclassical differential equation of growth* we make use of Itô's lemma. We are given that

$$\mathrm{d}K = sF(K, L)\mathrm{d}t,$$
$$\mathrm{d}L = nL\,\mathrm{d}t + \sigma L\,\mathrm{d}z.$$

Letting $k = K/L$, Itô's lemma yields

$$\mathrm{d}k = \frac{\partial k}{\partial t}\,\mathrm{d}t + \frac{\partial k}{\partial L}\,\mathrm{d}L + \frac{\partial k}{\partial K}\,\mathrm{d}K +$$

$$+ \frac{1}{2}\left[\frac{\partial^2 k}{\partial K^2}(\mathrm{d}K)^2 + 2\frac{\partial^2 k}{\partial K \partial L}\,(\mathrm{d}K)(\mathrm{d}L) + \frac{\partial^2 k}{\partial L^2}(\mathrm{d}L)^2\right]$$

$$= -\frac{K}{L^2}(nL\,\mathrm{d}t + \sigma L\,\mathrm{d}z) + \frac{1}{L}\,sF(K,L)\mathrm{d}t + \frac{1}{2}\left(\frac{2}{L^3}\,K\sigma^2 L^2\,\mathrm{d}t\right)$$

$$= [sf(k) - (n-\sigma^2)k]\mathrm{d}t - k\sigma\mathrm{d}z. \tag{2.4}$$

Note that if $\sigma = 0$ for all $k \in [0, \infty)$, then eq. (2.4) yields as a special case the certainty differential equation of neoclassical growth in (2.1). Comparing eqs. (2.1) and (2.4) we see that because of the new specification of the labor growth in (2.3), Itô's lemma has enabled us to obtain random fluctuations in the changes of the output per labor unit ratio $\mathrm{d}k$. These random fluctuations are due, of course, to the random fluctuations of the labor growth. Thus, uncertainty with respect to labor growth via Itô's lemma is translated into uncertainty with respect to output per labor unit, which is consistent with our intuition that fluctuations in an input are expected to cause fluctuations in output.

## 3. Growth in an open economy under uncertainty

Let $F(K, L)$ be a homogeneous production function of degree 1 and let saving in such an economy be

$$\text{saving} = sF(K,L)\mathrm{d}t + \rho K\mathrm{d}z, \tag{3.1}$$

where, as before, $s$ is the marginal propensity to save and $\rho K\mathrm{d}z$ indicates a random inflow of capital from the rest of the world. Observe that we do not explain the causes of $\rho K\mathrm{d}z$ because our purpose in this application is to illustrate Itô's lemma and how uncertainty in the rest of the world causes uncertainty in the domestic economy. Here we assume that there are *ad hoc* random inflows of capital but we hasten to add that one possible explanation of $\rho K\mathrm{d}z$ may be the random fluctuations in the differential between domestic interest rates and the average interest rate prevailing in the rest of the world.

Next, as in the previous application, we assume that the behavior of labor growth is given by

$$\mathrm{d}L = nL\,\mathrm{d}t + \sigma L\,\mathrm{d}z.$$

The model consists of the labor growth equation and the equilibrium condition equation

$$\mathrm{d}K = sF(K,L)\mathrm{d}t + \rho K\mathrm{d}z. \tag{3.2}$$

Note that (3.2) says that over a short period of time the proportionate rate of change of the capital stock is normally distributed with mean $sF(K, L)dt$ and variance $\rho^2 dt$. Thus, in this application we have two sources of uncertainty, i.e. uncertainty due to fluctuations in the labor force and uncertainty due to fluctuations in the inflow of foreign capital. Both such fluctuations are expected to influence the changes in the domestic output per labor unit and the precise formulation is obtained by computing $dk$. To obtain $dk$, $k = K/L$, we use Itô's lemma which yields

$$dk = -\frac{K}{L^2}[nL\,dt + \sigma L\,dz] + \frac{1}{L}[sF(K, L)dt + \rho K\,dz]$$

$$+ \frac{1}{2}\left[-2\left(\frac{1}{L^2}\right)\rho K\sigma L\,dt + \frac{2K}{L^3}\sigma^2 L^2\,dt\right]$$

$$= -\frac{K}{L^2}nL\,dt - \frac{K}{L^2}\sigma L\,dz + \frac{sF(K, L)}{L}dt + \rho\frac{K}{L}dz -$$

$$- \frac{1}{L^2}\rho K\sigma L\,dt + \frac{K}{L^3}\sigma^2 L^2\,dt$$

$$= [sf(k) - (n-\sigma^2 + \rho\sigma)k]dt - (\sigma-\rho)k\,dz. \tag{3.3}$$

Obviously, if $\rho = 0$ and there is no inflow of foreign capital, the last equation reduces to eq. (2.4).

## 4. Growth under uncertainty: Properties of solutions

Consider the stochastic differential equation of economic growth derived in section 2 of this chapter:

$$dk = [sf(k) - (n-\sigma^2)k]dt - \sigma k\,dz, \tag{4.1}$$

with initial random condition $k(0, \omega) = k(0) = k_0 > 0$. Suppose that (4.1) has a unique solution $k(t, \omega) = k(t)$ for $t \in [0, \infty)$. What are the properties of such a solution? The answer is provided by the following theorems.

**Theorem 4.1** (The Markov property). Suppose that the stochastic differential eq. (4.1), with initial random condition $k_0 > 0$, has a unique solution. Then its unique solution $k(t)$, $t \in [0, \infty)$ is a Markov process whose initial probability distribution at $t = 0$ is $k_0$ and whose transition probability is $P(s, k, t, B) = P[k(t) \in B \mid k(s) = k]$.

The proof follows immediately from theorem 7.3 of the previous chapter.

This theorem is a useful result, particularly for economic policy considerations. Suppose that for an economy the process of capital per worker is described by the Markov process $k(t)$. Given that the economy has capital per worker $k$ at time $s$, the economic policy makers may be interested in knowing the probability that at some future time $t$ the capital per worker will fall within the interval

$$(b_1, b_2), 0 < b_1 < b_2 < \infty.$$

We now establish that $k(t)$ is a diffusion process.

**Theorem 4.2** (The diffusion property). Suppose that the stochastic differential eq. (4.1), with initial condition $k_0 > 0$, has a unique solution. Then its unique solution $k(t)$, $t \in [0, \infty)$, is a diffusion process with drift coefficient $[sf(k(t)) - (n-\sigma^2)k(t)]$ and diffusion coefficient $\sigma^2 k^2(t)$.

The proof of this theorem follows from theorem 7.4 of the previous chapter if we note that eq. (4.1) is an autonomous stochastic differential equation and therefore the continuity with respect to $t$ is vacuously satisfied.

The economic significance of this result may be made clear by a comparison of various systems. For example, in the deterministic model of economic growth we obtain a differential equation whose present state determines its future evolution. The Markov property of the stochastic differential equation of growth is richer in content because the current state completely determines the probability of occupying various states at all future times. The diffusion property goes even further: it describes changes in the process of capital accumulation per worker during a small unit of time, say $\Delta t$, as the sum of two factors. The first factor, i.e. the drift coefficient, is the macroeconomic average velocity of the random motion of capital accumulation when $k(s) = k$. The second factor, i.e. the diffusion coefficient, measures the local magnitude of the fluctuation of $k(t) - k(s)$ about the average value, which is caused by collisions of the process of capital accumulation with economic and noneconomic variables undergoing a random movement.

A straightforward generalization of eq. (4.1) is when $s$ and $\sigma$ are functions of $k$, written as $s(k)$ and $\sigma(k)$, instead of being nonrandom and non-negative constants. Assuming this to be the case we may rewrite (4.1) as

$$dk = [s(k)f(k) - (n-\sigma^2(k))k]dt - \sigma(k)kdz. \tag{4.2}$$

Similar analysis as the one just completed can show that if a unique solution of (4.2) exists then it will satisfy the Markov and the diffusion properties.

## 5. Growth under uncertainty: Stationary distribution

The tools of stochastic calculus can be used to further study neoclassical growth under uncertainty. In this section we are interested in the existence of a stationary distribution of the stochastic process $k(t)$ which is assumed to be the solution of eq. (2.4). As discussed in section 9 of the previous chapter, the stochastic process for $k$ is completely characterized by the drift and diffusion coefficients. Using the results of section 9 of Chapter 2, the stationary distribution for $k$, denoted by $\pi(k)$, is given by

$$\pi(k) = \frac{m}{\sigma^2 k^2} \exp\left[2\int^k \frac{sf(y) - (n-\sigma^2)y}{\sigma^2 y^2} dy\right]$$

$$= mk^{-2n/\sigma^2} \exp\left[\frac{2}{\sigma^2}\int^k \frac{sf(y)}{y^2} dy\right]. \tag{5.1}$$

Note that (5.1) for the special case when the production function is given by a Cobb–Douglas function, i.e. $f(k) = k^\alpha$, $0 < \alpha < 1$, becomes

$$\pi(k) = mk^{-2n/\sigma^2} \exp\left[\frac{-2s}{(1-\alpha)\sigma^2} k^{-(1-\alpha)}\right]. \tag{5.2}$$

In eqs. (5.1) and (5.2) $m$ is determined so that $\int_0^\infty \pi(y)dy = 1$. For further analysis of (5.1) and (5.2) see Merton (1975a).

The remainder of this section follows Merton (1975a) in comparing the expected stationary value of per capita output with the steady state certainty value. To achieve this goal we need a brief analysis and a technical lemma.

Consider (2.4) and suppose that it has a stationary distribution denoted by $\pi$. Let $g(k)$ be a twice continuously differentiable function and use Itô's lemma to compute $dg(k)$, i.e.

$$dg(k) = g'(k)dk + \tfrac{1}{2} g''(k)(dk)^2$$

$$= (g'(k)[sf(k) - (n-\sigma^2)k] + \tfrac{1}{2} g''(k)\sigma^2 k^2)dt - g'(k)\sigma^2 k^2 dz.$$

The following lemma is useful.

**Lemma 5.1.** Suppose $g(k)$ is twice continuously differentiable and that (2.4) has a stationary distribution $\pi$. Also, assume that

$$\lim_{k \to 0} [g'(k)\sigma^2 k^2 \pi(k)] = \lim_{k \to \infty} [g'(k)\sigma^2 k^2 \pi(k)] = 0.$$

Then

$$E(g'(k) [sf(k) - (n-\sigma^2)k] + \tfrac{1}{2} g''(k)\sigma^2 k^2) = 0. \tag{5.3}$$

For a proof see Merton (1975a, p. 392).

We use lemma 5.1 with $g(k) = k$ as a special case to compute the expected stationary value of per capita output, i.e. $E(f(k))$. We use (5.3) as follows:

$$E(1 \cdot [sf(k) - (n-\sigma^2)k] + 0) = 0. \tag{5.4}$$

From (5.4) we obtain

$$E(sf(k)) = (n-\sigma^2)E(k)$$

and thus

$$E(f(k)) = \frac{n-\sigma^2}{s} E(k). \tag{5.5}$$

The result in (5.5) is of particular interest because it allows for a comparison with the certainty case. For example, let $f(k) = k^\alpha$, $0 < \alpha < 1$. From growth theory we know that the certainty estimate of the steady state per capita output is

$$\left(\frac{s}{n}\right)^{\alpha/(1-\alpha)}. \tag{5.6}$$

How do (5.5) and (5.6) compare? Obviously

$$E(f(k)) = \frac{n-\sigma^2}{s} E(k) > \left(\frac{s}{n}\right)^{\alpha/(1-\alpha)}. \tag{5.7}$$

Eq. (5.7) illustrates that the certainty estimate is biased since (5.5) is larger than

(5.6) and that therefore care must be taken in using the certainty analysis, even as an approximation of stochastic analysis.

## 6. The stochastic Ramsey problem

In this section we follow Merton (1975a) in determining the *optimal saving* policy function under uncertainty. The problem is to find a saving policy $s*(k, T-t)$ such that we

$$\text{maximize } E_0 \int_0^T u(c)dt \tag{6.1}$$

subject to

$$dk = (sf(k) - (n-\sigma^2)k)dt - \sigma k\,dz$$

and $k(t) \geqslant 0$ for each $t$ w.p.1, and in particular $k(T) \geqslant 0$. Here, $u$ is a strictly concave, von Neumann–Morgenstern utility function of per capita consumption $c$ for the representative consumer. Note that

$$c = (1-s)f(k). \tag{6.2}$$

To solve this stochastic maximization problem we use Bellman's Optimality Principle, as in section 10 of Chapter 2. Let

$$J(k(t), t, T) = \max_s E_t \int_t^T u[(1-s)f(k)]dt. \tag{6.3}$$

The Hamilton–Jacobi–Bellman equation for (6.3) is given by

$$0 = \max_s \left\{ u[(1-s)f(k)] + \frac{\partial J}{\partial t} + \frac{\partial J}{\partial k}[sf(k) - (n-\sigma^2)k] + \right.$$
$$\left. + \frac{1}{2}\frac{\partial^2 J}{\partial k^2}\sigma^2 k^2 \right\}. \tag{6.4}$$

The first-order condition to be satisfied by the optimal policy $s*$ from (6.4) is

$$0 = u'[(1-s*)f(k)](-f(k)) + \frac{\partial J}{\partial k}f(k),$$

which becomes

$$u'[(1-s^*)f(k)] = \frac{\partial J}{\partial k}. \tag{6.5}$$

Note that $u'$ means $du/dc$. To solve for $s^*$, in principle, one solves (6.5) for $s^*$ as a function of $k$, $T-t$ and $\partial J/\partial k$, and then substitutes this solution into (6.4) which becomes a partial differential equation for $J$. Once (6.4) is solved then its solution is substituted back into (6.5) to determine $s^*$ as a function of $k$ and $T-t$.

The nonlinearity of the Hamilton–Jacobi–Bellman equation causes difficulties in finding a closed form solution. One way of overcoming this difficulty is by letting $T \to \infty$, in which case the partial differential equation is reduced to an ordinary differential equation. This is done next.

Observe that $k$ is a time-homogeneous process and that $u$ is not a function of time; thus from (6.3) we deduce that

$$\frac{\partial J}{\partial t} = - \,\mathrm{E}_t\, \{u[(1-s^*(k, T-t))f(k(T-t))]\}. \tag{6.6}$$

Suppose that an optimal policy exists, $f$ is a well-behaved production function, and $n-\sigma^2 > 0$, then as $T \to \infty$

$$\lim s^*(k, T-t) = s^*(k, \infty) = s^*(k),$$

there will exist a stationary distribution for $k$ associated with the optimal policy $s^*(k)$ and denoted by $\pi^*$. Let $T \to \infty$ in (6.6) to obtain

$$\lim \left\{ \frac{\partial J}{\partial t} \right\} = -\mathrm{E}^*(u[(1-s^*)f(k)]) = -B, \tag{6.7}$$

where $\mathrm{E}^*$ is the expectation operator over the stationary distribution $\pi^*$ and $B$ is the level of expected utility of per capita consumption in the Ramsey optimal stationary distribution. Use (6.7) in (6.4) to write as $T \to \infty$,

$$0 = u[(1-s^*)f(k)] - B + \frac{\partial J}{\partial k}\,[s^*f(k) - (n-\sigma^2)k] + \frac{1}{2}\,\frac{\partial^2 J}{\partial k^2}\,\sigma^2 k^2. \tag{6.8}$$

Next we differentiate (6.5) with respect to $k$:

$$\frac{\partial^2 J}{\partial k^2} = u''[(1-s^*)f(k)]\left( (1-s^*)f'(k) - \frac{ds^*}{dk}\,f(k) \right). \tag{6.9}$$

Finally, substitute (6.5) and (6.9) into (6.8) and rearrange terms to conclude

$$0 = (-\tfrac{1}{2}\,\sigma^2 k^2 f u'') \frac{\mathrm{d}s^*}{\mathrm{d}k} + (f u' - \tfrac{1}{2}\,\sigma^2 k^2 u'' f')s^* +$$

$$+ \tfrac{1}{2}\,\sigma^2 k^2 u'' f' - u'(n - \sigma^2)k + u - B. \tag{6.10}$$

Note that (6.10) is a first-order differential equation for $s^*$ with boundary condition as $t \to \infty$,

$$\lim E_0\,\{J(k(t), t)\} = 0.$$

In (6.10), if $\sigma = 0$, the classical Ramsey rule of the certainty case is yielded, namely

$$u's^*f - u'nk + u - B = 0,$$

which is usually rewritten as

$$s^*f - nk = \frac{B-u}{u'},$$

where $B$ is the bliss level of utility associated with maximum steady state consumption and $\dot{k} = s^*f - nk$ along the optimal certainty path.


## 7. Bismut on optimal growth

In this section we give an application of Bismut's approach to optimal stochastic control presented in section 11 of Chapter 2. Here we follow Bismut (1975).

Consider a one-sector optimal growth model with the usual notation i.e. $k$ is capital per worker, $f(k)$ is a well-behaved production function, $s$ is the marginal propensity to save, $u$ is a concave utility function and $\rho$ is the discount rate. The problem is to maximize the expected discounted intertemporal utility assuming $u'(0) = \infty$,

$$\max_s E_0 \int_0^\infty e^{-\rho t}\, u((1-s)f(k))\mathrm{d}t \tag{7.1}$$

subject to the constraints

$$dk = sf(k)dt + \sigma(k, sf(k))dz, \tag{7.2}$$

$$k(0) = k_0 > 0.$$

The transformed Hamiltonian function, i.e. (11.5) of Chapter 2, is written as

$$\mathcal{H} = u((1-s)f(k)) + psf(k) + H\sigma(k, sf(k)). \tag{7.3}$$

Maximize $\mathcal{H}$ in terms of $s$, where $0 < s < 1$, to get

$$-u'(c)f + pf + H\sigma_I f = 0,$$

which becomes, after dividing by $f$,

$$u'(c) = p + H\sigma_I. \tag{7.4}$$

Note that $c$ is per capita consumption and $\sigma_I$ is the partial derivative of $\sigma$ with respect to investment where investment equals $sf(k)$.

Next, we write eq. (11.7) of Chapter 2 as it applies to our case. We have

$$dp = -[(1-s)f'(k)u'(c) + psf'(k) - p\rho + H(\sigma_k + sf'(k)\sigma_I)]dt +$$

$$+ Hdz + dM, \tag{7.5}$$

where $\sigma_k$ denotes the partial derivative of $\sigma$ with respect to $k$. Eq. (7.5) may be rewritten as

$$dp = \{-(p + H\sigma_I)f'(k) - H\sigma_k + p\rho\}dt + Hdz + dM. \tag{7.6}$$

Eqs. (7.4) and (7.6) uncover important economic reasoning. Eq. (7.4) indicates that the consumer will consume up to the point where the marginal utility of his consumption is equal to the expected marginal value of capital in terms of utility, minus the marginal risk of investment valued at its cost. If, for example, the consumer is a risk-averter with $-H > 0$, this will tend to make the consumer consume more than he would with the same $p$ when no risk is involved.

Let $R$ denote the cost of capital. The consumer pays $p + H\sigma_I$ to the producer and the cost of the producer is $Rk - H\sigma$. The equation of profits is

$$(p + H\sigma_I)f(k) + H\sigma(k, I) - Rk, \tag{7.7}$$

which once maximized in $k$ gives

$$(p + H\sigma_I)f'(k) + H\sigma_k - R = 0. \tag{7.8}$$

Then the profit rate in terms of the marginal value of capital $p$, denoted by $r$, is

$$r = \frac{R}{p} = \left(1 + \frac{H}{p}\sigma_I\right)f'(k) + \frac{H}{p}\sigma_k. \tag{7.9}$$

Now look for a moment at (7.6) and rewrite it as

$$\frac{dp}{dt} = \{-(p + H\sigma_I)f'(k) - H\sigma_k + p\rho\} + H\frac{dz}{dt} + \frac{dM}{dt}. \tag{7.10}$$

Multiply both sides of (7.10) by $(1/p)$ and take its expected value to conclude

$$E\left(\frac{dp}{dt}\frac{1}{p}\right) = -\left(1 + \frac{H}{p}\sigma_I\right)f'(k) - \frac{H}{p}\sigma_k + \rho. \tag{7.11}$$

Use (7.11) to rewrite (7.9) as

$$\rho = r + E\left(\frac{dp}{dt}\frac{1}{p}\right), \tag{7.12}$$

which is the neoclassical relation between interest rate $\rho$, rate of return $r$, and the expected inflation rate.

Finally, compare $f'(k)$ and $r$ from (7.9) to obtain

$$f'(k) = \frac{r - (H/p)\sigma_k}{1 + (H/p)\sigma_I},$$

where the *instantaneous risk premium* $f'(k) - r$ is equal to

$$-\frac{(H/p)(\sigma_k + r\sigma_I)}{1 + (H/p)\sigma_I}.$$

Having presented several applications of stochastic calculus to economic growth we next discuss the concept of rational expectations which recently has received attention from several economists. Rational expectations use the techniques of stochastic analysis and they are incorporated in several applications in this chapter and the next.

## 8. The rational expectations hypothesis

In this section we present the *rational expectations hypothesis* postulated by Muth (1961) which has found many applications in stochastic economic and financial models. Theorists realize that to make stochastic models complete, an expectations hypothesis is needed. What kind of information is used by agents and how it is put together to frame an estimate of future conditions is important because the character of dynamic processes is sensitive to the way expectations are influenced by the actual course of events. Muth (1961) suggests that expectations, since they are informed predictions of future events, are essentially the same as the predictions of the relevant economic theory. This hypothesis may be restated: the subjective probability distributions of outcomes tend to be distributed, for the same information set, about the objective probability distributions of outcomes. This hypothesis asserts that: (1) information is scarce and the economic system generally does not waste it; (2) the way expectations are formed depends specifically on the structure of the relevant system describing the economy; and (3) a prediction based on general information will have no substantial effect on the operation of the economic system. The hypothesis does not assert that predictions of economic agents are perfect or that their expectations are all the same.

From a theoretical standpoint there are good reasons for assuming rational expectations because: (1) it is a hypothesis applicable to all dynamic problems, and expectations in different markets would not have to be treated in different ways; (2) if expectations were not moderately rational there would be opportunities to make profits; and (3) rational expectations is a hypothesis that can be modified with its analytical methods remaining applicable in systems with incomplete or incorrect information.

As an illustration of the ideas above, we present Muth's (1961) model of price fluctuations in an isolated market, with a fixed production lag, of a commodity which cannot be stored. The model is given by:

$$C(t) = -\beta p(t) \qquad \text{demand,} \tag{8.1}$$

$$P(t) = \gamma p^{e}(t) + u(t) \quad \text{supply,} \tag{8.2}$$

$$P(t) = C(t) \qquad \text{market equilibrium,} \tag{8.3}$$

where $P(t)$ represents the number of units produced in a period lasting as long as the production lag, $C(t)$ is the amount consumed, $p(t)$ is the market price in the $t$th period, $p^{e}(t)$ is the market price expected to prevail during the $t$th period on the basis of information available through the $(t-1)$st period, and finally $u(t)$ is an error term. All the variables used here are deviations from equilibrium values.

Put (8.1) and (8.2) in (8.3) to get

$$p(t) = -(\gamma/\beta)p^e(t) - (1/\beta)u(t). \tag{8.4}$$

Note that $u(t)$ is unknown at the time the production decisions are made but it is known and relevant at the time the commodity is purchased in the market.

Suppose that the errors have no serial correlation and that $Eu(t) = 0$. Then the prediction of the model in (8.4) is

$$Ep(t) = -(\gamma/\beta)p^e(t). \tag{8.5}$$

In (8.5) $Ep(t)$ denotes the prediction of the model or the theory and it is objective, while $p^e(t)$ denotes the subjective prediction of the firms. If the prediction of the model were different from the expectations of the firms, there would be opportunities for profit. Note that we do not use parentheses with the expectation operation to simplify the notation. The rationality assumption given by

$$Ep(t) = p^e(t) \tag{8.6}$$

states that such profit opportunities could no longer exist. If $(\gamma/\beta) \neq -1$ in (8.5), then the rationality assumption implies that

$$p^e(t) = 0, \tag{8.7}$$

i.e. the expected price equals the equilibrium price.

Let us now introduce more realism into our illustration by allowing for effects in demand and alternative costs in supply. We assume that part of the shock variable may be predicted on the basis of prior information. From (8.4), taking conditional expectation, we write

$$Ep(t) = -(\gamma/\beta)p^e(t) - (1/\beta)Eu(t), \tag{8.8}$$

and using the rationality assumption in (8.6) we obtain

$$p^e(t) = -(\gamma/\beta)p^e(t) - (1/\beta)Eu(t), \tag{8.9}$$

which yields

$$p^e(t) = -[1/(\beta + \gamma)]Eu(t). \tag{8.10}$$

If the shock is observable, then the conditional expected value may be found directly. If the shock is not observable, it must be estimated from the past history of variables that can be measured. In this latter case we shall write the $u$'s as a linear combination of the past history of normally and independently distributed random variables $x(t)$ with zero mean and variance $\sigma^2$, i.e.

$$u(t) = \sum_{i=0}^{\infty} w(i)x(t-i), \tag{8.11}$$

$$Ex(i) = 0, \tag{8.12}$$

$$Ex(i)x(j) = \begin{cases} \sigma^2 & \text{if } i = j, \\ 0 & \text{if } i \neq j. \end{cases} \tag{8.13}$$

The price will be a linear function of the same independent disturbances and will be written as

$$p(t) = \sum_{i=0}^{\infty} W(i)x(t-i). \tag{8.14}$$

Similarly, from (8.12)

$$p^e(t) = W(0)Ex(t) + \sum_{i=1}^{\infty} W(i)x(t-i) = \sum_{i=1}^{\infty} W(i)x(t-i). \tag{8.15}$$

Putting (8.14) and (8.15) in (8.1) and (8.2), respectively, and solving using (8.3) we have

$$W(0)x(t) + \left(1 + \frac{\gamma}{\beta}\right)\sum_{1}^{\infty} W(i)x(t-i) = -\frac{1}{\beta}\sum_{0}^{\infty} w(i)x(t-i). \tag{8.16}$$

Eq. (8.16) is an identity in the $x$'s and therefore the relation between $W(i)$ and $w(i)$ is as follows:

$$W(0) = -\frac{1}{\beta}w(0), \tag{8.17}$$

$$W(i) = -\frac{1}{\beta + \gamma}w(i) \quad \text{for } i = 1, 2, 3 \dots . \tag{8.18}$$

Note that (8.17) and (8.18) give the parameters of the relation between the price function and the expected price function in terms of the past history of independent shocks. The next step is that of writing the expected price in terms of the history of observable variables, i.e.

$$p^e(t) = \sum_{j=1}^{\infty} V(j)p(t-j).$$ (8.19)

Use (8.14), (8.15) and (8.19) to obtain

$$
\begin{aligned}
p^e(t) &= \sum_{i=1}^{\infty} W(i)x(t-i) = \sum_{j=1}^{\infty} V(j)p(t-j) \\
&= \sum_{j=1}^{\infty} V(j)\left[ \sum_{i=0}^{\infty} W(i-j)x(t-i-j) \right] \\
&= \sum_{i=1}^{\infty} \left[ \sum_{j=1}^{i} V(j)W(i-j) \right] x(t-i).
\end{aligned}
$$ (8.20)

As before, we again conclude from (8.20) that the coefficients must satisfy

$$W(i) = \sum_{j=1}^{i} V(j)W(i-j)$$ (8.21)

since the equality in (8.20) must hold for all shocks. In (8.21) we have a system of equations with a triangular structure which may be solved successively for $V_1$, $V_2$, ... .

As a particular illustration suppose that in (8.11) $w(i) = 1$ for all $i = 0, 1, 2, ...$, which means that an exogenous shock, say a technological change, affects all future conditions of supply. Then using (8.17) and (8.18) eq. (8.21) yields

$$p^e(t) = \frac{\beta}{\gamma} \sum_{j=1}^{\infty} \left( \frac{\gamma}{\beta+\gamma} \right)^j p(t-j),$$ (8.22)

which expresses the expected price as a geometrically weighted moving average of past prices.

## 9. Investment under uncertainty

In this application we follow Lucas and Prescott (1971) who introduced an uncertain future into an adjustment-cost type model of the firm to study the time series behavior of investment, output and prices. Their paper is rich in methodological ideas and techniques and in this section we plan just to describe the model and state an existence theorem.

Consider an industry consisting of many small firms each producing a single output, $q_t$, by using one single input capital, $k_t$, under constant returns to scale. With an appropriate choice of units we may use $k_t$ to also denote production at full capacity and denote the production function as

$$0 \leqslant q_t \leqslant k_t. \tag{9.1}$$

Gross investment, denoted by $x_t$, is related to capacity in a nonlinear way:

$$k_{t+1} = k_t h(x_t/k_t), \tag{9.2}$$

where $h$ is assumed to be bounded, increasing, $h(0) > 0$, continuously differentiable, strictly concave, and that there exists a $\delta$, $0 < \delta < 1$, such that $\delta = h^{-1}$ (1). The last assumption means that $\delta k_t$ is the investment rate that is needed to maintain the capital stock $k_t$. The assumption of strict concavity is made because it gives rise to adjustment costs of investment and the model thus reflects gradual changes in capital stock as opposed to immediate passage to a long-run equilibrium level.

Let $p_t$ denote the product price and $r$ the cost of capital with $r > 0$. Using the standard discount factor, $\beta$, where $\beta = 1/(1 + r)$, then *ex post* the present value of the firm, $V$, is given by

$$V = \sum_{t=0}^{\infty} \beta^t [p_t q_t - x_t]. \tag{9.3}$$

We use the notation $k_t$, $x_t$ and $p_t$ interchangeably for industry and firm variables. The objective of the firm is the maximization of the mean value of (9.3) with the stochastic behavior of $p_t$ somehow specified. However, because we have omitted variable factors of production in (9.1) the firm will choose to produce at full capacity and the only nontrivial decision of the firm is the choice of an investment level. This investment level is decided, as usual, by comparing the known cost of a unit of investment to an expected marginal return. To simplify the firm's investment decision we place the burden of evaluating the income stream changes due to a given investment to the traders in the firm's securities. We denote by $w_t^*$ the undiscounted value per unit of capital expected to prevail next period. The firm's problem can now be stated as

$$\max_{x \geqslant 0} [-x + \beta k_t h(x/k_t) w_t^*], \tag{9.4}$$

where $x$ is the cost of investment and $\beta k_{t+1} w_t^*$ is the next period value resulting from $x$. Using (9.2) we obtain (9.4). The maximization problem of (9.4) is subject to the two constraints

$$w_t k_t = p_t k_t - x + \beta k_t h(x/k_t) w_t^* \tag{9.5}$$

and

$$-1 + \beta h'(x/k_t) w_t^* \leqslant 0 \quad \text{with equality, if } x > 0. \tag{9.6}$$

Since (9.5) and (9.6) are solved jointly for $x_t$ and $w_t^*$ as functions of $k_t, p_t$ and $w_t$, we can write the investment function as

$$x_t = k_t g(w_t - p_t), g' > 0. \tag{9.7}$$

The industry demand function is assumed to be subject to random shifts and is written as

$$p_t = D(q_t, u_t), \tag{9.8}$$

where $\{u_t\}$ is a Markov process with a transition function $p(\cdot, \cdot)$ defined on $R^2$. For given $u_t$, $D$ is a continuous, strictly decreasing function of $q_t, p_t = D(0, u_t) < \infty$, and with

$$\int_0^q D(z, u_t) dz \tag{9.9}$$

bounded uniformly in $u_t$ and $q$. We also assume that $D$ is continuous and increasing in $u_t$ so that an increase in $u_t$ causes a shift to the right of the demand function.

For given $(k_0, u_0)$ an anticipated price process is defined to be a sequence $\{p_t\}$ of functions of $(u_1, u_2, ..., u_t)$, or functions with domain $R^t$. Similarly, an investment–output plan is defined as a sequence $\{q_t, x_t\}$ of functions on $R^t$. We restrict the sequences $\{p_t\}$, $\{q_t\}$ and $\{x_t\}$ to belong to $L^+$, i.e. to be elements of the class $L$ with non-negative terms for all $(t, u_1, u_2, ..., u_t)$. Here $L$ denotes the set of all sequences $x = \{x_t\}$, $t = 0, 1, 2, ...$, where $x_0$ is a number and for $t \geq 1$, $x_t$ is a bounded measurable function on $R^t$, bounded in the sense that norm is finite, i.e.

$$\| x \| = \sup_t \sup_{(u_1, ..., u_t) \in E^t} |x_t(u_1, ..., u_t)| < \infty.$$

Therefore, for any sequences $\{p_t\}$, $\{q_t\}$ and $\{x_t\}$, elements of $L^+$, the present value $V$ in (9.3) is a well-defined random variable with a finite mean. The objective of the firm then becomes the maximization of the mean value of $V$ with respect to the investment—output policy, given an anticipated price sequence.

To link the anticipated price sequence to the actual price sequence we assume that expectations of the firms are rational or that the anticipated price at time $t$ is the same function of $(u_1, ..., u_t)$ as the actual price. We are now ready to define an industry equilibrium for fixed initial state $(k, u)$ as an element $\{q_t^0, x_t^0, p_t^0\}$ of $L^+ \times L^+ \times L^+$ such that (9.8) is satisfied for all $(t, u_1, ..., u_t)$ and such that

$$\mathrm{E}\left( \sum_{t=0}^{\infty} \beta^t [p_t^0 q_t^0 - x_t^0] \right) \geq \mathrm{E}\left( \sum_{t=0}^{\infty} \beta^t [p_t^0 q_t - x_t] \right) \tag{9.10}$$

for all $\{q_t, x_t\} \in L^+ \times L^+$ satisfying (9.1) and (9.2). Note that the expectation of the $t$th term in (9.10) is taken with respect to the joint distribution of $(u_1, ..., u_t)$.

Having defined industry equilibrium the question naturally arises of whether a unique equilibrium exists. Lucas and Prescott (1971) first show that a competitive equilibrium leads the industry to maximize a certain "consumer surplus" expression, and then they show that the latter maximum problem can be solved using the techniques of dynamic programming.

Define the function $s(q, u)$, $q \geq 0$, $u \in R$ by

$$s(q, u) = \int_0^q D(z, u)\mathrm{d}z, \tag{9.11}$$

so that for given $u$, $s(q, u)$ is a continuously differentiable, increasing, strictly concave, positive and bounded function of $q$, and for given $q$, $s$ is increasing in $u$. Note that $s(q_t, u_t)$ is the area under the industry's demand curve at an output of $q_t$ and with the state of demand $u_t$. Let the discounted consumer surplus, $S$, for the industry be

$$S = \mathrm{E}\left( \sum_{t=0}^{\infty} \beta^t [s(q_t, u_t) - x_t] \right). \tag{9.12}$$

We are interested in using the connection between the maximization of $S$ and competitive equilibrium in order to determine the properties of the latter. Associated with the maximization of $S$ is the functional equation

$$v(k,u) = \sup_{x \geq 0} \left( s(k,u) - x + \beta \int v[kh(x/k),z\,]p(dz,u) \right). \qquad (9.13)$$

We now state a basic result of Lucas and Prescott (1971).

**Theorem 9.1.** The functional eq. (9.13) has a unique, bounded solution $v$ on $(0,\infty) \times R$; and for all $(k,u)$ the right-hand side of (9.13) is attained by a unique $x(k,u)$. In terms of $x(k,u)$, the unique industry equilibrium, given $k_0$ and $v_0$, is given by

$$x_t = x(k_t, u_t), \qquad (9.14)$$

$$k_{t+1} = k_t h[x(k_t, u_t)/k_t], \qquad (9.15)$$

$$q_t = k_t,$$

$$P_t = D(q_t, u_t)$$

for $t = 0, 1, 2, \ldots$, and all realizations of the process $\{u_t\}$.

*Proof.* See Lucas and Prescott (1971, pp. 666–671).

With the question of existence and uniqueness being settled Lucas and Prescott study the long-run equilibrium assuming (1) independent errors and (2) serially dependent errors. Below we state the results for the first case and refer the reader to the Lucas and Prescott paper for the results of the second case.

Suppose that the shifts $u_t$ and $u_s$ are independent for $s \neq t$, i.e. the transition function $p(z,u)$ does not depend on $u$. It follows in this case that $x(k,u)$ will not depend on $u$ and from (9.14) and (9.15) it follows that the time path of the capital stock will be deterministic, given by

$$k_{t+1} = k_t h[x(k_t)/k_t], \qquad (9.16)$$

where $x(k_t)$ is the unique investment rate obtaining $v(k_t, u)$. We define a capital stock $k^c > 0$ to be a stationary solution of (9.16) if and only if it is a solution to $x(k) = \delta k$, since $h(\delta) = 1$.

**Theorem 9.2.** Under the hypothesis of independence of the process $\{u_t\}$ there are two possibilities for the behavior of the optimal stock $k_t$. First, if $k_0 > 0$ and if

$$\int D(0, u)p(\mathrm{d}u) > \delta + [r/h'(\delta)] \tag{9.17}$$

holds, then $k_t$ will converge monotonically to the stationary value $k^c$, given implicitly by

$$\int D(k, u)p(du) = \delta + [r/h'(\delta)].$$

Or, secondly, if (9.17) fails to hold, or if $k_0 = 0$, then $k_t$ will converge monotonically to zero.

*Proof.* See Lucas and Prescott (1971, pp. 671–673).

## 10. Competitive processes, the transversality condition and convergence

The methods of Bismut (1973, 1975), briefly presented earlier, have been used by Brock and Magill (1979) in an attempt to develop a general approach to the continuous time stochastic processes that arise in dynamic economics. In this section we follow Brock and Magill (1979) who show that under a concavity assumption, to be specified, a competitive process which satisfies a transversality condition is optimal under a discounted catching-up criterion.

Let $(\Omega, \mathscr{F}, P)$ denote a complete probability space, $\mathscr{F}$ a $\sigma$-field on $\Omega$, and $P$ a probability measure on $\mathscr{F}$. Let $I = [0, \infty)$ denote the non-negative time interval and $(I, \mathscr{M}, \mu)$ the complete measure space of Lebesgue measurable sets $\mathscr{M}$, with Lebesgue measure $\mu$. Let $(\Omega \times I, \mathscr{H}, P \times \mu)$ denote the associated complete product measure space with complete measure $P \times \mu$ and $\sigma$-field $\mathscr{H} \supset \mathscr{F} \times \mathscr{M}$. Let $(R^n, \mathscr{M}^n)$, with $n \geqslant 1$, denote the measurable space formed from the $n$-dimensional real Euclidean space $R^n$ with $\sigma$-field of Lebesgue measurable sets $\mathscr{M}^n$. Let

$$k(\omega, t): (\Omega \times I, \mathscr{H}) \to (R^n, \mathscr{M}^n)$$

be an $\mathscr{H}$-measurable function (random process) induced by the following stochastic control problem. Find an $\mathscr{H}$-measurable control $v(\omega, t) \in U \subset R^s, s \geqslant 1$, such that for $\delta > 0$

$$\sup_{v \in U} \int_{\Omega} \int_{I} e^{-\delta t} u(\omega, t, k(\omega, t), v(\omega, t)) \mathrm{d}t \mathrm{d}P(\omega), \tag{10.1}$$

$$k(\omega, t) = k_0 + \int_0^t f(\omega, \tau, k(\omega, \tau), v(\omega, \tau)) \mathrm{d}\tau +$$

$$+ \int_0^\tau \sigma(\omega, \tau, k(\omega, \tau), v(\omega, \tau)) \, dz(\omega, \tau), \qquad (10.2)$$

where

$$u \in R^1; \quad f = (f^1, ..., f^n) \in R^n; \sigma = \begin{bmatrix} \sigma^{11} & ... & \sigma^{1m} \\ \cdot & & \cdot \\ \cdot & & \cdot \\ \cdot & & \cdot \\ \sigma^{n1} & ... & \sigma^{nm} \end{bmatrix} \in R^{nm} \quad \text{and} \quad k_0 \in K \subset R^n$$

is a nonrandom initial condition; $u(\cdot, k, v), f(\cdot, k, v), \sigma(\cdot, k, v)$ are $\mathscr{H}$-measurable random processes for all $(k, v)$ in $k \times U \subset R^n \times R^s$ and $u(\omega, \cdot), f(\omega, \cdot),$ $\sigma(\omega, \cdot)$, are continuous on $I \times K \times U$ for almost all $\omega$, while $z(\omega, t) \in R^m$, $m \geqslant 1$, is a Brownian motion process. Let

$$\mathscr{F}_t = \mathscr{F}(z(\omega, t)), \tau \in [0, t])$$

denote the smallest complete $\sigma$-field on $\Omega$ relative to which the random variaables $\{z(\omega, \tau), \tau \in [0, t]\}$ are measurable. We require that $k(\omega, t)$ be $\mathscr{F}_t$-measureable for all $t \in I$, so that $f(\cdot)$ and $\sigma(\cdot)$ are nonanticipating with respect to the family of $\sigma$-fields $\{\mathscr{F}_t, t \in I\}$. To ensure the existence of a unique random process $k(\omega, t)$ as a solution of (10.2) we assume that the Lipschitz and growth conditions of Chapter 2 are satisfied, i.e. we make the following assumption.
**Assumption 10.1.** Lipschitz and growth conditions: there exist positive constants, $\alpha$ and $\beta$, such that

(i) $\| f(\omega, t, k, v) - f(\omega, t, \overline{k}, v) \| + \| \sigma(\omega, t, k, v) - \sigma(\omega, t, \overline{k}, v) \| \leqslant \alpha \| k - \overline{k} \|$

for all $(k, v), (\overline{k}, v) \in K \times U$, for almost all $(\omega, t) \in \Omega \times I$, and

(ii) $\| f(\omega, t, k, v) \|^2 + \| \sigma(\omega, t, k, v) \|^2 \leqslant \beta(1 + \| k \|^2)$

for all $(k, v) \in K \times U$, for almost all $(\omega, t) \in \Omega \times I$. Here, as in previous sections, double bars denote vector norms.

Recall from Chapter 2 that assumption 10.1 is sufficient for the existence and uniqueness of a solution of a stochastic differential equation.

We will exhibit a sufficient condition for a random process to be a solution of the problem (10.1) and (10.2) in terms of a certain price support property, the nature of which is most clearly revealed by restricting this stochastic control in the manner of Bismut (1973, p. 393) and Rockafellar (1970, p. 188) as follows. Consider the new integrand

$$L(\omega, t, k, \dot{k}, \sigma) = \begin{cases} \sup_{v \in U} u(\omega, t, k, v) \mid f(\omega, t, k, v) = \dot{k}, \sigma(\omega, t, k, v) = \sigma \\ \\ -\infty \quad \text{if there is no } v \in U \text{ such that} \\ \qquad f(\omega, t, k, v) = \dot{k}, \sigma(\omega, t, k, v) = \sigma. \end{cases}$$

Note that $L(\omega, t, \cdot)$ is upper semicontinuous for all $(\omega, t) \in \Omega \times I$ and $L(\omega, t, k(\omega, t), \dot{k}(\omega, t), \sigma(\omega, t))$ is $\mathcal{H}$-measurable whenever $k(\omega, t)$, $\dot{k}(\omega, t)$ and $\sigma(\omega, t)$ are $\mathcal{H}$-measurable.

We impose indirect concavity and boundedness conditions on the functions $u(\omega, t, \cdot)$, $f(\omega, t, \cdot)$ and $\sigma(\omega, t, \cdot)$, and a convexity condition on the domain $K \times U$ by the following assumption.

**Assumption 10.2.** Concavity-boundedness: $L(\omega, t, \cdot)$ is concave in $(k, \dot{k}, \sigma)$ for all $(k, \dot{k}, \sigma) \in R^n \times R^n \times R^{nm}$ for all $(\omega, t) \in \Omega \times I$ and there exists $\gamma \in R, |\gamma| < \infty$, such that $L(\cdot) < \gamma$ for all $(\omega, t, k, \dot{k}, \sigma) \in \Omega \times I \times R^n \times R^n \times R^{nm}$.

Let $(k, \dot{k}, \sigma) = (k(\omega, t), \dot{k}(\omega, t), \sigma(\omega, t))$ denote the $\mathcal{H}$-measurable random process defined by the equation

$$k(\omega, t) = k_0 + \int_0^t \dot{k}(\omega, \tau) d\tau + \int_0^t \sigma(\omega, \tau) dz(\omega, \tau), \tag{10.3}$$

where $k_0 \in K \subset R^n$ is a nonrandom initial condition, and where there exists an $\mathcal{H}$-measurable control $v(\omega, t) \in U$ such that

$$\dot{k}(\omega, \tau) = f(\omega, \tau, k(\omega, \tau), v(\omega, \tau));$$

$$\sigma(\omega, \tau) = \sigma(\omega, \tau, k(\omega, \tau), v(\omega, \tau))$$

for almost all $(\omega, t) \in \Omega \times I$. In view of assumption 10.1

$$\int_\Omega \left( \int_0^t \| \dot{k}(\omega, \tau) \|^2 \, d\tau + \int_0^t \| \sigma(\omega, \tau) \|^2 \, d\tau \right) dP(\omega) < \infty \quad \text{for all } t \in I. \tag{10.4}$$

We let $\mathcal{P}$ denote the class of random processes satisfying (10.3) and (10.4), where $\dot{k}(\omega, \tau)$, $\sigma(\omega, \tau)$ are $\mathcal{H}$-measurable and nonanticipating with respect to the family of $\sigma$-fields $\{\mathscr{F}_t, t \in I\}$. The control problem (10.1) and (10.2) then reduces to the following.

*Stochastic variational problem:* Let $L$ satisfy assumption 10.2, let $L(\omega, t, \cdot)$ be upper semicontinuous for all $(\omega, t) \in \Omega \times I$, and let $L(\cdot, x, v, s)$ be $\mathscr{H}$-measurable for all $(x, v, s) \in R^n \times R^n \times R^{nm}$. Find an $\mathscr{H}$-measurable random process $(k, \dot{k}, \sigma) \in \mathscr{P}$ such that

$$\sup_{(k,\dot{k},\sigma) \in \mathscr{P}} \int_\Omega \int_I e^{-\delta t} L(\omega, t, k(\omega, t), \dot{k}(\omega, t), \sigma(\omega, t)) \mathrm{d}t \mathrm{d}P(\omega). \qquad (10.5)$$

In order to give (10.5) a broad interpretation we introduce the following definition. Let $\mathscr{K} \subset \mathscr{P}$ denote a class of $\mathscr{H}$-measurable random processes $(k, \dot{k}, \sigma)$. A random process $(\bar{k}, \bar{\dot{k}}, \bar{\sigma}) \in \mathscr{K}$ is optimal (in $\mathscr{K}$) if

$$\liminf_{T \to \infty} \int_\Omega \int_0^T e^{-\delta \tau} (L(\omega, \tau, \bar{k}, \bar{\dot{k}}, \sigma) - L(\omega, \tau, k, \dot{k}, \sigma)) \mathrm{d}\tau \mathrm{d}P(\omega) \geqslant 0 \qquad (10.6)$$

for all random processes $(k, \dot{k}, \sigma) \in \mathscr{K}$. Next, let

$$p(\omega, t): (\Omega \times I, \mathscr{H}) \to (R^n, \mathscr{M}^n)$$

denote an $\mathscr{H}$-measurable random price process dual to $k(\omega, t)$. We let

$$(\dot{p} - \delta p, p, \pi) = (\dot{p}(\omega, t) - \delta p(\omega, t), p(\omega, t), \pi(\omega, t))$$

denote the $\mathscr{H}$-measurable random price process defined by the equation

$$p(\omega, t) = p_0 + \int_0^t \dot{p}(\omega, \tau) \mathrm{d}\tau + \int_0^t \pi(\omega, \tau) \mathrm{d}z(\omega, \tau), \qquad (10.7)$$

where $p_0 \in R^n$ is nonrandom and where $\dot{p}(\omega, \tau)$ and $\pi(\omega, \tau)$ are $\mathscr{H}$-measurable random processes, nonanticipating with respect to the family of $\sigma$-fields $\{\mathscr{F}_t, t \in I\}$, with values in $(R^n, \mathscr{M}^n)$ and $(R^{nm}, \mathscr{M}^{nm})$, respectively, and which satisfy

$$\int_\Omega \left( \int_0^t \| \dot{p}(\omega, \tau) \|^2 \mathrm{d}\tau + \int_0^t \| \pi(\omega, \tau) \|^2 \mathrm{d}\tau \right) \mathrm{d}P(\omega) < \infty \qquad (10.8)$$

for all $t \in I$. Let $\mathscr{P}^*$ denote the class of random processes defined in this way.

The following concept is fundamental to all the analysis that follows. We define a random process $(\overline{k}, \dot{\overline{k}}, \overline{\sigma}) \in \mathscr{P}$ to be *competitive* if there exists a dual random price process $(\dot{\overline{p}} - \delta\overline{p}, \overline{p}, \overline{\pi}) \in \mathscr{P}^*$ such that

$$(\dot{\overline{p}} - \delta\overline{p})'\overline{k} + \overline{p}'\dot{\overline{k}} + \mathrm{tr}(\overline{\pi}\,\overline{\sigma}') + L(\omega, t, \overline{k}, \dot{\overline{k}}, \overline{\sigma})$$

$$\geqslant (\dot{\overline{p}} - \delta\overline{p})'k + \overline{p}'\dot{k} + \mathrm{tr}(\overline{\pi}\sigma') + L(\omega, t, k, \dot{k}, \sigma) \tag{10.9}$$

for all $(k, \dot{k}, \sigma) \in R^n \times R^n \times R^{nm}$, for almost all $(\omega, t) \in \Omega \times I$.

The economic interpretation of this concept is this: a competitive random process is a random process $(\overline{k}, \dot{\overline{k}}, \overline{\sigma}) \in \mathscr{P}$ that has associated with it a dual random price process $(\dot{\overline{p}} - \delta\overline{p}, \overline{p}, \overline{\pi}) \in \mathscr{P}^*$ under which it maximizes profit almost surely, at almost every instant. For $-(\dot{\overline{p}} - \delta\overline{p})$ denotes the vector of unit rental costs, $-\overline{\pi}$ denotes the matrix of unit risk costs induced by the disturbance matrix $\overline{\sigma}$, while $(1, \overline{p})$ is the vector of unit output prices, so that

$$L + \overline{p}'\dot{\overline{k}} + (\dot{\overline{p}} - \delta\overline{p})'\overline{k} + \mathrm{tr}(\overline{\pi}\overline{\sigma}')$$

is the (imputed) profit which is maximized almost surely, at almost every instant, by a competitive random process. We also give a geometric interpretation which is this: the random process $(\dot{\overline{p}} - \delta\overline{p}, \overline{p}, \overline{\pi}) \in \mathscr{P}^*$ generates supporting hyperplanes to the epigraph of $-L(\omega, t, k, \dot{k}, \sigma)$ at the point $(\overline{k}, \dot{\overline{k}}, \overline{\sigma})$ for almost all $(\omega, t) \in \Omega \times I$. The hyperplanes parallel to a given supporting hyperplane indicate hyperplanes of constant profit, so that the supporting hyperplanes are precisely the hyperplanes of maximum profit at each instant.

Note that under assumption 10.2 a random process $(k, \dot{k}, \sigma) \in \mathscr{P}$ is competitive if and only if

$$(\dot{p}(\omega, t) - \delta p(\omega, t), p(\omega, t), \pi(\omega, t)) \in -$$

$$- \partial L(\omega, t, k(\omega, t), \dot{k}(\omega, t)\sigma(\omega, t)) \tag{10.10}$$

for almost all $(\omega, t) \in \Omega \times I$, where $\partial L$ denotes the subdifferential of $L(\omega, t, \cdot)$. Eq. (10.10) is a generalization of the standard Euler–Lagrange equation.

The Fenchel conjugate of $-L(\omega, t, k, \dot{k}, \sigma)$ with respect to $(\dot{k}, \sigma)$ will be called the *generalized Hamiltonian*

$$G(\omega, t, k, p, \pi) = \sup_{(\dot{k},\sigma) \in R^n \times R^{nm}} \{p'\dot{k} + \mathrm{tr}(\pi\sigma') +$$

$$+ L(\omega, t, k, \dot{k}, \sigma)\}. \tag{10.11}$$

Observe that $G(\omega, t, k, p, \pi)$ is concave in $k$ and convex in $(p, \pi)$ for all $(\omega, t) \in \Omega \times I$ and is defined for all $(k, p, \pi) \in R^n \times R^n \times R^{nm}$.

Under assumption 10.2 if $G(\omega, t, \cdot)$ is differentiable, a random process $(k, \dot{k}, \sigma) \in \mathscr{P}$ is competitive if and only if

$$k(\omega, t) = k_0 + \int\limits_0^t G_p(\omega, \tau) \mathrm{d}\tau + \int\limits_0^t G_\pi(\omega, \tau) \mathrm{d}z(\omega, \tau), \qquad (10.12)$$

and also

$$p(\omega, t) = p_0 + \int\limits_0^t [-G_k(\omega, \tau) + \delta p(\omega, \tau)] \mathrm{d}\tau +$$

$$+ \int\limits_0^t \pi(\omega, \tau) \mathrm{d}z(\omega, \tau). \qquad (10.13)$$

Eqs. (10.12) and (10.13), which will be called the *stochastic Hamiltonian equations*, are a generalization of the standard Hamiltonian canonical equations for a discounted stochastic variational problem.

Assume that for all $(x, v, s) \in R^n \times R^n \times R^{nm}$ that

$$L(\omega, t, x, v, s) = L(x, v, s) \quad \text{for all } (\omega, t) \in \Omega \times I,$$

so that $L$ is nonrandom and time independent. When (10.5) is finite we define the current value function $W(k): R^n \to R$

$$W(k(t)) = \sup_{(k, \dot{k}, \sigma) \in \mathscr{P}} E_t \int\limits_t^\infty e^{-\delta(\tau - t)} L(k(\omega, \tau), \dot{k}(\omega, \tau), \sigma(\omega, \tau)) \mathrm{d}\tau, \qquad (10.14)$$

where $E_t$ denotes the conditional expectation given $k$ at time $t$, and where $k(t)$ replaces $k_0$ as the initial condition in (10.3).

Note that under assumption 10.2, $W(k)$ is a concave function for all $k \in K$. In establishing convergence properties, the following class of McKenzie competitive processes is of special importance. A random process $(\bar{k}, \dot{\bar{k}}, \bar{\sigma}) \in \mathscr{P}$ is *McKenzie competitive* if it is competitive and if the dual random price process $\bar{p}(\omega, t)$ supports the value function

$$W(\bar{k}(\omega, t)) - \bar{p}(\omega, t)' \bar{k}(\omega, t) \geqslant W(k) - \bar{p}(\omega, t)' k \qquad (10.15)$$

for all $k \in R^n$, for almost all $(\omega, t) \in \Omega \times I$.

If $(k, \dot{k}, \sigma) \in \mathscr{P}$ is McKenzie competitive then $p_0$ in (10.8) is determined by the condition $p_0 \in \partial W(k_0)$. We are now ready to state

**Theorem 10.1.** (Transversality condition). A competitive random process $(\bar{k}, \dot{\bar{k}}, \bar{\sigma}) \in \mathscr{K}$ with dual price process $(\dot{\bar{p}} - \delta \bar{p}, \bar{p}, \bar{\pi}) \in \mathscr{P}^*$, which satisfies the transversality condition

$$\limsup_{T \to \infty} E_0 e^{-\delta T} \bar{p}(\omega, T)' \bar{k}(\omega, T) \leqslant 0$$

is optimal in the class $\mathscr{K}$ of random processes for which

$$\liminf_{T \to \infty} E_0 e^{-\delta T} \bar{p}(\omega, T)' k(\omega, T) \geqslant 0. \tag{10.16}$$

*Proof.* See Brock and Magill (1979).

The sample paths of a McKenzie competitive process starting from nonrandom initial conditions have a remarkable convergence property. Consider a point $k_0 \in K$ and a McKenzie competitive process emanating from this point. Under assumptions, which include a strict concavity assumption on the basic integrand $L$, a McKenzie competitive process emanating from any other point $k_0 \in K$ converges almost surely to the first process. This result, which has its origin in the dual relationship between the prices and quantities of a McKenzie competitive process, may be stated as follows.

**Theorem 10.2** (Almost sure convergence). Let assumption 10.2 be satisfied and let the function $L$ be time independent and nonrandom as in (10.14). If two McKenzie competitive random processes

$$(k, \dot{k}, \sigma) \in \mathscr{P} \quad \text{and} \quad (\bar{k}, \dot{\bar{k}}, \bar{\sigma}) \in \mathscr{P}, \tag{10.17}$$

with associated dual price processes

$$(\dot{p} - \delta p, p, \pi) \in \mathscr{P}^* \quad \text{and} \quad (\dot{\bar{p}} - \delta \bar{p}, \bar{p}, \bar{\pi}) \in \mathscr{P}^*, \tag{10.18}$$

starting from the nonrandom initial conditions

$$(k_0, p_0) \quad \text{and} \quad (\bar{k}_0, \bar{p}_0)$$

satisfy the following conditions:

(i)   there exists a compact convex subset $M \subset R^n \times R^n$ such that for all $t \in I$

$$(k(\omega,t),p(\omega,t)) = (k(\omega,t;k_0),p(\omega,t;p_0)) \in M,$$
$$(\bar{k}(\omega,t),\bar{p}(\omega,t) = (\bar{k}(\omega,t;\bar{k}_0),\bar{p}(\omega,t;\bar{p}_0) \in M$$

for almost all $\omega \in \Omega$;

(ii)   there exists $\mu > 0$ such that the function

$$V(k-\bar{k},p-\bar{p}) = -(p-\bar{p})'(k-\bar{k}) \tag{10.19}$$

satisfies

$$\mathscr{D}\, V(k-\bar{k},p-\bar{p}) \leqslant -\mu\,\|\,(k-\bar{k},p-\bar{p})\,\|^2 \tag{10.20}$$

for all $(k-\bar{k},p-\bar{p}) \in Y = \{(k-\bar{k},p-\bar{p})\,|\,(k,p),(\bar{k},\bar{p}) \in M\}$;

(iii)   the value function is (a) strictly concave, (b) differentiable, and (c) strictly concave and differentiable, for all $k$ in the interior of $K$, $K = \{k\,|\,(k,p) \in M\}$; then (i), (ii) and (iii) (a) imply

$$k(\omega,t) - \bar{k}(\omega,t) \to 0 \quad \text{w.p.l. as } t \to \infty;$$

(i), (ii) and (iii) (b) imply

$$p(\omega,t) - \bar{p}(\omega,t) \to 0 \quad \text{w.p.l. as } t \to \infty;$$

(i), (ii) and (iii) (c) imply

$$(k(\omega,t) - \bar{k}(\omega,t),p(\omega,t) - \bar{p}(\omega,t)) \to 0 \quad \text{w.p.l. as } t \to \infty.$$

*Proof.*   See Brock and Magill (1979, pp. 852–855).

## 11.  Rational expectations equilibrium

In this section we follow Brock and Magill (1979) to show how the concept of a competitive process, in conjunction with the stochastic Hamiltonian equations (10.12) and (10.13), provides a useful framework for the analysis of rational expectations equilibrium. We examine in particular a rational expectations equilibrium for a competitive industry in which a fixed finite number of firms behave according to a stochastic adjustment cost theory by creating an extended integrand problem analogous to that of Lucas and Prescott (1971).

Consider therefore an industry composed of $N \geqslant 1$ firms, each producing the same industry good with the aid of $n \geqslant 1$ capital goods. All firms have identical expectations regarding the industry product's price process, which is an $\mathscr{H}$-measurable, nonanticipating process

$$r(\omega, t): (\Omega \times I, \mathscr{H}) \to (R^+, \mathscr{M}). \tag{11.1}$$

The instantaneous flow of profit of the $i$th firm is the difference between its revenue $r(\omega, t)f^i(k^i(\omega, t))$ and its costs $C^i(v^i(\omega, t))$, where $k^i = (k^{i1}, ..., k^{in})$ and $v^i = (v^{i1}, ..., v^{in})$ denote the capital stocks and investment rates of the $i$th firm, and where $f^i(k^i)$ and $-C^i(v^i)$ are the standard strictly concave production and adjustment cost functions. If $\delta > 0$ denotes the nonrandom interest rate, then each firm seeks to maximize its expected discounted profit by selecting an $\mathscr{H}$-measurable, nonanticipating investment processes

$$v^i(\omega, t): (\Omega \times I, \mathscr{H}) \to (R^n, \mathscr{M}^n), \quad i = 1, ..., N,$$

such that

$$\sup_{v^i(\omega,t)} E_0 \int_0^\infty e^{-\delta \tau} [r(\omega, \tau)f^i(k^i(\omega, \tau)) - C^i(v^i(\omega, \tau))]d\tau,$$

$$k^i(\omega, t) = k_0^i + \int_0^t v^i(\omega, \tau)d\tau + \int_0^t \sigma^i(k^i(\omega, \tau))dz^i(\omega, \tau), \tag{11.2}$$

$$\sigma^i(k^i)dz^i = \sum_{j=1}^m (H^{ij}k^i + \sigma_0^{ij})dz^{ij}, \tag{11.3}$$

where $H^{ij}$, $\sigma_0^{ij}$ are $n \times n$ and $n \times 1$ matrices with constant coefficients and $z^i(\omega, \tau)$ is an m-dimensional Brownian motion process. This model is a simple stochastic version of the basic Lucas (1967) and Mortensen (1973) adjustment cost model, with the standard additional neoclassical assumption that the investment and output process of the $i$th firm have no direct external effects on the investment and output processes of the $k$th firm, for $i \neq k$.

On the product market the total market supply, given by

$$Q_S(\omega, t) = \sum_{i=1}^N f^i(k^i(\omega, t)),$$

depends in a complex way through the maximizing behavior of firms on the price process (11.1). On the demand side of the product market we make the simplifying assumption that the total market demand depends only on the current market price

$$Q_D(\omega, t) = \psi^{-1}(r(\omega, t)), \quad r \geqslant 0,$$

where $\psi(Q) > 0$, $\psi'(Q) < 0$ and $Q \geqslant 0$.

A rational expectations equilibrium for the product market of the industry is an $\mathcal{H}$-measurable, nonanticipating random process (11.1) such that

$$Q_D(\omega, t) = Q_S(\omega, t) \quad \text{for almost all } (\omega, t) \in \Omega \times I. \tag{11.4}$$

The firms' expectations are rational in that the anticipated price process coincides almost surely with the actual price process generated on the market by their maximizing behavior.

Consider the integral of the demand function

$$\Psi(Q) = \int_0^Q \psi(y)dy, \quad Q \geqslant 0,$$

so that $\Psi'(Q) = \psi(Q)$, $\Psi''(Q) = \psi'(Q) < 0$ and $Q \geqslant 0$.

We call the problem of finding $N$ $\mathcal{H}$-measurable, nonanticipating investment processes

$$(v^1(\omega, t), ..., v^N(\omega, t)) \colon (\Omega \times I, \mathcal{H}) \to (R^{nN}, \mathcal{M}^{nN})$$

such that

$$\sup_{(v^1(\omega,t), ..., v^N(\omega,t))} E_0 \int_0^\infty e^{-\delta\tau} \left[ \Psi\left( \sum_{i=1}^N f^i(k^i(\omega, \tau)) \right) - \right.$$

$$\left. - \sum_{i=1}^N C^i(v^i(\omega, \tau)) \right] d\tau, \tag{11.5}$$

where $(k^1(\omega, t), ..., k^N(\omega, t))$ satisfy (11.2) and (11.3), almost surely, the extended integrand problem.

**Theorem 11.1.** (Rational expectations equilibrium). If the generalized Hamiltonian of the extended integrand problem (11.5) is differentiable, if

$$(\bar{k}(\omega, t), \bar{p}(\omega, t)) = (\bar{k}^1(\omega, t), ..., \bar{k}^N(\omega, t), \bar{p}^1(\omega, t), ..., \bar{p}^N(\omega, t)$$

is a competitive process for (11.5) which satisfies the transversality condition

$$\lim_{T \to \infty} \sup E_0\, e^{-\delta T}\, \bar{p}(\omega, T)'\bar{k}(\omega, T) \leqslant 0, \tag{11.6}$$

and if for any alternative random process $k(\omega, t)$ with $k_0 = \bar{k}_0$

$$\lim_{T \to \infty} \inf E_0\, e^{-\delta T}\, \bar{p}(\omega, T)'k(\omega, T) \geqslant 0, \tag{11.7}$$

then the $\mathcal{H}$-measurable, nonanticipating random process

$$\bar{r}(\omega, t) = \psi\left[ \sum_{i=1}^{N} f^i(\bar{k}^i(\omega, t)) \right] \tag{11.8}$$

is a rational expectations equilibrium for the product market of the industry.

*Proof.* Since the generalized Hamiltonian for the extended integrand problem is differentiable, $(\bar{k}(\omega, t), \bar{p}(\omega, t))$ is competitive if and only if, writing (10.12) and (10.13) in shorthand form,

$$\left.\begin{aligned}
dk^i &= h^i(p^i)dt + o^i(k^i)dz^i \\
dp^i &= [\delta p^i - \Psi'\left( \sum_{i=1}^{N} f^i(k^i) \right) f^i_{k^i} - \sum_{j=1}^{m} \pi^{ij}o^{ij}_{k^i}]dt + \pi^i dz^i \\
&\qquad\qquad\qquad\qquad\qquad i = 1, ..., N,
\end{aligned}\right\} \tag{11.9}$$

where $h^i = (C^i_{v^i})^{-1}$. Eqs. (11.8) and (11.9) imply

$$\left.\begin{aligned}
dk^i &= h^i(p^i) + o^i(k^i)dz^i \\
dp^i &= \left[ \delta p^i - r(\omega, t)f^i_{k^i} - \sum_{j=1}^{m} \pi^{ij}o^{ij}_{k^i} \right]dt + \pi^i dz^i.
\end{aligned}\right\}, i = 1, ..., N. \tag{11.10}$$

Eqs. (11.6), (11.7) and (11.8) are sufficient conditions for each firm to maximize expected discounted profit by theorem 10.1. Eq. (11.8) implies that (11.4) is satisfied and the proof is complete.

## 12.  Linear quadratic objective function

In this section and the following one we present the analysis of particular stochastic control problems which have found applicability in economics and finance.

Consider the one-dimensional state and control problem

$$-W(x(t)) = \min_{v} E_t \int_{s=t}^{\infty} e^{-\rho(s-t)} \{a(x(s))^2 + b(v(s))^2\} ds$$

subject to $x(t)$ known and

$$dx(t) = v(t)dt + \sigma x(t)dz(t).$$

Note that $\sigma > 0$, $a > 0$, $b > 0$ and $\rho > 0$. Since the objective function is convex we try as a solution

$$W(x) = -Px^2, \tag{12.1}$$

which yields

$$W_x = -2Px \quad \text{and} \quad W_{xx} = -2P. \tag{12.2}$$

We now use the Hamilton–Jacobi–Bellman equation for the case with discounting, i.e.

$$\rho W(k(t)) = \max_{v} \{u(k,v) + W_k T + \tfrac{1}{2} \sigma^2 W_{kk}\}$$

to write in our specific case

$$-\rho Px^2 = \max_{v} \{-ax^2 - bv^2 - 2Pxv + \tfrac{1}{2}(-2P\sigma^2 x^2)\}. \tag{12.3}$$

From (12.3) we obtain that $-2bv - 2Px = 0$, or that

$$v^\circ = -Px/b. \tag{12.4}$$

Substitute (12.4) into (12.3); then

$$-\rho Px^2 = \left\{ -ax^2 - b\left(-\frac{Px}{b}\right)^2 - 2Px\left(-\frac{Px}{b}\right) - P\sigma^2 x^2 \right\},$$

which after simple algebraic manipulations is reduced to

$$\frac{1}{b} P^2 + (\rho - \sigma^2) P - a = 0. \tag{12.5}$$

Choose the largest root because of the convexity of the objective function and denote it by $P_+$. A candidate solution that is optimum is given by

$$dx(t) = \left[ -\frac{1}{b} P_+ x(t) \right] dt + \sigma x(t) dz(t), \tag{12.6}$$

the solution of which is

$$x(t) = x_0 \exp \left\{ \left( -\frac{1}{b} P_+ - \frac{\sigma^2}{2} \right) t + \sigma z(t) \right\}. \tag{12.7}$$

## 13. State valuation functions of exponential form

In the previous section we saw that the state valuation function of the linear quadratic example turned out to be quadratic in the state variable. This is a general principle in linear quadratic problems. We turn now to a type of problem where the state valuation function turns out to be of exponential form. In examining this type of problem we hope that the reader will pick up the common thread of technique that is used in searching for closed form solutions for the state valuation functions when they exist. We also hope that this will illustrate a technique of showing that a state valuation function of a particular closed form cannot satisfy the Hamilton–Jacobi–Bellman equation as well. In this way the search for closed form solutions to the Hamilton–Jacobi–Bellman partial differential equations is systematic instead of random guesswork. It is well to learn how to prove that solutions of a particular form do not exist as well as formulating hypotheses upon the objective and the constraints of the problem so that a closed form solution of a particular type does exist.

Consider the problem

$$J(x(t), t, N) = \max_{} E_t \left\{ \int_t^N D(s) u(c(s), s) ds + B(x(N), N) \right\} \tag{13.1}$$

subject to

$$dx(s) = (b(s) x(s)^{\beta(s)} - c(s)) ds + \sigma(x(s), s) dz(s). \tag{13.2}$$

Here $D(s)$, $c(s)$, $u(c(s), s)$, $x(s)$, $B(x(N), N)$, $b(s)$, $\beta(s)$, $\sigma(x(s), s)$, and $dz(s)$ denote discount rate at time $s$, consumption at time $s$, utility of consumption at time $s$, state variable at time $s$ which in economic applications is usually capital, bequest function of the state variable at time $N$, efficiency multiplier of production function at time $s$, output elasticity at time $s$, standard deviation function at time $s$, and Wiener process that is standardized at each moment of time, respectively.

Notice that the problem in (13.1) and (13.2) is quite general. We shall see how general we can make it, and still get a closed form solution for the state valuation function $J$. The stochastic maximum principle applied to (13.1) and (13.2) gives

$$0 = \underset{c \geq 0}{\text{maximum}} \{D(t)u(c, t) + \mathcal{L}(J)\}, \tag{13.3}$$

where

$$\mathcal{L}(J) \equiv \lim_{\Delta t \to 0} E_t\{[J(x(t + \Delta t), t + \Delta t, N) - J(x(t), t, N)]/\Delta t\}. \tag{13.4}$$

The operator $\mathcal{L}(\cdot)$ is just the conditional expectation of the instantaneous rate of change of $J$ and is a general case of the operator defined in eq. (7.9) of Chapter 2. See also eq. (4.15) of Chapter 4.

Let us calculate the partial differential equation given in eq. (13.3) for the constant relative risk-aversion class of utility functions given by

$$u(c(s), s) = (k(s)(c(s))^{1-\alpha(s)})/(1-\alpha(s)). \tag{13.5}$$

Here $k(s)$ is just a constant independent of $c$. Assume that the optimal consumption given by eq. (13.3) is positive. Then we can write

$$kc^{-\alpha} = J_x. \tag{13.6}$$

In eq. (13.6) it is understood that everything is a function of time, and subscripts denote the obvious partial differentiation. Insert the solution for $c$ from eq. (13.6) into eq. (13.3) to get the partial differential equation

$$0 = J_x^{(\alpha-1)/\alpha} k^{\alpha-1} \alpha(\alpha-1)^{-1} + J_t + bJ_x x^\beta + \tfrac{1}{2} J_{xx} \sigma^2. \tag{13.7}$$

Eq. (13.7) is obtained by the following steps. First, evaluate the operator in eq. (13.4) to get

$$\mathcal{L}(J) = J_t + J_x(bx^\beta - c) + \tfrac{1}{2} J_{xx}\, \sigma^2. \tag{13.8}$$

Eq. (13.8) is obtained either by a direct application of Itô's lemma, or by expanding $J$ into a formal Taylor series discarding all terms of higher order than $\Delta t$, and taking conditional expectation of the remaining conditional on information received at time $t$. Secondly, insert the optimal value for $c$ from eq. (13.6) into eq. (13.8). Insert all of this into eq. (13.3) and rearrange terms to get eq. (13.7). Notice that everything in eq. (13.7) is a function of time, but we have suppressed notation of this dependence in order to simplify the expression.

Now look carefully at eq. (13.7), especially the last three terms. If we tried as a candidate for $J$ a function of the form

$$J(x, t, N) = g(t, N)x^{e(t)} + f(t, N) \tag{13.9}$$

then we should attempt to determine the unknown exponent $e(t)$ by equating exponents and testing for consistency by examining the possibility of identical exponents on the $x$ variable of all four terms of eq. (13.7).

The exponents of $x$ in eq. (13.7) for each of the four terms reading from left to right are:

$$(e-1)\,(\alpha-1)/\alpha = e + \text{exponent on } (x^e \log x) = (e-1) + \beta$$
$$= (e-2) + \text{exponent on } x \text{ from } \sigma^2. \tag{13.10}$$

Eq. (13.10) follows immediately by inspecting the partial derivatives of $J$, which we calculate by using eq. (13.9), and list below in equations

$$J_t = g_t x^e + f_t + gx^e \log x\, e_t, \tag{13.11}$$

$$J_x = eg\, x^{e-1}; \quad J_{xx} = e(e-1)gx^{e-2}. \tag{13.12}$$

We see immediately from eqs. (13.11) and (13.12) that in our search for the most general problem that will give us a solution for the state valuation function of the form (13.9), we will have to assume the conditions listed below in order to get the same exponent for the last three equalities of eq. (13.10). First assume that

$$e_t = 0. \tag{13.13}$$

Assumption (13.13) is needed to get rid of the term $x^e \log x$ that obstructs the validity of eq. (13.11). Secondly, assumption

$$\sigma^2(x, s) = h(s)x^2 \qquad\qquad (13.14)$$

on the variance function is needed so that the last equality of eq. (13.10) is equal to $e$.

Finally, assumption

$$\beta(s) = 1 \qquad\qquad (13.15)$$

is needed for identity with the other terms of eq. (13.10). Thus, assumptions (13.13)–(13.15) allow us to assert that it is possible to find a function of the form in eq. (13.9) that will satisfy eq. (13.7).

To solve for the unknown exponent $e$, which by eq. (13.13) must be constant in time, we solve

$$(e-1)(\alpha-1)/\alpha = e. \qquad\qquad (13.16)$$

Thus,

$$e = 1-\alpha. \qquad\qquad (13.17)$$

We see now that $\alpha$ must be independent of time in order to obtain a solution of the form of (13.9).

We have now solved for the exponent, $e$, in eq. (13.9). What remains is to solve for the functions $g$ and $f$.

In order to solve for the functions $g$ and $f$, write eq. (13.7) using eq. (13.10) as follows:

$$0 = (eg)^{(\alpha-1)/\alpha} k^{\alpha-1} \alpha(\alpha-1)^{-1} x^e + g_t x^e + f_t + beg\, x^e +$$

$$+ \tfrac{1}{2} e(e-1)ghx^e. \qquad\qquad (13.18)$$

Now, eq. (13.18) must hold for all $x$ and therefore $f_t = 0$ must hold. Hence, $f$ is independent of time. Now cancel $x^e$ off both sides of (13.18) to get

$$0 = (eg)^{(\alpha-1)/\alpha} k^{\alpha-1} \alpha(\alpha-1)^{-1} + g_t + beg + \tfrac{1}{2} e(e-1)gh = 0. \qquad (13.19)$$

Notice that (13.19) can be written in the form

$$0 = a_0 g^{(\alpha-1)/\alpha} + a_1 g + \dot{g}, \qquad\qquad (13.20)$$

where $\dot{g} \equiv g_t$.

We see that (13.20) is a differential equation that looks formidable to solve. However, such a differential equation can be transformed. Experiment with transformations of the form

$$y = g^\gamma \tag{13.21}$$

to discover that if $\gamma = 1/\alpha$, then eq. (13.20) can be transformed into the form

$$0 = a_0 y^{\alpha - 1} + a_1 y^\alpha + \alpha \dot{y}\, y^{\alpha - 1}. \tag{13.22}$$

Cancel $y^{\alpha - 1}$ off of both sides of (13.22) to get

$$0 = a_0 + a_1 y + \alpha \dot{y}, \tag{13.23}$$

which is a differential equation linear in $y, \dot{y}$. This is a standard form which can be solved regardless of whether the coefficients $a_0$ and $a_1$ are dependent upon time or not. Here the coefficients $a_0$ and $a_1$ are defined in the obvious manner by eq. (13.19). A boundary condition is needed before eq. (13.23) can be solved. This is obtained from the restriction

$$J(x, N, N) = B(x, N). \tag{13.24}$$

Of course, we will not be able to solve eqs. (13.23) and (13.24) for arbitrary bequest functions. To see how things go put

$$B(x, N) \equiv 0. \tag{13.25}$$

Look now at eq. (13.9). From eqs. (13.9) and (13.25) we infer immediately that $f(t, N) = 0$. This is so because $f_t(t, N) = 0$ for all $t$ and $f(N, N) = 0$. Furthermore, eqs. (13.25) and (13.19) imply

$$g(N, N) = 0 \quad \text{and} \quad y(N, N) = 0. \tag{13.26}$$

Thus, we have our boundary condition on eq. (13.23) and it can be solved for an explicit solution $y(t, N)$, from which $g$ can be calculated from eq. (13.21).

What about the case of more general bequest functions than that given in eq. (13.25)? We see immediately by inspection of eq. (13.24) that if there is any hope of a solution for $J$ of the form given by eq. (13.9), then the exponent on $x$ in the function $B(x, N)$ must be the same as $e$ in eq. (13.9). Thus, a more general class of bequest functions where $x$ enters with the same exponent as that of the utility function may be treated without much extra effort. Thus, closed form solutions exist in this case as well.

Now, retrace through our derivation using the Hamilton–Jacobi–Bellman equation, eq. (13.7), and recall how much had to be assumed on the structure of the utility function, the production function, and the bequest function in order to get a closed form solution for $J$. We saw from eqs. (13.13) and (13.15) that the exponent on the utility function had to be independent of time and the exponent on the production function had to be unity at all points in time. Furthermore, the variance had to be proportional to the square of the state variable, but the constant of proportionality could vary in time. Also, all other coefficients could vary in time.

It is worthwhile to note that other sources of randomness may be introduced into problem (13.1) and (13.2) above and beyond the randomness in the change of the state variable, provided that these sources of randomness are independent of the Wiener process $dz$; that is to say the coefficients $k$, $b$ and $h$ may be generated by Itô equations as well.

We may summarize this application by noticing that it is fairly illustrative of the method of searching for closed form solutions when they exist and determining assumptions that are necessary to put on the problem in order to get the existence of a closed form solution. Furthermore, we have approached the problem of searching for a closed form solution in such a way that illustrates the maximum amount of generality that one can have and still get the existence of a closed form solution. Once the solution has been found, then the optimal controls may be solved for explicitly, and the optimal law of motion that describes the system may be written down in closed form as well. Quadratic objectives with linear dynamics or exponential objectives with linear dynamics are by far the most common examples where closed form solutions to the HJB equation are available.

## 14. Money, prices and inflation

In this section we follow Gertler (1979) to present a rational expectations macro-economic model to illustrate the convergence of the state variables to a stable distribution over time. We begin by stating the deterministic system of structural relationships:

$$\ln Y = b_1 \ln K - b_2 r, \qquad b_1, b_2, > 0 \tag{14.1}$$

$$\ln (M/P) = c \ln Y - mi, \qquad c, m > 0, \tag{14.2}$$

$$\pi^d = \lambda \,(\ln Y - a \ln K) + \pi^*, \quad \lambda, a > 0, \tag{14.3}$$

$$\pi = \phi \,(\pi^d - \tilde{\pi}) + \tilde{\pi}, \qquad\qquad 0 \leqslant \phi \leqslant 1, \tag{14.4}$$

$$\pi = \pi^*, \tag{14.5}$$

$$i = r + \pi^*. \tag{14.6}$$

Equation (14.1) is a reduced form IS function relating the logarithm of real output positively to the logarithm of the capital stock and negatively to the real interest rate. It is assumed that the capital stock is constant. Eq. (14.2) is the LM function relating the logarithm of real money balances positively to the logarithm of real income and negatively to the nominal interest rate. Eq. (14.3) relates the desired rate of inflation, $\pi^d$, to the aggregate excess effective demand and the anticipated inflation rate $\pi^*$. Note that aggregate excess effective demand is the difference between output and the natural full employment level of output. Eq. (14.4) describes the price velocity constraint where actual current inflation $\pi$ is the sum of the *ad hoc* inertia inflation $\tilde{\pi}$ and a convex combination of $\pi^d$ and $\tilde{\pi}$. Eq. (14.5) denotes myopic perfect foresight and it is the equivalent of rational expectations in the deterministic context. Finally, eq. (14.6) is a definitional identity where the nominal interest rate $i$ is the sum of the real interest rate $r$ and the anticipated inflation rate $\pi^*$.

Next we introduce uncertainty into the deterministic system by postulating

$$\ln Y \, \mathrm{d}t = b_1 \ln K \mathrm{d}t - b_2 r \mathrm{d}t + \mathrm{d}z. \tag{14.7}$$

Note that this last equation generalizes eq. (14.1) since the random term $\mathrm{d}z$ is being added. In (14.7), $\mathrm{d}z$ equals $sy(t)\mathrm{d}t$, where $s$ is a parameter and $y(t)$ is a normally distributed random variable with mean zero, unit variance, and the $y(t)$'s are serially uncorrelated.

The introduction of uncertainty also affects eq. (14.5). Assuming rational expectations instead of (14.5) we now postulate

$$\pi^*(t) = \lim_{u \to t} \, \mathrm{E}[\pi(t) \,|\, \mathcal{F}_u], \tag{14.8}$$

where $\mathcal{F}_u$ is the $\sigma$-field incorporating all information available at time $u$. Such information includes the structure of the model and both the initial values and the past behavior of the state variables.

Using eqs. (14.2), (14.3) and (14.7) we obtain:

$$\pi dt = \phi\lambda w \ln (M/P)dt + \phi(\lambda wm + 1)\pi^* dt$$
$$+ (1-\phi)\bar{\pi} dt + \phi\lambda [(wmb_1/b_2) - a] \ln K dt$$
$$+ [(\phi\lambda mw)/b_2] dz, \tag{14.9}$$

where for convenience we have $w \equiv b_2/(m + cb_2)$. Eq. (14.9) describes the dynamic random behavior of actual inflation as a function of various parameters, of the logarithm of real money supply, the anticipated inflation rate, the inertia inflation of the system, the logarithm of the capital stock and uncertainty. From the assumption of rational expectations in (14.8) and (14.9) we conclude that

$$\pi^* = [mw/(1-m\delta w)] \ln (M/P) + [1/(1-m\delta w)]\bar{\pi}$$
$$- [\delta/(1-m\delta w)]a \ln K + [b_1 m\delta w/b_2(1-m\delta w)] \ln K. \tag{14.10}$$

In this last equation we let $\delta \equiv \phi\lambda(1-\phi)$. Note that $\delta$ is the coefficient on excess demand in the price adjustment equation. Substituting (14.10) into (14.9) we have

$$\pi dt = [\delta w/(1-m\delta w)] \ln (M/P) dt + [1/(1-m\delta w)]\bar{\pi} dt$$
$$- [\delta/(1-m\delta w)]a \ln K dt + [b_1 m\delta w/b_2(1-m\delta w)] \ln K dt$$
$$+ (\phi\lambda mw/b_2)dz. \tag{14.11}$$

At this point we assume that the money growth rate is deterministic and fixed at $\mu$. Using Itô's lemma we obtain

$$d \ln (M/P) = \mu dt - \pi dt + (\phi\lambda mw/b_2)^2 dt, \tag{14.12}$$

where $\pi dt$ is as in (14.11). Finally, we write

$$d\bar{\pi} = \beta^*(\pi dt - \bar{\pi} dt), \tag{14.13}$$

which is an assumption about the evolution of the system's inertia inflation $\bar{\pi}$.

We are now ready to study the problem of convergence. Substituting $\pi dt$ from (14.11) into (14.12) and (14.13) we obtain the system of linear stochastic differential equations

$$dh = Ah dt + f' dt + dv, \tag{14.14}$$

where

$$h = \begin{pmatrix} \ln{(M/P)} \\ \tilde{\pi} \end{pmatrix},$$

$$A = \begin{pmatrix} -\delta w/(1-m\delta w) & -1/(1-m\delta w) \\ \beta^*\delta w/(1-m\delta w) & \beta^* m\delta w/(1-m\delta w) \end{pmatrix},$$

$$f = \begin{pmatrix} \mu + [\delta/(1-m\delta w)]a \ln K - [b_1 m\delta w/b_2(1-m\delta w)] \ln K \\ - [\beta^*\delta/(1-m\delta w)]a \ln K + [\beta^* b_1 m\delta w/b_2(1-m\delta w)] \ln K \end{pmatrix},$$

$$f' = f + \begin{pmatrix} -(\phi\lambda mw/b_2)^2 \\ 0 \end{pmatrix},$$

$$dv = \begin{pmatrix} -(\phi\lambda mw/b_2)dz \\ \beta^*(\phi\lambda mw/b_2)dz \end{pmatrix}.$$

The solution of (14.14) is given by

$$h(t) = e^{At} h(t) + \int_0^t e^{A(t-s)} f \, ds + \int_0^t e^{A(t-s)} \, dv(s) \, ds. \tag{14.15}$$

Take the expected value of $h(t)$ in this last equation to yield

$$Eh(t) = e^A h(0) + \int_0^t e^{A(t-s)} f \, ds. \tag{14.16}$$

Assume that $A$ is negative-definite. Then from (14.16) we conclude that the mean, $Eh(t)$, converges to a stable path. The necessary and sufficient conditions for stability are

$$\beta^* m < 1 \tag{14.17}$$

and

$$\delta < 1/mw. \tag{14.18}$$

Eq. (14.17) restricts the adjustment speed of the price inertia and requires that the demand for real balances cannot be too sensitive to the nominal interest rate. Eq. (14.18) restricts the size of $\delta$, the coefficient on excess demand in the price adjustment mechanism.

Gertler (1979, p. 232) also shows that if $A$ is negative-definite the variance–covariance matrix of $h(t)$ will converge to a stable value. The above discussion concludes our illustration and we refer the reader to Gertler (1979) for further analysis of specific aspects.

## 15. An *N*-sector discrete growth model

In this application we follow Brock (1979) to present an *n*-process discrete optimal growth model which generalizes the Brock and Mirman (1972, 1973) model. This section and sections 11–16 of Chapter 4 attempt to put together ideas from the modern theory of finance and the literature on stochastic growth models. Here we develop the growth theoretic part of an intertemporal general equilibrium theory of capital asset pricing. Basically, what is done is to modify the stochastic growth model of Brock and Mirman (1972) in order to put a nontrivial investment decision into the asset pricing model of Lucas (1978). The finance side of the theory is presented in sections 11–16 of Chapter 4 and they derive their inspiration from Merton (1973b). However, Merton's (1973b) intertemporal capital asset pricing model (ICAPM) is not a general equilibrium theory in the sense of Arrow–Debreu, that is to say, the technological sources of uncertainty are not related to the equilibrium prices of the risky assets in Merton (1973b). To make Merton's ICAPM a general equilibrium model, first the Brock and Mirman (1972) stochastic growth model is modified, and secondly, Lucas' (1978) asset pricing model is extended to include a nontrivial investment decision. This is done in such a way as to preserve the empirical tractibility of the Merton formulation and at the same time determine endogenously the risk prices derived by Ross (1976) in his arbitrage theory of capital asset pricing. The model is given by

$$\text{maximize } E_1 \sum_{t=1}^{\infty} \beta^{t-1} u(c_t) \tag{15.1}$$

such that

$$c_{t+1} + x_{t+1} - x_t = \sum_{i=1}^{N} [g_i(x_{it}, r_t) - \delta_i x_{it}], \tag{15.2}$$

$$x_t = \sum_{i=1}^{N} x_{it}, x_{it} \geqslant 0, \quad i = 1, 2, ..., N, \ t = 1, 2, ..., \tag{15.3}$$

$$c_t \geqslant 0, \qquad\qquad\qquad t = 1, 2, ..., \qquad\qquad\qquad (15.4)$$

$$x_1, x_{i1}, \qquad\qquad i = 1, 2, ..., N, r_1 \text{ historically given}, \qquad (15.5)$$

where $E_1$, $\beta$, $u$, $c_t$, $x_t$, $g_i$, $x_{it}$, $r_t$ and $\delta_i$ denote mathematical expectation conditioned at time 1, discount factor on future utility, utility function of consump-
tion, consumption at date $t$, capital stock at date $t$, production function of process $i$, capital allocated to process $i$ at date $t$, random shock which is common to
all processes $i$, and depreciation rate for capital installed in process $i$, respectively.

The space of $\{c_t\}_{t=1}^{\infty}$, $\{x_t\}_{t=1}^{\infty}$ over which the maximum is being taken in
(15.1) needs to be specified. Obviously, decisions at date $t$ should be based only
upon information at date $t$. In order to make the choice space precise some for-
malism is needed which is developed in what follows.

The environment will be represented by a sequence $\{r_t\}_{t=1}^{\infty}$ of real vector
valued random variables which will be assumed to be independently and identically distributed. The common distribution of $r_t$ is given by a measure $\mu$: $\mathcal{R}(R^m)$
$\rightarrow [0, 1]$, where $\mathcal{R}(R^m)$ is the Borel $\sigma$-field of $R^m$. In view of a well-known one-
to-one correspondence (see, for example, Loeve, 1977), we can adequately rep-
resent the environment as a measure space $(\Omega, \mathcal{F}, \nu)$, where $\Omega$ is the set of all
sequences of real $m$ vectors, $\mathcal{F}$ is the $\sigma$-field generated by cylinder sets of the
form $\Pi_{t=1}^{\infty} A_t$, where $A_t \in \mathcal{R}(R^m)$, $t = 1, 2, ...$, and $A_t = R^m$ for all but a finite
number of values of $t$. Also $\nu$, the stochastic law of the environment, is simply
the product probability induced by $\mu$, given the assumption of independence.

The random variables $r_t$ may be viewed as the $t$th coordinate function on $\Omega$,
i.e. for any $\omega = \{\omega_t\}_{t=1}^{\infty} \in \Omega, r_t(\omega)$ is defined by $r_t(\omega) = \omega_t$.

We shall refer to $\omega$ as a possible state of the environment, or an environment
sequence, and shall refer to $\omega_t$ as the environment at date $t$. In what follows, $\mathcal{F}_t$
is the $\sigma$-field guaranteed by partial histories up to period $t$ (i.e. the smallest
$\sigma$-field generated by cylinder sets of the form $\Pi_{\tau=1}^{\infty} A_\tau$, where $A_\tau$ is in $\mathcal{R}(R^m)$
for all t, and $A_\tau = R^m$ for all $\tau > t$). The $\sigma$-field $\mathcal{F}_t$ contains all of the information about the environment which is available at date t.

In order to express precisely the fact that decisions $c_t, x_t$ only depend upon
information that is available at the time the decisions are made, we simply require that $c_t, x_t$ be measurable with respect to $\mathcal{F}_t$.

Formally the maximization in (15.1) is taken over all stochastic processes
$\{c_t\}_{t=1}^{\infty}$, $\{x_t\}_{t=1}^{\infty}$ that satisfy (15.2)$-$(15.5) and such that for each $t = 1, 2, ...$,
$c_t, x_t$ are measurable $\mathcal{F}_t$. Call such processes *admissible*.

Existence of an optimum $\{c_t\}_{t=1}^{\infty}$, $\{x_t\}_{t=1}^{\infty}$ may be established by imposing
an appropriate topology $\mathcal{T}$ on the space of admissible processes such that the
objective (15.1) is continuous in this topology and the space of admissible pro-

*Stochastic methods in economics and finance*

cesses is $\mathcal{F}$-compact. While it is beyond the scope of this application to discuss existence, presumably a proof can be constructed along the lines of Bewley (1977).

The notation almost makes the working of the model self-explanatory. There are $N$ different processes. At date $t$ it is decided how much to consume and how much to hold in the form of capital. It is assumed that capital goods can be costlessly transformed into consumption goods on a one-for-one basis. After it is decided how much capital to hold then it is decided how to allocate the capital across the $N$ processes. After the allocation is decided, nature reveals that $r_t$ and $g_i(x_{it}, r_t)$ units of new production are available from process $i$ at the end of period $t$. But $\delta_i x_{it}$ units of capital have evaporated at the end of $t$. Thus, net new output is $g_i(x_{it}, r_t) - \delta_i x_{it}$ from process $i$. The total output available to be divided into consumption and capital stock at date $t + 1$ is given by

$$\sum_{i=1}^{N} [g_i(x_{it},r_t) - \delta_i x_{it}] + x_t = \sum_{i=1}^{N} [g_i(x_{it},r_t) + (1-\delta_i)x_{it}]$$

$$= \sum_{i=1}^{N} f_i(x_{it},r_t) \equiv y_{t+1},$$ 

(15.6)

where

$$f_i(x_{it},r_t) \equiv g_i(x_{it},r_t) + (1-\delta_i)x_{it}$$ 

(15.7)

denotes the total amount of output emerging from process $i$ at the end of period $t$. The output $y_{t+1}$ is divided into consumption and capital stock at the beginning of date $t + 1$ and so on it goes.

Note that we assume that it is costless to install capital into each process $i$ and it is costless to allocate capital across processes at the beginning of each date $t$.

The objective is to maximize the expected value of the discounted sum of utilities over all consumption paths and capital allocations that satisfy (15.2)–(15.4).

In order to obtain sharp results we place restrictive assumptions on this problem. We collect the basic working assumptions below.

**A.1.** The functions $u(\cdot)$, $f_i(\cdot)$ are all concave, increasing, and are twice continuously differentiable.

**A.2.** The stochastic process $\{r_t\}_{t=1}^{\infty}$ is independently and identically distributed. Each $r_t\colon (\Omega, \mathcal{R}, \mu) \to R^m$, where $(\Omega, \mathcal{R}, \mu)$ is a probability space. Here $\Omega$ is the space of elementary events, $\mathcal{R}$ is the sigma field of measurable sets with respect to $\mu$, and $\mu$ is a probability measure defined on subsets $B \subset \Omega, B \in \mathcal{R}$. Furthermore, the range of $r_t, r_t(\Omega)$, is compact.

**A.3.** For each $\{x_{i1}\}_{i=1}^N, r_1$ the problem in (15.1) has a unique optimal solution (unique up to a set of realizations of $\{r_t\}$ of measure zero).

Notice that A.3 is implied by A.1 and strict concavity of $u$ and $\{f_i\}_{i=1}^N$. Rather than try to find the weakest possible assumptions sufficient for uniqueness of solutions to (15.1), it seemed simpler to assume it. Furthermore, since we are not interested in the study of existence of optimal solutions in this application we have simply assumed that also. Since the case $N = 1$ has been dealt with by Brock and Mirman (1972, 1973) and Mirman and Zilcha (1975, 1976, 1977), we shall be brief where possible.

By A.3 we see that to each output level $y_t$ the optimum $c_t, x_t, x_{it}$, given $y_t$, may be written

$$c_t = g(y_t); \quad x_t = h(y_t); \quad x_{it} = h_i(y_t). \tag{15.8}$$

The optimum policy functions $g(\cdot)$, $h(\cdot)$ and $h_i(\cdot)$ do not depend upon $t$ because the problem is time stationary.

Another useful optimum policy function may be obtained. Given $x_t$ and $r_t$, A.3 implies that the optimal allocation $\{x_{it}\}_{i=1}^N$ and next period's optimal capital stock $x_{t+1}$ are unique. Furthermore, these may be written in the form

$$x_{it} = a_i(x_t, r_{t-1}) \tag{15.9}$$

and

$$x_{t+1} = H(x_t, r_t). \tag{15.10}$$

Eqs. (15.9) and (15.10) contain $r_{t-1}$ and $r_t$, respectively, because the allocation decision is made after $r_{t-1}$ is known but before $r_t$ is revealed, while the capital-consumption decision is made after $y_{t+1}$ is revealed, i.e. after $r_t$ is known.

Equation (15.10) looks very much like the optimal stochastic process studied by Brock–Mirman and Mirman–Zilcha. It was shown in Brock and Mirman (1972, 1973) for the case $N = 1$ that the stochastic difference equation (15.10) converges in distribution to a unique limit distribution independent of initial conditions. We shall show below that the same result may be obtained for our $N$ process model by following the argument of Mirman and Zilcha (1975). Some lemmas are needed.

**Lemma 15.1.** Assume A.1. Let $U(y_1)$ denote the maximum value of the objective in (15.1) given initial resource stock $y_1$. Then $U(y_1)$ is concave, nondecreasing in $y_1$ and, for each $y_1 > 0$, the derivative $U'(y_1)$ exists and is nonincreasing in $y_1$.

*Proof.* Mirman and Zilcha (1975) prove that

$$U'(y_1) = u'(g(y_1)) \quad \text{for } y_1 > 0,$$

for the case $N = 1$. The same argument may be used here. The details are left to the reader.

Note that $g(y_1)$ in the last equation is nondecreasing since $u''(c) < 0$ and $U'(y)$ is nonincreasing in $y$ owing to the concavity of $U(\cdot)$.

**Lemma 15.2.** Suppose that A.2 holds and $u(c) \geqslant 0$ for all $c$. Furthermore, assume that along optima

$$E_1 \beta^{t-1} U(y_t) \to 0 \quad \text{as } t \to \infty.$$

Then if $\{c_t\}_{t=1}^\infty$, $\{x_t\}_{t=1}^\infty$, $\{x_{it}\}_{i=1}^N$, $t = 1, 2, ...$, is optimal then the following conditions must be satisfied. For each $i, t$

$$u'(c_t) \geqslant \beta E_t \{u'(c_{t+1}) f_i'(x_{it}, r_t)\}, \tag{15.11}$$

$$u'(c_t) x_{it} = \beta E_t \{u'(c_{t+1}) f_i'(x_{it}, r_t) x_{it}\} \tag{15.12}$$

and

$$\lim_{t \to \infty} E_1 \{\beta^{t-1} u'(c_t) x_t\} = 0. \tag{15.13}$$

*Proof.* The proof of (15.11) and (15.12) is an obvious application of calculus to (15.1) with due respect to the constraints $c_t \geqslant 0$ and $x_t \geqslant 0$. An argument analogous to that of Benveniste and Scheinkman (1977) establishes (15.13). By concavity of $f_i$, $i = 1, 2, ... N$, $U(\cdot)$ and by lemma 15.1 we have for any constant $\gamma$, $0 < \gamma < 1$

$$E_1 \beta^{t-1} \left\{ U\left[ \sum_{i=1}^N f_i(x_{i,t-1}, r_{t-1}) \right] - U\left[ \sum_{i=1}^N f_i(\gamma x_{i,t-1}, r_{t-1}) \right] \right\}$$

$$\geqslant E_1\beta^{t-1}U'(y_t)\left[\sum_{i=1}^N f'_i(x_{i,t-1},r_{t-1})(1-\gamma)x_{i,t-1}\right]$$

$$= (1-\gamma)E_1\beta^{t-1}u'(c_t)\left[\sum_{i=1}^N f'_i(x_{i,t-1},r_{t-1})(x_{i,t-1})\right]. \tag{15.14}$$

But since $U$ is nondecreasing in $y_t$ and each $f_i$ is increasing in $x_i$, the l.h.s. of (15.14) is bounded above by $E_1\beta^{t-1}U(y_t)$ which goes to zero as $t\to\infty$. Since $u'\geqslant 0$ and $y_t\geqslant 0$, the r.h.s. of (15.14) must go to zero as well. But by (15.12) as $t\to\infty$

$$E_1\beta^{t-1}u'(c_t)\left[\sum_i f'_i(x_{i,t-1},r_{t-1})(x_{i,t-1})\right] = E_1\beta^{t-2}\left(\sum_i u'(c_{t-1})x_{i,t-1}\right)$$

$$= E_1\beta^{t-2}u'(c_{t-1})x_{t-1}\to 0,$$

as was to be shown.

**Lemma 15.3.** Assume that $u'(c) > 0$, $u''(c) < 0$ and $u'(0) = \infty$. Furthermore, assume that $f_j(0,r) = 0$, $f''_j(x,r) > 0$ and $f'_j(x,r) < 0$ for all values of $r$. Also suppose that there is a set of $r$-values with positive probability such that $f_j$ is strictly concave in $x$. Then the function $h(y)$ is continuous in $y$, increasing in $y$, and $h(0) = 0$.

*Proof.* See Brock (1979).

Now by A.3 and (15.8)–(15.10) it follows that $y_{t+1}$ may be written

$$y_{t+1} = F(x_t,r_t). \tag{15.15}$$

Following Mirman and Zilcha (1975) define

$$\underline{F}(x) \equiv \min_{r\in R} F(x,r), \quad \overline{F}(x) \equiv \max_{r\in R} F(x,r), \tag{15.16}$$

where $R$ is the range of the random variable $r: (\Omega,\mathscr{R},\mu)\to R^m$ which is compact by A.2. The following lemma shows that $\underline{F}$ and $\overline{F}$ are well defined.

**Lemma 15.4.** Assume the hypotheses of lemma 15.3 and suppose that each $f_i(x,r)$ is continuous in $r$ for each $x$. Then $F(x,r)$ is continuous in $r$.

*Proof.*   This is straightforward because

$$y_{t+1} \equiv \sum_j f_j(x_{jt}, r_t) = \sum_j f_j(\eta_j(x_t)x_t, r_t) \equiv F(x_t, r_t).$$

Since $\eta_j(x_t)$ is continuous in $x_t > 0$ and each $f_j(x, r)$ is continuous in $r$ we conclude that $F(x, r)$ is continuous in $x$ and $r$. This concludes the proof.

Let $\underline{x}, \bar{x}$ be any two fixed points of the functions

$$H(x) \equiv h(\underline{F}(x)); \quad \bar{H}(x) \equiv h(\bar{F}(x)), \tag{15.17}$$

respectively. Then

**Lemma 15.5.**   Any two fixed points of the pair of functions defined in (15.17) must satisfy $\underline{x} < \bar{x}$.

*Proof.*   See Brock (1979).

Finally, applying arguments similar to Brock and Mirman (1972) we obtain the following:

**Theorem 15.1.**   There is a distribution $F(x)$ of the optimum aggregate capital stock $x$ such that $F_t(x) \to F(x)$ uniformly for all $x$. Furthermore, $F(x)$ does not depend on the initial conditions $(x_1, r_1)$. Here $F_t(x) \equiv P[x_t < x]$.

*Proof.*   See Brock (1979).

Theorem 15.1 shows that the distribution of optimum aggregate capital stock at date $t$, $F_t(x)$, converges pointwise to a limit distribution $F(x)$. Theorem 15.1 is important because we will use the optimal growth model to construct equilibrium asset prices and risk prices. Since these prices will be time stationary functions of $x_t$, and since $x_t$ converges in distribution to $F$, we will be able to use the mean ergodic theorem and stationary time series methods to make statistical inferences about these prices on the basis of time series observations. More will be said about this in chapter 4.

## 16.   Competitive firm under price uncertainty

In this application we follow Sandmo (1971) to illustrate the use of stochastic techniques in the theory of the competitive firm under price uncertainty. These techniques are rather elementary and make use of concepts introduced in Chapter 1.

Consider a competitive firm in the short run whose output decisions are dominated by a concern to maximize the expected utility of profits. The sales price, $p$, is a non-negative random variable whose distribution is subjectively determined by the firm's beliefs. The density function of the sales price is $f(p)$ and we denote the expected value of the sales price by $\mu$, i.e. $E(p) = \mu$. We assume that the firm is a price taker in the sense that it is unable to influence the sales price distribution.

Let $u$ denote the von Neumann–Morgenstern utility function of the firm and $\pi(x)$ the profits function, where $x$ is output. We assume that $u$ is a bounded, concave, continuous and differentiable function such that

$$u'(\pi) > 0 \quad \text{and} \quad u''(\pi) < 0. \tag{16.1}$$

Thus, the firm is assumed to be risk averse. The total cost function of the firm, $F(x)$, consists of total variable cost, $C(x)$, and fixed cost $B$. We write

$$F(x) = C(x) + B, \tag{16.2}$$

where

$$C(0) = 0 \quad \text{and} \quad C'(x) > 0. \tag{16.3}$$

In the usual way we define the firm's profit function by

$$\pi(x) = px - C(x) - B, \tag{16.4}$$

and the firm's objective to maximize the expected utility of profits can be written as

$$E(u(px - C(x) - B)). \tag{16.5}$$

The necessary and sufficient conditions for a maximum of (16.5) are obtained by differentiating (16.5) with respect to $x$; they are

$$E(u'(\pi)(p - C'(x))) = 0 \tag{16.6}$$

and

$$E(u''(\pi)(p - C'(x))^2 - u'(\pi)C''(x)) < 0. \tag{16.7}$$

Suppose that eqs. (16.6) and (16.7) determine a positive, finite and unique solution to the maximization problem (16.5). For our analysis the basic question is: How does the optimal output under uncertainty compare with the well-known competitive solution under certainty, where price is equated with marginal cost? To provide an answer we proceed as follows. Rewrite (16.6) as

$$E(u'(\pi)p) = E(u'(\pi)C'(x)) \tag{16.8}$$

and subtract $E(u'(\pi)\mu)$ on each side of (16.8) to get

$$E(u'(\pi)(p - \mu)) = E(u'(\pi)[C'(x) - \mu]). \tag{16.9}$$

Note that taking the expectation of (16.4) we obtain

$$E(\pi) = E(p)x - C(x) - B = \mu x - C(x) - B.$$

Therefore, $\pi(x) - E(\pi) = px - \mu x = (p - \mu)x$, or equivalently $\pi(x) = E(\pi) + (p - \mu)x$. If $p \geqslant \mu$, then from the last sentence we obtain that $\pi(x) \geqslant E(\pi)$ and therefore $u'(\pi) \leqslant u'(E(\pi))$. In general, for all $p$ we have that

$$u'(\pi)(p - \mu) \leqslant u'(E(\pi))(p - \mu). \tag{16.10}$$

Take expectations on both sides of (16.10) to get

$$E(u'(\pi)(p - \mu)) \leqslant u'(E(\pi)) E(p - \mu) = 0. \tag{16.11}$$

Observe that the zero in the right-hand side of (16.11) comes from the fact that $u'(E(\pi))$ is a constant and $E(p) = \mu$. Combine the result in (16.11), namely that $E(u'(\pi)(p - \mu)) \leqslant 0$ with eq. (16.9) to conclude that

$$E(u'(\pi)[C'(x) - \mu]) \leqslant 0, \tag{16.12}$$

which finally implies

$$C'(x) \leqslant \mu \tag{16.13}$$

because $u'(\pi) > 0$ by (16.1). Our result in (16.13) says that optimal output for a competitive firm under price uncertainty is characterized by marginal cost being less than the expected price. If we characterize the certainty output as that quantity where $C'(x) = \mu$, then we may conclude that under price uncertainty, output is smaller than the certainty output. This result is a generalization of McCall's (1967) theorem for the special case of a constant absolute risk aversion utility function.

Next, suppose that $x^*$ denotes the positive, finite and unique optimum output which is the solution to (16.6) and satisfies (16.7). Then $x^*$ will give a global utility maximum provided

$$E(u(px^* - C(x^*) - B)) \geqslant u(-B), \tag{16.14}$$

where $-B$ is the level of profit when $x = 0$. Consider the left-hand side of (16.14) and approximate it by a Taylor series around the point $p = \mu$ to rewrite (16.14) as

$$E(u(\mu x^* - C(x) - B) + u'(\mu x^* - C(x^*) - B)x^*(p - \mu)$$
$$+ \tfrac{1}{2} u''(\mu x^* - C(x^*) - B)x^{*2}(p - \mu)^2) \geqslant u(-B). \tag{16.15}$$

Note that higher-order terms in the Taylor series have been neglected and also note that by definition the second term on the left-hand side of (16.15) is zero. Rearranging the remaining terms in (16.15) and dividing through by $u'(\mu x^* - C(x^*) - B)$ so as to make the expressions invariant under linear transformations of the utility function, we then obtain

$$\frac{u(\mu x^* - C(x^*) - B) - u(-B)}{u'(\mu x^* - C(x^*) - B)} \geqslant -\frac{1}{2}\frac{u''(\mu x^* - C(x^*) - B)}{u'(\mu x^* - C(x^*) - B)} x^{*2} E(p - \mu)^2. \tag{16.16}$$

Observe that the factor $-u''/u'$ on the right-hand side of (16.16) is the risk aversion function evaluated at the expected level of profit for optimum output $x^*$. The factor $x^{*2} E(p - \mu)^2$ denotes the variance of sales. Each of these two factors is positive and therefore from (16.16) we conclude that

$$\mu x^* - C(x^*) - B > -B, \tag{16.17}$$

given that the utility function is strictly increasing. Finally from (16.17) we obtain

$$\frac{C(x^*)}{x^*} < \mu, \tag{16.18}$$

which says that at the optimum level of output of a competitive firm under uncertainty, the expected price is greater than average cost. This fact further implies that the firm requires strictly positive expected profit in order to operate in a competitive environment under price uncertainty. Therefore, price uncertainty leads to a modification of the standard results of the microeconomic theory of the competitive firm in an environment of certainty.

## 17. Stabilization in the presence of stochastic disturbances

In this section we follow Brainard (1967) and Turnovsky (1977) to illustrate the effects of uncertainty in stabilization policy. To fully demonstrate the role of uncertainty we use the simple case of one target, denoted by $y$, and one instrument, denoted by $x$, related linearly as follows:

$$y = ax + u. \tag{17.1}$$

Here we assume that $y$ and $x$ are scalars while $a$ and $u$ are random variables having expectations and variances denoted by $E(a) \equiv \bar{a}$, $E(u) \equiv \bar{u}$, $\text{var}(a) \equiv \sigma_a^2$, and $\text{var}(u) \equiv \sigma_u^2$, respectively.

Suppose that the policy-maker has chosen a target value $y^*$. To fix the ideas involved, suppose that (17.1) is a reduced form equation between GNP, $y$, and the money supply, $x$, subject to additive, $u$, and multiplicative, $a$, disturbances. Given $y^*$, the stabilization problem is to choose $x$ so that the policy-maker will maximize his expected utility. Brainard (1967) uses a quadratic utility function, $U$, of the form

$$U = -(y - y^*)^2. \tag{17.2}$$

The problem then becomes

$$\max_x E(U) = \max_x E(-(y - y^*)^2) \tag{17.3}$$

subject to (17.1)

Substituting (17.1) to (17.3) and taking expectations we obtain

$$E(U) = -[(\bar{a}x + \bar{u} - y^*)^2 + \sigma_a^2 x^2 + \sigma_u^2 + 2\rho \sigma_a \sigma_u x], \tag{17.4}$$

where $\rho$ denotes the correlation coefficient between $a$ and $u$. Put (17.4) into (17.3), differentiate with respect to $x$, and set this derivative equal to zero to obtain the optimal value of $x$; it is given by

$$x_{\mathrm{u}} = \frac{\overline{a}(y^* - \overline{u}) - \rho \sigma_a \sigma_u}{\overline{a}^2 + \sigma_a^2} \, . \tag{17.5}$$

In (17.5) $x_{\mathrm{u}}$ denotes the optimal value under uncertainty. Note that if $\sigma_a = \sigma_u = 0$, which means that certainty prevails, then the certainty optimal value of the instrument variable, denoted $x_{\mathrm{c}}$, is given by

$$x_{\mathrm{c}} = \frac{y^* - \overline{u}}{\overline{a}} \, . \tag{17.6}$$

Comparing (17.5) and (17.6) we observe that the difference between the values of $x_{\mathrm{u}}$ and $x_{\mathrm{c}}$ depends on $\sigma_a$. If $\sigma_a = 0$ then $x_{\mathrm{u}} = x_{\mathrm{c}}$. This shows that for an additive random disturbance the values of $x_{\mathrm{u}}$ and $x_{\mathrm{c}}$ are the same; this is called the *certainty equivalence* result. Therefore, to understand the role of uncertainty we must study the role of the multiplicative disturbances arising from the random variable $a$. To do so, suppose that $\sigma_u = 0$; then (17.5) becomes

$$x_{\mathrm{u}} = \frac{\overline{a}(y^* - \overline{u})}{\overline{a}^2 + \sigma_a^2} = \frac{\overline{a}^2}{\overline{a}^2 + \sigma_a^2} \, x_{\mathrm{c}}, \tag{17.7}$$

from which we obtain that

$$|x_{\mathrm{u}}| < |x_{\mathrm{c}}|. \tag{17.8}$$

This last equation means that the policy-maker is more conservative when uncertainty prevails. What are the implications of such a conservative policy? Inserting $x_{\mathrm{u}}$ from (17.7) into (17.1) we solve for the target value achieved from $x_{\mathrm{u}}$ in the special case under discussion, i.e. when $\sigma_u^2 = 0$ and $u = \overline{u}$; the result is

$$y = \frac{a\overline{a}(y^* - \overline{u})}{\overline{a}^2 + \sigma_a^2} + \overline{u}. \tag{17.9}$$

Without loss of generality we assume in (17.9) that $y^* > \overline{u}$; this assumption simply says that the target value $y^*$ is greater than the expectation of the additive disturbance term. From (17.9) we are now able to discover the implications of uncertainty in economic policy. Observe that

$$E(y) - y^* = E\left(\frac{a\overline{a}(y^* - \overline{u})}{\overline{a}^2 + \sigma_a^2} + \overline{u}\right) - y^*$$

$$= \frac{\overline{a}^2(y^* - \overline{u})}{\overline{a}^2 + \sigma_a^2} + \frac{(\overline{a}^2 + \sigma_a^2)(\overline{u} - y^*)}{\overline{a}^2 + \sigma_a^2}$$

$$= \frac{\sigma_a^2(\overline{u} - y^*)}{\overline{a}^2 + \sigma_a^2} < 0. \tag{17.10}$$

From (17.10), under the assumption that $y^* > u$, we conclude that $E(y) < y^*$, which means that the target variable will on the average undershoot its desired value. If we assume that $\sigma_a^2 = 0$ and $\sigma_u^2 \neq 0$, it can be shown in a similar way that $E(y) = y^*$, so that $y$ will fluctuate randomly about its target value. Thus even in the simplest possible case with one target on one instrument variable, the results of stochastic analysis are more general and richer than those of the certainty analysis.

## 18. Stochastic capital theory in continuous time

In this section we conduct an analysis similar to that of section 8 of Chapter 1 for discrete time except that now we work with the case of continuous time. Consider the Markov process $\{X_t\}_{t=0}^{\infty}$ with $t \in [0, \infty)$. Most of the time we suppose that $\{X_t\}_{t=0}^{\infty}$ is given by the Itô stochastic differential equation

$$dX = f(X, t)dt + \sigma(X, t)dz, \tag{18.1}$$

where $dz$ is normal with mean zero and variance $dt$.
   Consider the problem

$$\gamma(t, X, T) \equiv \sup_{t \leq \tau \leq T} E[e^{-r\tau} X_\tau \mid X(t) = X]. \tag{18.2}$$

Here the supremum is taken over the set of measurable stopping times, i.e. the events $\{\tau \leq s\}$ depend only upon $\{X_r\}$ for $r \leq s$.
   The existence of optimal stopping times and of critical boundaries becomes a very technical matter in continuous time as the analysis of section 13 of Chapter 2 has shown. Hence, we shall proceed heuristically making use of some ideas from section 13 of Chapter 2. Actually the basic ideas are simple, intuitive and quite pretty when unencumbered by technicalities.
   We define the *continuation region* $C(t, T)$ for problem (18.2) by

$$C(t, T) = \{(s, X) \mid \gamma(s, X, T) > e^{-rs}X, t \leqslant s \leqslant T\} \qquad (18.3)$$

Note that solutions $X(s, T)$ to

$$\gamma(s, X, T) = X(s, T)e^{-rs} \qquad (18.4)$$

play the role that the critical numbers $\{\bar{X}_t\}$ played in the discrete time case Just as in the discrete time case we would like the critical numbers to exist and to be unique so that $X(s, T)$ is a function and not a point to set mapping. We want to describe the boundary of $C(t, T)$ by a function. We will be particularly interested in $C(0, T) \equiv C(T)$ and $C(0, \infty) \equiv C$.

Fortunately a theorem by Miroshnichenko (1975, p. 388) gives us what we need under mild assumptions. In order to motivate Miroshnichenko's theorem we proceed heuristically as follows.

Put

$$R(s, X) \equiv e^{-rs}X. \qquad (18.5)$$

Since $(t, X) \in C(T)$ therefore $\gamma(t, X, T) > R(t, X)$. Now sample paths are continuous. Therefore if $\Delta t$ is small enough, it will always be worthwhile to continue on from $(t + \Delta t, X(t + \Delta t))$. Since the value at $(t, X)$ is the maximum of the value of stopping before $t + \Delta t$ and continuing on after $t + \Delta t$, it follows that $(t, X) \in C(T)$ implies

$$\gamma(t, X, T) = E[\gamma(t + \Delta t, X(t + \Delta t), T) \mid X(t) = X]. \qquad (18.6)$$

In order to shorten notation, put

$$LH \equiv H_t + H_X f + \tfrac{1}{2} H_{XX}\sigma^2 \qquad (18.7)$$

for any function $H(t, X)$. Let

$$A(t, T) \equiv \{(s, X) \mid LR(s, X) > 0, t \leqslant s \leqslant T\}.$$

Then we claim that $A(t, T) \subset C(t, T)$.

To prove our claim let

$$(s, X) \in \gamma(t, T) = \{(s, X) \mid \gamma(s, X, T) = R(s, X)\}.$$

Hence,

$$R(s, X) = \sup_{s \leqslant \tau \leqslant T} E[R(\tau, X(\tau)) \mid X(s) = X]. \tag{18.8}$$

In particular, for any $\Delta s \geqslant 0$,

$$R(s, X) \geqslant E[R(s + \Delta s, X(s + \Delta s)) \mid X(s) = X]. \tag{18.9}$$

Expand the r.h.s. of (18.9) and let $\Delta s \to 0$, to obtain

$$R(s, X) \geqslant R(s, X) + LR(s, X). \tag{18.10}$$

Thus, $LR(s, X) \leqslant 0$ and

$$\gamma(t, T) \subset \{(s, X) \mid LR(s, X) \leqslant 0\}, \tag{18.11}$$

from which we conclude that $A(t, T) \subset C(t, T)$. This ends the proof of the claim.

Clearly for $(s, X) \in C(t, T)$ we have $R(s, X) = \gamma(s, X, T)$. Furthermore, for $(s, X) \in C(t, T)$, (18.6) holds. Hence, for $(s, X) \in C(t, T)$, for $\Delta t$ small, we must have by expanding the r.h.s. of (18.6) in a formal Taylor series about $(t, X)$ and taking expected values

$$\gamma(t, X, T) = \gamma(t, X, T) + \gamma_t \Delta t + \gamma_X f \Delta t + \tfrac{1}{2} \gamma_{XX} \sigma^2 \Delta t + o(\Delta t). \tag{18.12}$$

Cancel $\gamma(t, X, T)$ off of both sides of (18.12), divide by $\Delta t$, and take $\Delta t \to 0$ to obtain

$$0 = \gamma_t + \gamma_X f + \tfrac{1}{2} \gamma_{XX} \sigma^2 \equiv L\gamma(t, X, T). \tag{18.13}$$

Equation (18.13) is the *fundamental partial differential equation of optimal stopping theory*. It only holds on the continuation set. It is so important that it is worthwhile to repeat how it was obtained. For $(t, X)$ in the continuation region, by continuity of sample paths,

$$E[R(t + \Delta t, X(t + \Delta t)) \mid X] < E[\gamma(t + \Delta t, X(t + \Delta t), T) \mid X]$$

for $\Delta t$ small. Hence, the value $\gamma(t, X, T)$ must equal $E[\gamma(t + \Delta t, X(t + \Delta t), T) \mid X]$ for $\Delta t$ small. Thus,

$$\gamma(t, X, T) = E[\gamma(t + \Delta t, X(t + \Delta t), T) \mid X].$$

Expanding the r.h.s. of this last equation in a formal Taylor series we get (18.13).

We are now at the point where Miroshnichenko's theorem (1975, lemma 1, p. 388) should have meaning for the reader.

**Theorem 18.1** (Miroshnichenko).   Under mild regularity conditions on $f$ and $\sigma$ the following obtains:

(1)   $A(t, T) \subset C(t, T)$, and

(2)   $C(t, T)$ is open and each connected component of $C(t, T)$ contains at least one connected component of the set $A(t, T)$. In particular, if the set $A$ is connected, then it consists of a single connected component and the set $C(t, T)$ also consists of a single connected component, i.e. it is connected.

We gave an heuristic proof of part (1) when we established the claim. The reader is referred to Miroshnichenko (1975, pp. 388–389) for the proof of part (2).

Theorem 18.1 is useful in obtaining a sufficient condition for uniqueness of critical numbers in the time independent case. To find such a condition, work with current values. Put

$$W(t, X, T) = e^{rt}\gamma(t, X, T). \tag{18.14}$$

Note that $W$ and $\gamma$ increase as $T$ increases. Suppose that finite limits exist as $T \to \infty$ and that these limits are values for the infinite horizon problems.

From (18.14) we obtain

$$W_t = re^{rt}\gamma + e^{rt}\gamma_t; \quad W_X = e^{rt}\gamma_X; \quad W_{XX} = e^{rt}\gamma_{XX}.$$

The fundamental equation (18.13) becomes

$$0 = W_t - rW + W_X f + \tfrac{1}{2} W_{XX}\sigma^2. \tag{18.15}$$

In the *time independent* case the term $W_t = 0$, so that the fundamental partial differential equation becomes an ordinary differential equation. This is pleasant and will be exploited heavily in what follows.

What does the connectedness of $A \equiv \{(t, X) \mid Lg(t, X) > 0\}$ boil down to in special cases? For example, if $g(t, X) = e^{-rt}X$, then

$$Lg = -re^{-rt}X + e^{-rt}f > 0 \quad \text{if and only if } f/X > r. \tag{18.16}$$

Hence, if $f$ is time independent, we may state.

**Theorem 18.2.**   If $f$ and $\sigma$ are time independent, $R(t, X) = e^{-rt}X, A(X) \equiv f/X$

is decreasing in $X$ with $A(0) > r$, $A(\infty) < r$, then $C(0, \infty) \equiv C = (0, \bar{X})$ for some $\bar{X}$.

*Proof.* The set $A = \{(t, X) \mid Lg(t, X) > 0\}$ is connected, nonempty and time independent. Therefore by theorem 18.1, $C$ is connected and nonempty. $C$ is time independent because the problem is time independent. Therefore, the projection of $C$ on the $X$-axis is a connected open set in the real line. The only such sets are open intervals. This ends the proof.

Theorem 18.2 drastically simplifies the time independent problem. The value $\gamma(0, X_0, \infty) \equiv \gamma(0, X_0)$ must be the form:

$$\gamma(0, X_0) = E[e^{-r\tau} X(\tau) \mid X(0) = X_0]$$
$$= X E[e^{-r\tau} \mid X(0) = X_0]$$
$$\equiv XM(X; X_0), \tag{18.17}$$

where $\tau$ is the time of first passage from $X_0$ to some barrier $X$, and $M(X; X_0)$ is the *Laplace transform* of the time of first passage from $X_0$ to $X$.

Much is known about $M(X; X_0)$. Furthermore, it is clear that if we find $\bar{X}$ that solves

$$\max_{X} XM(X; X_0), \tag{18.18}$$

then we have found the optimum barrier and the value function. This makes our problem much easier to solve. For if the tree is of size $X$ and we plan to cut it down when it reaches size $Y$, then its expected present discounted value is

$$H(X, Y) = Y E[e^{-rT_Y} \mid X(0) = X] \equiv YM(Y; X). \tag{18.19}$$

For the moment put $f(X, t) = a(X, t)$, $\sigma(X, t) = \sqrt{b(X, t)}$. We want to use $f$ for something else now. Notice that $H$ is the product of $Y$ and the Laplace transform of first passage time from $X$ to $Y$. Let us suppress $Y$ for a moment and consider the value of the tree as a function of the tree's current size alone. We saw above that this valuation function, $f(X)$, satisfies the linear second-order differential equation

$$rf(X) - a(X)f'(X) - \tfrac{1}{2} b(X)f''(X) = 0. \tag{18.20}$$

Put $M'(Y; X) = \partial M / \partial X$. $M$ also satisfies (18.20) in the $X$ argument. Furthermore, $H(X, Y)$ satisfies (18.20) even if $Y$ is not optimal. Fix $Y$ and suppose

$X(t) = X$ is strictly less than $Y$. Then if $\Delta t$ is small, $X(t + \Delta t) < Y$ w.p.l. and the tree will not be cut down before $t + \Delta t$. Thus,

$$f(X) = e^{-r\Delta t}Ef(X(t + \Delta t)).$$

Use stochastic calculus to expand the r.h.s. of this last equation in a Taylor's series to obtain

$$f(X) = e^{-r\Delta t}E[f(X) + f'(X)\Delta X + \tfrac{1}{2}f''(X)\Delta X^2 + ...]$$
$$= [1 - r\Delta t + ...][f(X) + f'(X)a(X)\Delta t + \tfrac{1}{2}f''(X)b(X)\Delta t + ...].$$

Rearrange and discard all terms of order $(\Delta t)^2$ to get (18.20).

Since (18.20) is a second-order differential equation, two boundary conditions are required to determine a particular solution. One condition is given by observing that if a tree were cut down and sold for $Y$ when its size is $Y$, then its value at $Y$ must be $Y$ or

$$f(Y) = Y. \tag{18.21}$$

The other boundary condition is not so easy to pin down. We discuss some alternative specifications, all of which have different economic interpretations. The first possibility is that there is a particular size $Q$ such that if the tree ever becomes as small as $Q$, it stays at $Q$. In this case a tree of size $Q$ is worth $Q$, so our second boundary condition is simply that

$$f(Q) = Q. \tag{18.22a}$$

We call (18.22a) an *absorbing barrier* at $Q$. Another possibility is that there is no natural lower bound to a tree's size. In this case we require that $f(X)$ continues to make mathematical sense when $X$ approaches $-\infty$. Since

$$f(X) = YE[e^{-rTY} \mid X(0) = X],$$

and $e^{-rTY}$ is always less than unity, it must be that $f(X)$ remains bounded or that

$$\lim_{X \to -\infty} |f(X)| < \infty. \tag{18.22b}$$

A third possibility is to suppose that there is, as with the absorbing barrier, a size $Q$ below which the tree cannot sink, but that, in contrast to the absorbing barrier, when it reaches size $Q$ it does not stay at $Q$ but instead bounces back to life – and away from $Q$. This specification, called a *reflecting barrier* at $Q$, is, admittedly, not very plausible for trees, but it is a useful case for other problems. A fourth boundary condition is when $f(Q) = 0$, but when $Q \neq 0$ this discontinuity in value causes technical problems.

Cox and Miller (1965, pp. 231–232) show that if a diffusion process has a reflecting boundary at $Q$, then its Laplace transform must satisfy

$$M'(Y;Q) \equiv \partial M/\partial X = 0. \tag{18.22c}$$

From (18.22c) and $f(X) = YM(Y;Q)$ we conclude that $f'(Q) = 0$.

Notice that if $\phi(X)$ satisfies any of the three boundary conditions (18.22) so does $K\phi(X)$, where $K$ is a constant. Hence it is natural to search for solutions $H(X, Y)$ to (18.20) of the form $H(X, Y) = K(Y)\phi(X)$. We use the boundary condition $H(Y, Y) = Y$ to determine $K$. Let $\phi(X)$ be a solution to (18.20) satisfying (18.22), then $H(X, Y) = K(Y)\phi(X)$ satisfies (18.21) only if $H(Y, Y) = Y$ or if $K(Y) = Y/\phi(Y)$. Thus, the value of a tree of size $X$ which will be cut down when it reaches $Y$ is $H(X, Y) = (Y/\phi(Y))\phi(X)$ and clearly the optimum cutting size should be chosen to maximize $Y/\phi(Y)$. Let $X^*$ be the optimal cutting size and define

$$V(X) = H(X, X^*) = X^* \left( \frac{\phi(X)}{\phi(X^*)} \right). \tag{18.23}$$

Notice that $V(X)$ also satisfies (18.20) because $X^*$ is independent of $X$. Notice also that the smooth pasting condition $V'(X^*) = 1$ is satisfied.

To see how eq. (18.20) and these boundary conditions determine $X^*$ and $V(X)$ it is instructive to begin by analyzing the case where $a(X)$ and $b(X)$ are constant and equal to $a$ and $b$, respectively. Most techniques and results carry over to the more general variable coefficient case. With constant coefficients the tree's growth process is simply Brownian motion with drift $a$ and infinitesimal variance $b$. Then solutions of (18.20) are of the form

$$f(X) = Ae^{\lambda X} + Be^{\mu X} \tag{18.24}$$

where

$$\lambda = \frac{-a + (a^2 + 2rb)^{\frac{1}{2}}}{b} > 0 \tag{18.25}$$

and

$$\mu = \frac{-a - (a^2 + 2rb)^{\frac{1}{2}}}{b} < 0 \qquad (18.26)$$

are the roots of the characteristic equation of (18.20). If $B = -A\,e^{(\lambda - \mu)Q}$, then $f(Q) = 0$, so all solutions with absorbing barriers for $Q = 0$ are of the form

$$A(e^{\lambda X} - e^{(\lambda - \mu)Q}e^{\mu X})$$

and the optimal cutting size $X^*$ maximizes

$$\frac{X}{(e^{\lambda X} - e^{(\lambda - \mu)Q}e^{\mu X})} .$$

It is not easy to analyze the effects of parameter changes on $X^*$ and $V(X)$. For general $Q$, the absorbing case is complicated. In the remaining we discuss only the case $Q = 0$.

If $f(\cdot)$ is bounded at $-\infty$, things are much simpler. Note that $f(\cdot)$ can remain finite only if $B = 0$. Thus, $X^*$ maximizes $Xe^{-\lambda X}$ or $X^* = \lambda^{-1}$, and $V(X) = \lambda^{-1}e^{\lambda X - 1}$. Effects of changes in the process work entirely through $\lambda$. Since $X < X^* = \lambda^{-1}$, $dV(X)/d\lambda = (X - \lambda^{-1})V(X) < 0$. It is straightforward to calculate that $d\lambda/db < 0$. Thus, increasing the instantaneous variance of the process governing the tree's growth increases both the tree's value and the time at which the tree is cut down, since $d\lambda/da < 0$ and $d\lambda/dr > 0$. The effects of changes in the growth rate and the interest rate are similarly straightforward.

A reflecting barrier is not so simple. For a solution to (18.24) to satisfy $f'(Q) = 0$ it must be that $B = -A(\lambda/\mu)e^{(\lambda - \mu)Q}$, so that $X^*$ must be chosen to maximize $X/(e^{\lambda X} - (\lambda/\mu)e^{(\lambda - \mu)Q}e^{\mu X})$. It is difficult to analyze the effects of changes in parameters on cutting size and value.

We show below in proposition 18.4, summarizing our results, that the reflecting barrier case is qualitatively the same as the case when $f(\cdot)$ is bounded at $-\infty$. The absorbing barrier case is considerably more subtle.

To analyze more completely the constant coefficient case for reflecting and absorbing barriers, it is helpful to examine the function

$$g(X) = \frac{f'(X)}{f(X)} = \frac{\phi'(X)}{\phi(X)} . \qquad (18.27)$$

One should note that $g$ is the rate of change of the logarithm of the Laplace transform of first passage to a *fixed* barrier $Y$ with respect to $X$. It is *independent* of $Y$. Call $g$ the logarithmic rate of change of the Laplace transform with respect to $X$. The value of a tree of size $X$ which will be cut down at $Y > X$ is

$$H(X, Y) = Y \frac{\phi(X)}{\phi(Y)} = Y \exp\left(-\int_X^Y g(u)\mathrm{d}u\right).$$

It follows from (18.20) that $g$ satisfies a first-order differential equation

$$g'(X) = r/b - (a/b)g(X) - g^2(X). \tag{18.28}$$

Since $g$ is a first-order differential equation, only one boundary condition is required to identify a particular solution. Thus, in principle, to calculate $H(X, Y)$ in a particular case, use a boundary condition of the form (18.22) to identify a particular solution of (18.28) and integrate along that solution from $X$ to $Y$ to get $-\log(H(X, Y)/Y)$.

This observation suggests a way of getting comparative statics results. Suppose $g$ and $\hat{g}$ are two solutions to (18.28) corresponding to different parameters. For simplicity we call the tree which grows according to the process which determines $g(\cdot)$ the $g$ tree and that which grows according to the process which determines $\hat{g}(\cdot)$ the $\hat{g}$ tree. Suppose also that

$$g(X) < \hat{g}(X) \quad \text{for all } X. \tag{18.29}$$

Then

$$\exp\left(-\int_X^Y g(u)\mathrm{d}u\right) > \exp\left(-\int_X^Y \hat{g}(u)\mathrm{d}u\right) \quad \text{and} \quad H(X, Y) > \hat{H}(X, Y).$$

This means that for any cutting size $Y$ the $g$ tree is worth more than the $\hat{g}$ tree. If $X^*$ is the optimal cutting size for the $g$ tree and $\hat{X}^*$ is the optimal cutting size for the $\hat{g}$ tree, then in an obvious notation

$$V(X) = H(X, X^*) \geqslant H(X, \hat{X}^*) > \hat{H}(X, X^*) = \hat{V}(X),$$

so that the $g$ tree is worth more than the $\hat{g}$ tree. It also follows that $X^* > \hat{X}^*$. The optimal cutting size for the $g$ tree, $X^*$, must maximize $X/\phi(X)$. First-order conditions for maximization imply $X^*$ must satisfy $\phi(X^*) = X^*\phi'(X^*)$ or that

$$g(X^*) = 1/X^*. \tag{18.30}$$

Second-order conditions are that $g(X)$ intersect $1/X$ from below. Thus, if $g(X) < \hat{g}(X)$, it must be that if $\hat{g}(\hat{X}^*) = \hat{X}^{*-1}$, $g(\hat{X}^*) < \hat{X}^{*-1}$; at $\hat{X}^*$ the value of the $g$ tree is still increasing. Thus, $X^* > \hat{X}^*$.

To use this method of analysis it is necessary to translate the boundary conditions on $f$ given by (18.22a, b, c) into boundary conditions on $g$. Since solutions to (18.20) are of the form (18.24), in general

$$g(X) = \frac{A\lambda e^{\lambda X} + B\mu e^{\mu X}}{Ae^{\lambda X} + Be^{\mu X}} .$$

The solution $\gamma_Q(X)$ corresponding to a reflecting barrier at $Q$ for $f$, satisfies $\gamma_Q(Q) = 0$ and is given by

$$\gamma_Q(X) = \frac{\exp[(\lambda - \mu)(X - Q)] - 1}{\frac{1}{\lambda} \exp[(\lambda - \mu)(X - Q)] - \frac{1}{\mu}} . \tag{18.31}$$

The solution $\alpha_Q(X)$ corresponding to an absorbing barrier at $Q = 0$ for $f$ is given by

$$\alpha_Q(X) = -\frac{\lambda \exp[(\lambda - \mu)X] - \mu}{\exp[(\lambda - \mu)X] - 1} . \tag{18.32}$$

The solution to (18.20) which is bounded at $-\infty$ is of the form $f(X) = Ae^{\lambda X}$; the corresponding $g = f'/f$ is given by

$$\beta(X) = \lambda. \tag{18.33}$$

We now have enough information to draw a phase diagram for g. Eq. (18.28) can be rewritten as

$$g' = (\lambda - g)(g - \mu), \tag{18.34}$$

where $\lambda$ and $\mu$ are the roots of the characteristic equation of $f(\cdot)$ and are given by (18.25) and (18.26). For $g > \lambda$, $g'(X) < 0$; for $\lambda > g > \mu$, $g'(X) > 0$, and for $g < \mu$, $g'(X) < 0$, so the phase diagram looks as in fig. 18.1. In this figure we have drawn $\alpha_Q(X)$, $\gamma_Q(X)$ and $\beta(X)$ solutions corresponding to an absorbing barrier at $Q$, a reflecting barrier at $Q$, and bounded behavior at $-\infty$, respectively.

Since solutions to (18.28) cannot cross, $\alpha_Q(X) > \beta(X) > \gamma_Q(X)$. It is easy to calculate from (18.31) and (18.32) that

$$\lim_{X \to \infty} \alpha_Q(X) = \lim_{X \to \infty} \gamma_Q(X) = \beta(X) = \lambda \tag{18.35}$$

and

$$\lim_{Q \to -\infty} \alpha_Q(X) = \lim_{Q \to -\infty} \beta_Q(X) = \beta(X) = \lambda. \qquad (18.36)$$

Also, if $Q > \hat{Q}$, it is easy to see that $\alpha_Q(X) > \alpha_{\hat{Q}}(X) > \beta_{\hat{Q}}(X) > \beta_Q(X)$.
We may sum up these observations in two propositions.

**Proposition 18.1.** As $Q \to -\infty$, trees with reflecting barriers at $Q$ become indistinguishable from trees with absorbing barriers at $Q$ or trees which are simply bounded at $\infty$.

**Proposition 18.2.** Trees with reflecting barriers are worth more than trees with absorbing barriers. The value and the optimal cutting size of a tree with a reflecting barrier is an increasing function of the barrier. For trees with absorbing barriers, the opposite is true.

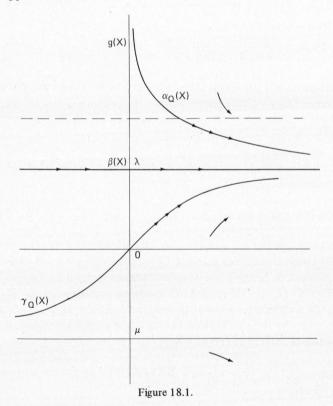

Figure 18.1.

The phase diagram makes it easy to do comparative statics, to analyze the effects of changes in parameters on value and cutting size. The parameters of our problem are $a$, $b$ and $r$. Differentiating (18.28) we see that $dg'(X)/dr = 1/b > 0$ and $dg'(X)/da = -(1/b)g(X)$ which is negative since we are only interested in positive $g(X)$. This proves the rather obvious

**Proposition 18.3.** Increasing the infinitesimal mean growth rate of a tree or decreasing the interest rate increases the value and the optimal cutting time of a tree.

The effects of increasing the variance are only slightly more complicated. Note that $dg'(X)/db = -(1/b^2)[r - ag(X)]$ which is negative whenever $g(X) < r/a$. It is straightforward to calculate that $\lambda < r/a$. If $g(\cdot)$ corresponds to a reflecting barrier at $-\infty$, $g(X) \leqslant \lambda$ for all $X$, it follows that increasing the variance increases the cutting size and value of the tree.

**Proposition 18.4.** If a tree's growth process is a diffusion with constant coefficients which is bounded at infinity or which has reflecting barriers, then increasing the variance increases the value and cutting size of the tree.

The effects of increasing the variance on the absorbing case are more complex and the reader is referred to Brock, Rothschild and Stiglitz (1979). We conclude this analysis of the stationary case by observing that under uncertainty the optimal cutting size is greater than it would be under certainty.

## 19. Miscellaneous applications and exercises

(1) Consider the stochastic differential equation of economic growth derived in section 2,

$$dk = [sf(k) - (n-\sigma^2)k]dt - \sigma k\,dz,\qquad(19.1)$$

with initial random condition $k(0) = k_0 > 0$. Find a set of sufficient conditions for the existence of a unique solution $k(t)$, $t \in [0, \infty)$ and use theorem 6.1 of Chapter 2 to establish the existence and uniqueness of $k(t)$.

(2) Equation (19.1) may be generalized by assuming that $s$ and $\sigma$ are no longer constants but instead are functions of $k$, written as $s(k)$ and $\sigma(k)$. With the new functions $s(k)$ and $\sigma(k)$ the generalized stochastic differential equation of growth becomes

$$dk = [s(k)f(k) - (n-\sigma^2(k))k]dt - \sigma(k)k\,dz, \qquad (19.2)$$

with initial condition $k(0) = k_0 > 0$. Find a set of sufficient conditions for the existence of a unique solution $k(t)$, $t \in [0, \infty)$ and use theorem 6.1 of Chapter 2 to establish the existence and uniqueness of $k(t)$.

(3) Suppose that eq. (19.2) has a unique solution $k(t)$. Discover sufficient conditions such that $k(t)$ is bounded w.p.1, i.e. such that

$$P[\omega: k(t, \omega) < \infty] = 1.$$

(4) Assume that the coefficients of (19.2) vanish for the equilibrium solution $k^*$, where $k^*$ is a nonrandom, nonzero constant; that is, assume that

$$[s(k^*)f(k^*) - (n-\sigma^2(k^*))k^*] = \sigma(k^*)k^* = 0.$$

Also, assume that there exists an $\eta > 0$ such that $\sigma(k)k > 0$ for $0 < |k-k^*| < \eta$. Is the equilibrium solution k* stable?

(5) Money, growth and uncertainty: Tobin (1965) was one of the first economists to introduce money in an economic growth model. The Tobin model has been studied by several authors. In this application we will first present briefly the deterministic Tobin model following Hadjimichalakis (1971) and then proceed to propose an extension by introducing uncertainty. We begin the model by describing its equations:

homogeneous production
functions of degree 1

$$Y = F(K, L), \qquad (19.3)$$

perfect foresight

$$\frac{\dot{P}}{P} = q, \qquad (19.4)$$

investment function

$$I = \dot{K} + \frac{d}{dt}\left(\frac{M}{P}\right), \qquad (19.5)$$

saving function

$$S = s\left[Y + \frac{d}{dt}\left(\frac{M}{P}\right)\right], \qquad (19.6)$$

labor growth

$$L(t) = L(0)\,e^{nt}, \quad L(0) > 0, \qquad (19.7)$$

money supply growth

$$M(t) = M(0)\,e^{\theta t}, \quad M(0) > 0. \qquad (19.8)$$

Note that $\dot{P}/P$ denotes actual inflation and $q$ denotes expected inflation. It is assumed that the two are equal under the assumption of perfect foresight.

Under equilibrium, saving must equal investment. Thus,

$$\dot{K} + \frac{d}{dt}\left(\frac{M}{P}\right) = s\left[Y + \frac{d}{dt}\left(\frac{M}{P}\right)\right],$$

which yields

$$\dot{K} = s\left[Y + \frac{d}{dt}\left(\frac{M}{P}\right)\right] - \frac{d}{dt}\left(\frac{M}{P}\right).$$

This last equation is called *Tobin's fundamental equation*. On the basis of the above, the differential equation of money and growth, where we define $m = M/P \cdot 1/L$, i.e. $m$ is per capita real money balances, is given by

$$\dot{k} = sf(k) - (1 - s)(\theta - q)m - nk. \tag{19.9}$$

Note that if $\theta = 0 = q$, then we obtain as a special case the Solow equation (2.1).

Suppose that uncertainty is now introduced in Tobin's monetary growth model by postulating randomness in the growth of the money supply and described by the stochastic differential equation

$$dM = \theta M dt + \mu M dz. \tag{19.10}$$

Use eq. (19.10) and the appropriately modified model of eqs. (19.3)–(19.7) to study the effects of uncertainty in the money supply growth.

(6) Random demand functions: Let a consumer solve

$$\max U(X, Y) \tag{19.11}$$

subject to

$$p_1 X + p_2 Y = M. \tag{19.12}$$

Let

$$X = g(M, p_1, p_2) \quad \text{and} \quad Y = h(M, p_1, p_2)$$

solve (19.11) and (19.12). Now, let $M$, $p_1$ and $p_2$ be random and follow the Itô processes:

$$dM = \mu_M dt + \sigma_M dz_M, \tag{19.13}$$

$$dp_1 = \mu_{p_1} dt + \sigma_{p_1} dz_{p_1}, \tag{19.14}$$

$$dp_2 = \mu_{p_2} dt + \sigma_{p_2} dz_{p_2}. \tag{19.15}$$

Assume that each instantaneous mean and variance of (19.13)–(19.15) is a function of $(M, p_1, p_2)$. Assume that $\{z_M\}, \{z_{p_1}\}, \{z_{p_2}\}$ are Wiener processes that satisfy the formal rules

$$dz_M \cdot dz_{p_1} = \rho dt; \quad dz_M \cdot dz_{p_2} = \rho dt; \quad dz_{p_1} \cdot dz_{p_2} = \rho dt,$$

$$dz_{p_1} \cdot dz_{p_1} = dt; \quad dz_{p_2} \cdot dz_{p_2} = dt; \quad dz_M \cdot dz_M = dt.$$

Use Itô's lemma to write down the stochastic differential equations of the demands $X = g(M, p_1, p_2)$, and $Y = h(M, p_1, p_2)$ for the case

$$U(X, Y) = AX^\alpha Y^\beta, \quad 0 < \alpha < 1, 0 < \beta < 1, 0 < \alpha + \beta \leqslant 1.$$

This is a random world when the individual cannot store commodities, i.e. it is a stochastic world without stock variables just as standard static demand theory is a deterministic world with no stock variables.

(7) Edgeworth's random boxes: This is an Edgeworth Box with random endowment. Let consumer $i = 1, 2$ solve

$$\max X_i^\alpha Y_i^\beta$$

$$p_1 X_i + p_2 Y_i = p \cdot e_i,$$

where $e_1 = (a, 0), e_2 = (0, b), p \cdot e_i$ is a scalar product. Note that $\alpha$ and $\beta$ do not vary across consumers. Use problem (6) to solve for the demand functions

$$X_i = h_i(p_1, p_2, p \cdot e_i) \quad \text{and} \quad Y_i = g_i(p_1, p_2, p \cdot e_i).$$

Determine the relative price $p_1/p_2$ by supply equals demand:

$$h_1 + h_2 = a, \quad g_1 + g_2 = b.$$

Now let

$$da = \mu_a dt + \sigma_a dz_a,$$

$$db = \mu_b dt + \sigma_b dz_b,$$

$$dz_a dz_b = \rho dt$$

determine the endowments, $a$ and $b$, of the two consumers. Use Itô's lemma to find the stochastic process that equilibrium relative prices $p_1/p_2$ must follow.

(8) Marshall with $\sigma^2$: Consider the following algebraic version of the Marshallian Cross:

$$p = D(q) = Aq^{-\alpha} = S(q) = Bq^{\beta}, \quad A > 0, B > 0, \alpha > 0, \beta > 0.$$

Find the stochastic processes generating equilibrium $p(t)$ and $q(t)$ when

$$dA = \mu_A dt + \sigma_A dz_A,$$

$$dB = \mu_B dt + \sigma_B dz_B,$$

where we assume $dz_A dz_B = 0$ for simplicity. Does the old proposition about random supply and demand leading to high price variation and low quantity variation when supply and demand are inelastic hold up in this new framework?

(9) Monopoly vs. competition in exploring for oil: A monopolist solves

$$\max E_0 \int_0^\infty e^{-\rho t} (R(q) - I) dt \tag{19.16}$$

subject to

$$dE(t) = -q(t)dt + b(E(t))da(t), \tag{19.17}$$

where

$$da(t) = 1 \quad \text{with probability } \lambda(I(t))dt,$$

$$da(t) = 0 \quad \text{with probability } 1 - \lambda(I(t))dt.$$

Here $R(q) = D(q)q$ = total revenue, and $I(t)$ = investment in enlarging the stock of oil, $E(t)$. Investment takes the form of allocating money $I(t)$ to "dig" where with probability $\lambda(I(t))dt$ of "find" of size $b(E(t))$ will be discovered. Write down the Hamilton–Jacobi–Bellman equation for (19.16) and (19.17). Let competition solve the same problem with $R(q)$ replaced by $CS(q)$ where

$$CS(q) = \int_{q_0}^q D(y)dy,$$

and $q_0$ is a small positive lower limit. Note that $q_0$ could be taken to be zero provided that

$$\int\limits_{q_0}^{q} D(y)\,\mathrm{d}y \quad \text{for } q > 0, \text{ is finite.}$$

Write down the Hamilton–Jacobi–Bellman equation for competition. Point out the difference from monopoly. Also, attempt to say if monopoly or competition will invest more in finding new wells.

The above model may be looked upon as a proxy for the amount of innovation in the sense of divesting profits into projects to enlarge the stock of salable resources. One would expect that monopoly would invest less than competition. Any hints from this model? Finally, try to find a form of $D(q), b(E(t)), \lambda(I(t))$ that permits a closed form solution for the Hamilton–Jacobi–Bellman equation.

(10) Exhaustible resource problem: Consider the problem

$$\max \mathrm{E}_0 \int\limits_0^{\infty} e^{-rt} R(q)\,\mathrm{d}t$$

subject to

$$\mathrm{d}E = -q\,\mathrm{d}t + \sigma_0 E\,\mathrm{d}z,$$

for the special case where

$$R(q) \equiv D(q)q \equiv (Aq^{-\alpha})q = Aq^{1-\alpha}.$$

As in the previous problem $R(q)$ is total revenue for the firm, $q$ is quantity of output and $E(t)$ is the stock of the resource at time $t$ with $E(0) = E_0 > 0$.

Solve this problem for a closed form solution for the cases: (i) $\sigma_0 = 0$ and (ii) $\sigma_0 > 0$, and compare your solutions.

(11) Write the quantity theory of money equation

$$MV = PO,$$

where $M$ is money supply, $V$ is velocity of money, $P$ is price level, and $O$ is real output. Let

$$\mathrm{d}O = \alpha_1\,\mathrm{d}t + \beta_1\,\mathrm{d}z_1$$

and

$$\mathrm{d}M = \alpha_2\,\mathrm{d}t + \beta_2\,\mathrm{d}z_2,$$

where, $E_t dz_1^2 = dt$, $E_t dz_2^2 = dt$, $E_t dz_1 dz_2 = \rho dt$. Let $V$ be constant over time and nonrandom. Write out the formula for $dP/P$ in percentage terms.

(12) Consider the quadratic problem presented in section 12:

$$- W(x(t)) = \min_v E_t \int_{s=t}^{\infty} e^{-\rho(s-t)} \{a(x(s))^2 + b(v(s))^2\} ds.$$

Solve this problem for each of the following laws of motion:

(i)   $dx(s) = v(s)dt + \sigma_0 x(s)dz(s),$

where $\sigma_0$ is a constant independent of time and of x;

(ii)   $dx(s) = [v(s) + \ell_0]dt + \sigma_0 x(s)dz(s),$

where $\ell_0$ is a nonrandom constant; and

(iii)   $dx(s) = [ax(s) + v(s)]dt + \sigma_0 x(s)dz(s),$

where $a$ is a nonrandom constant.
Furthermore, in each case check that your solution satisfies the transversality condition,

$$\lim_{t \to \infty} E_0 e^{-\rho t} q(t) \cdot x(t) = 0.$$

(13) Exchange rate in a two-country stochastic monetary model: We follow Lau (1977) to formulate a simple international monetary model. Let there be two countries: country 1, say England, and country 2, say France. Let there be two goods: good 1, cloth, and good 2, wheat. This is not a production model, so we assume that both goods are perishable and that at the beginning of each period England is endowed with a constant amount of cloth, $y_1$, and France is endowed with a constant amount of wheat, $y_2$. We use $q_1$ to denote the price of good 1, cloth, and $q_2$ to denote the price of good 2, wheat. We choose $q_1 \equiv 1$, i.e. cloth is the numeraire. Let $c_{ij}$ be the real consumption by country $i$ of good $j$ and let $M_{ij}$ be country $i$'s holdings of money $j$. Denote by $\delta_i$ the discount rate for country $i$. Let $P_1$ be pounds per unit of cloth and $P_2$ francs per unit of cloth and let $E$ be the exchange rate. By purchasing power parity, $P_1 = EP_2$ or $E = P_1/P_2$, i.e. $E$ is pounds per franc. We assume that the money stocks, $M_i^s$ follow the stochastic differential equations.

$$dM_1^s = \mu_1 M_1^s dt + \sigma_1 M_1^s dz_1$$

and

$$dM_2^s = \mu_2 M_2^s dt + \sigma_2 M_2^s dz_2,$$

where $\mu_1, \mu_2, \sigma_1$ and $\sigma_2$ are constants, and $z_1$ and $z_2$ are normalized Wiener processes such that

$$(dz_1)^2 = (dz_2)^2 = dt \quad \text{and} \quad dz_1 \cdot dz_2 = \rho_{12} dt.$$

With the above information we now formulate the problem: England solves

$$\max E_0 \int_0^\infty e^{-\delta_1 t} [u_1(c_{11}, c_{12}) + v_1(M_{11}/P_1, M_{12}/P_2)] dt$$

subject to

$$P_1 c_{11} dt + EP_2 q_2 c_{12} dt + dM_{11}^d + E dM_{12}^d = P_1 y_1 dt + dM_1^s$$

for $\{c_{11}^d\}_{t=1}^\infty, \{c_{12}^d\}_{t=1}^\infty, \{M_{11}^d\}_{t=1}^\infty$ and $\{M_{12}^d\}_{t=1}^\infty$, while France solves

$$\max E_0 \int_0^\infty e^{-\delta_2 t} [u_2(c_{21}, c_{22}) + v_2(M_{21}/P_1, M_{22}/P_2)] dt$$

subject to

$$\frac{1}{E} P_1 c_{21} dt + P_2 q_2 c_{22} dt + \frac{1}{E} dM_{21}^d + dM_{22}^d = P_2 q_2 y_2 dt + dM_2^s$$

for $\{c_{21}^d\}_{t=1}^\infty, \{c_{22}^d\}_{t=1}^\infty, \{M_{21}^d\}_{t=1}^\infty$ and $\{M_{22}^d\}_{t=1}^\infty$. Then $\{P_1\}_{t=1}^\infty$ and $\{P_2\}_{t=1}^\infty$ is a monetary perfect foresight equilibrium if for every $t \in [0, \infty)$ all of the following hold:

$$c_{11}^d + c_{21}^d = y_1; \quad c_{12}^d + c_{22}^d = y_2, M_{11}^d + M_{21}^d = M_1^s; \quad M_{12}^d + M_{22}^d = M_2^s.$$

Observe that we postulate separable utility functions $u_i$ and $v_i$ and also note that the superscript d denotes quantities demanded. Make any additional assumptions that are needed to study in this model the behavior of the exchange rate $E$.

(14) Stochastic search theory: Consider a representative individual who searches for a higher wage. This search takes the form of allocating time to influence the probability of arrival of a Poisson event. The more time that he allo-

cates to search activity the more frequently this Poisson event will arrive. If the Poisson event arrives the searcher's wage increases by a jump. The searcher faces a stochastic differential equation that determines the evaluation of his wage. This equation consists of an exogenous rise in the nominal wage plus a Brownian motion term plus a Poisson term. Only the Poisson term is influenced by the searcher's search activity. Let us get into the details. Consider the following model:

$$\max E_0 \int_0^\infty e^{-\rho t} y(t) dt \equiv J(W(0), 0)$$

subject to

$$dW(t) = rW(t)dt + \sigma W(t)dz(t) + g(W(t))dq(t), W_0 \text{ given},$$

where $W(t)$, $y(t)$, $\rho$, $\{z(t)\}_{t=0}^\infty$ and $\{q(t)\}_{t=0}^\infty$ are nominal wage, flow income, discount rate on future income, standardized Wiener process, and Poisson process, respectively. The probability that the Poisson event occurs, i.e. $q(t + \Delta t) - q(t) = 1$, is given by $\lambda(\ell_1)\Delta t + o(\Delta t)$ and $y(t) = W(t)(1 - \ell_1(t))$, where $\ell_1(t)$ is the percentage of time devoted to wage augmentation activity, i.e. search activity. Furthermore, $E_t$ denotes expectation conditioned at $t$. The numbers $r$ and $\sigma$ do not depend on $t$ or $W$. Here $g(W(t))$ is the amount of the wage jump if the Poisson event occurs. The idea is that the amount of a better job offer should depend upon current wage. Form

$$\phi(\ell_1, W, t) = e^{-\rho t}(1 - \ell_1)W + J_t + J_W rW$$
$$+ \tfrac{1}{2} J_{WW}\sigma^2 W^2 + \lambda[J(W + g(W), t) - J(W, t)]. \tag{19.18}$$

Assume an interior solution for $\ell_1$:

$$\frac{\partial \phi}{\partial \ell_1} = 0; \quad 0 = -e^{-\rho t}W + \frac{d\lambda}{d\ell_1}[J(W + g(W), t) - J(W, t)]. \tag{19.19}$$

Notice that eqs. (19.18) and (19.19) are the standard partial differential equations of stochastic control.

Make any assumptions that seem economically reasonable and proceed to study the effects of search on nominal wages.

## 20. Further remarks and references

We would like to note that there is some arbitrariness on the part of the author in the selection of applications and in distinguishing them either as economics or finance applications. This chapter and the next are not intended to be exhaustive nor is it possible to have an empty intersection. An important book of readings and exercises which supplements this chapter is Diamond and Rothschild (1978).

The use of continuous time stochastic calculus in macroeconomic growth under uncertainty first appeared in the papers of Bourguignon (1974), Merton (1975a) and Bismut (1975). These papers build on several earlier papers on economic growth under uncertainty such as Brock and Mirman (1972), Levhari and Srinivasin (1969), Mirman (1973), Stigum (1972), Leland (1974) and Mirman and Zilcha (1975), among others. The main unresolved issue in continuous time economic growth under uncertainty is the stochastic stability of the stationary distribution. Stochastic point equilibrium is discussed in Malliaris (1978). For discrete time growth models under uncertainty several results are presented in Brock and Majumdar (1979) and section 15 of this chapter. Two recent contributions on optimal saving under uncertainty are Foldes (1978a, 1978b).

Our analysis of growth in an open economy under uncertainty is very limited. However, there has been great interest recently in introducing uncertainty in international economics with several papers, such as Batra (1975), Mayer (1976), Baron and Forsythe (1979), and the recent book by Helpman and Razin (1979). See also the international monetary model under uncertainty in exercise 13 of section 19 due to Lau (1977).

In section 8 we presented the concept of rational expectations to illustrate the use of stochastic methods and more importantly to familiarize the reader with the definitional aspects of this concept. Rational expectations as a concept has found numerous applications and for a survey article we suggest Shiller (1978) and Kantor (1979). We use rational expectations in sections 9 and 11.

The cost of adjustment type of investment theory developed by Lucas (1967b, 1967), Gould (1968) and Treadway (1969, 1970), among others, was extended under uncertainty in the paper of Lucas and Prescott (1971). Elements of the Lucas and Prescott (1971) paper are presented in section 9 to illustrate the use of stochastic methods in microeconomic theory. Another microeconomic application is found in section 16. Some basic references in the analysis of the firm under uncertainty, other than Sandmo (1971), are Mills (1959), Nelson (1961), Pratt (1964), McCall (1967), Stigum (1969a, 1969b), Baron (1970, 1971), Zabel (1970), Leland (1972), Ishii (1977), Wu (1979) and Perrakis (1980), just to mention a few. Note that there is a considerable bibliography on problems of general equilibrium under uncertainty. We have not presented any applications in this area primarily because the techniques are not similar to the ones presented in this book. Techniques of general equilibrium under uncertainty usually involve arguments of a topological nature and/or arguments of functional analysis. A representative paper in general equilibrium under uncertainty is Bewley (1978). However, we remark that stochastic stability techniques have been used in general equilibrium in some specific formulations such as Turnovsky and Weintraub (1971).

In sections 10 and 11 we attempt to develop a general approach to continuous time stochastic processes that arise in dynamic economics from maximizing behavior of agents, as in Brock and Magill (1979). The analysis considers a class of stochastic discounted infinite horizon maximum problems that arise in economics and uses Bismut's (1973) approach in solving these problems. It is shown that the idea of a competitive path, introduced in the continuous time deterministic case in Magill (1977b), generalizes in a natural way in the case of uncertainty to a competitive process. Theorem 10.1 shows that under a concavity assumption on the basic integrand of the problem a competitive process which satisfies a transversality condition is optimal under a discounted catching-up criterion. Next, we consider the sample path properties of a competitive process. If for almost every realization of a competitive process the associated dual price process generates a path of subgradients for the value function, we call the process McKenzie competitive, since it was McKenzie (1976) who first recognized the importance of this property in the deterministic case. Theorem 10.2 shows that two McKenzie competitive processes starting from distinct nonrandom initial conditions converge almost surely if the processes are bounded almost surely and if a certain curvature condition is satisfied by the Hamiltonian of the system. The problem of finding sufficient conditions for the existence of a McKenzie competitive process remains an open problem.

Business cycles and macroeconomic stabilization methods in an environment where uncertainty prevails are areas of research in which stochastic calculus techniques are quite appropriate. However, the research has just begun; we note that Lucas' (1975) paper represents a methodological advance in business cycle theory. In the Lucas (1975) paper, discrete time stochastic techniques are used to show that random monetary shocks and an accelerator effect interact to generate serially correlated cyclical movements in real output and procyclical movements in prices, in the ratio of investment to output, and in nominal interest rates. Also, note the Slutsky (1937) paper in which it is shown that a weighted sum of independent and identically distributed random variables with mean zero and finite variance leads to approximately regular cyclical motion. Slutzky's (1937) ideas have not yet been fully utilized. Magill (1977a) has a brief analysis in which he shows that the introduction of uncertainty imbeds the short-run study of the business cycle into the long-run process of optimal capital accumulation. Tinbergen's (1952) classic work on static stabilization has been extended to allow for uncertainty by Brainard (1967) and in section 17 we give a simple illustration. See also Poole (1970), Chow (1970, 1973) and Turnovsky (1973).

In section 14 we illustrate the use of stochastic calculus techniques in a macroeconomic model with rational expectations, following Gertler (1979). The Gertler (1979) paper explores the consequences for price dynamics of imperfect price flexibility and it demonstrates that the same condition which ensures stability in the deterministic system also ensures that the distribution of the state variables converges to a stable path in the stochastic case. In section 14 we illustrated the stochastic case; for a comparison between the deterministic and the stochastic case see Gertler (1979).

Section 18 follows Brock, Rothschild and Stiglitz (1979) to illustrate various methods of continuous optimal stopping to stochastic capital theory. This section treats the time independent case. Similar results continue to hold in the general case when the instantaneous mean and the instantaneous variance of the diffusion process are functions of the tree's current size. For details see the Brock, Rothschild and Stiglitz (1979) paper. See also the paper by Miller and Voltaire (1980) which treats the repeated or sequential stochastic tree problem, i.e. the problem of deciding when to harvest and replant trees given the knowledge of the process of each tree's growth history. Miller and Voltaire (1980) show that the results for the nonrepeated and for the repeated cases are qualitatively similar.

The reader who is interested in further applications of optimal stopping methods in economics should consult Samuelson and McKean (1965), Boyce (1970) and Jovanovic (1979a).

CHAPTER 4


# APPLICATIONS IN FINANCE

> There is no need to enlarge upon the importance of a realistic theory explaining how individuals choose among alternate courses of action when the consequences of their actions are incompletely known to them.
>
> Arrow (1971, p. 1).


## 1. Introduction

In this chapter we present several applications of stochastic methods in finance to illustrate the techniques discussed in Chapters 1 and 2. We also include some applications which use additional techniques to familiarize the reader with a sufficient sample of stochastic methods applied in modern finance.


## 2. Stochastic rate of inflation

In this application we illustrate the use of Itô's lemma in determining the solution of the behavior of prices and real return of an asset when inflation is described by an Itô process. The analysis follows Fischer (1975).

Suppose that the rate of inflation is stochastic and the price level is describable by the process

$$\frac{\mathrm{d}P}{P} = \Pi \, \mathrm{d}t + s \, \mathrm{d}z. \tag{2.1}$$

The stochastic part is $dz$ with $z$ being a Wiener process. The drift of the process, $\Pi$, is the expected rate of inflation per unit of time. It is defined by

$$\Pi = \lim_{h \to 0} \mathrm{E}_t \frac{1}{h} \left\{ \frac{P(t+h) - P(t)}{P(t)} \right\}, \tag{2.2}$$

where $\mathrm{E}_t$ is the expectation operator conditioned on the value of $P(t)$. The variance of the process per unit of time is defined by

$$s^2 = \lim_{h \to 0} \mathrm{E}_t \frac{1}{h} \left\{ \left[ \frac{P(t+h) - P(t)}{P(t)} - \Pi h \right] \right\}^2 . \tag{2.3}$$

A discrete time difference equation which satisfies $\Pi$ and $s^2$ as defined in (2.2) and (2.3) is

$$\frac{P(t+h) - P(t)}{P(t)} = \Pi h + s y(t) (h)^{1/2}, \tag{2.4}$$

where $y(t)$ is a normal random variable with zero mean and unit variance which is not temporally correlated. The limit as $h \to 0$ of (2.4) then describes a Wiener process for the variable $s y(t) (h)^{1/2}$ and eq. (2.1) can be written as

$$\frac{dP}{P} = \Pi dt + s y(t) (h)^{1/2} = \Pi dt + s dz,$$

where $dz = y(t) (h)^{1/2}$. Note that (2.1) says that over a short time interval the proportionate change in the price level is normal with mean $\Pi dt$ and variance $s^2 dt$. Rewrite (2.1) as

$$dP = P\Pi dt + Ps dz \tag{2.5}$$

and let

$$y(t) = P(0) \exp \left[ \left( \Pi - \frac{s^2}{2} \right) t + s \int_0^t dz \right] . \tag{2.6}$$

We use Itô's lemma to show that $y(t)$ satisfies eq. (2.5). Let

$$F(t, z) = P(0) \exp \left[ \left( \Pi - \frac{s^2}{2} \right) t + s \int_0^t dz \right]$$

and compute $\partial F/\partial t$, $\partial F/\partial z$ and $\partial^2 F/\partial z^2$ as below:

$$\frac{\partial F}{\partial t} = P(0) \exp \left[ \left( \Pi - \frac{s^2}{2} \right) t + s \int_0^t dz \right] \left( \Pi - \frac{s^2}{2} \right) = y(t) \left( \Pi - \frac{s^2}{2} \right),$$

$$\frac{\partial F}{\partial z} = P(0) \exp \left[ \left( \Pi - \frac{s^2}{2} \right) t + s \int_0^t dz \right] s = sy(t),$$

$$\frac{\partial^2 f}{\partial z^2} = P(0) \exp \left[ \left( \Pi - \frac{s^2}{2} \right) t + s \int_0^t dz \right] s^2 = s^2 y(t).$$

Thus, applying Itô's formula and by making the necessary substitutions we end up with

$$\begin{aligned}
dy &= \frac{\partial F}{\partial t} \, dt + \frac{\partial F}{\partial z} \, dz + \frac{1}{2} \frac{\partial^2 F}{\partial z^2} (dz)^2 \\
&= y(t) \left( \Pi - \frac{s^2}{2} \right) dt + sy(t) \, dz + \tfrac{1}{2} s^2 y(t) \, (dz)^2 \\
&= y(t) \Pi dt - y(t) \frac{s^2}{2} \, dt + sy(t) \, dz + \tfrac{1}{2} s^2 \, y(t) \, dt \\
&= y(t) \Pi dt + y(t) s \, dz.
\end{aligned}$$

Thus, (2.6) satisfies eq. (2.5), which is what we wanted to show.

We continue with a further application. Consider the two Itô processes

$$\frac{dP}{P} = \Pi dt + s \, dz \quad \text{and} \quad \frac{dQ}{Q} = r \, dt. \tag{2.7}$$

We use Itô's lemma to compute the stochastic process describing the variable $q = u(P, Q) = Q/P$. Recall that Itô's formula for this case is:

$$\begin{aligned}
dq &= \frac{\partial q}{\partial t} \, dt + \frac{\partial q}{\partial P} \, dP + \frac{\partial q}{\partial Q} \, dQ \\
&+ \frac{1}{2} \left( \frac{\partial^2 q}{\partial P^2} \, dP^2 + 2 \frac{\partial^2 q}{\partial P \partial Q} \, dP dQ + \frac{\partial^2 q}{\partial Q^2} \, dQ^2 \right).
\end{aligned}$$

Computing the various terms and using the multiplication rules in (4.12) in Chapter 2, the result is:

$$dq = -\frac{Q}{P^2}\,dP + \frac{1}{P}\,dQ + \frac{1}{2}\left(\frac{2Q}{P^3}\right)(Ps)^2\,dt$$

$$= -\frac{Q}{P^2}\,(\Pi P dt + Ps dz) + \frac{1}{P}\,(rQ dt) + \frac{Q}{P}\,s^2\,dt.$$

Finally,

$$\frac{dq}{q} = (r - \Pi + s^2)\,dt - s\,dz,$$

which describes the *proportional rate of change of the real return of an asset* having a nominal return as in (2.7).

### 3. The Black–Scholes option pricing model

In this section we follow Black and Scholes (1973) and Merton (1973a) to develop an option pricing model. Consider an asset, a stock option for example, denoted by $A$, the price of which at time $t$ can be written as

$$W(t) = F(S, t), \tag{3.1}$$

where $F$ is a twice continuously differentiable function. Here $S(t)$ is the price of some other asset, denoted by $B$, for example the stock upon which the option is written. The price of $B$ is assumed to follow the stochastic differential equation

$$dS(t) = f(S(t), t)\,dt + \eta(S(t), t)\,dz(t), \tag{3.2}$$

$$S(0) = S_0 \text{ given.}$$

Consider an investor who builds up a portfolio of three assets, $A$, $B$ and a riskless asset denoted by $C$. We assume that $C$ earns the competitive rate of return $r(t)$. The nominal value of the portfolio is

$$P(t) = N_1(t)S(t) + N_2(t)W(t) + Q(t), \tag{3.3}$$

where $N_1$ denotes the number of shares of $B$, $N_2$ the number of shares of $A$, and $Q$ is the number of dollars invested in the riskless asset $C$. Assume that $B$ pays no dividends or other distributions. By Itô's lemma we compute

$$\mathrm{d}W(t) = \mathrm{d}F(t) = F_t\mathrm{d}t + F_S\mathrm{d}S + \tfrac{1}{2}F_{SS}\mathrm{d}S^2 \tag{3.4}$$

$$\equiv a\mathrm{d}t + b\mathrm{d}z,$$

where

$$a \equiv F_t + F_S f + \tfrac{1}{2}F_{SS}\,\eta^2 \equiv \alpha_W\,W, \tag{3.5}$$

$$b \equiv F_S\eta \equiv \sigma_W W. \tag{3.6}$$

Here we follow Black and Scholes (1973) and assume as a simplifying special case that $f(S, t) = \alpha S$ and that $\eta(S, t) = \sigma S$, where $\alpha$ and $\sigma$ are constants. Next we write the dynamics for $S(t)$ in this special case of (3.2) in percentage terms as

$$\frac{\mathrm{d}S}{S} = \alpha\mathrm{d}t + \sigma\mathrm{d}z. \tag{3.7}$$

Now for a portfolio strategy where $N_1$ and $N_2$ are adjusted slowly relative to the change in $S$, $W$ and $t$ we may assume that $\mathrm{d}N_1 = \mathrm{d}N_2 = 0$ and proceed to study the change in the nominal value of the portfolio, $\mathrm{d}P$, as follows:

$$\mathrm{d}P = N_1(\mathrm{d}S) + N_2(\mathrm{d}W) + \mathrm{d}Q$$

$$= (\alpha\mathrm{d}t + \sigma\mathrm{d}z)N_1 S + (\alpha_W\mathrm{d}t + \sigma_W\mathrm{d}z)N_2 W + rQ\mathrm{d}t. \tag{3.8}$$

Set $W_1 = N_1 S/P$, $W_2 = N_2 W/P$ and $W_3 = Q/P = 1 - W_1 - W_2$. Then (3.8) becomes

$$\frac{\mathrm{d}P}{P} = (\alpha\mathrm{d}t + \sigma\mathrm{d}z)W_1 + (\alpha_W\mathrm{d}t + \sigma_W\mathrm{d}z)W_2 + (r\mathrm{d}t)W_3. \tag{3.9}$$

Design the proportions $W_1$ and $W_2$ so that the position is *riskless* for all $t \geqslant 0$:

$$\mathrm{var}_t\left(\frac{\mathrm{d}P}{P}\right) = \mathrm{var}_t(W_1\sigma\mathrm{d}z + W_2\sigma_W\mathrm{d}z) = 0, \tag{3.10}$$

where $\mathrm{var}_t$ denotes variance conditioned on $(S(t), W(t), Q(t))$. In other words, choose $(W_1, W_2) = (\bar{W}_1, \bar{W}_2)$ so that

$$\bar{W}_1\sigma + \bar{W}_2\sigma_W = 0. \tag{3.11}$$

Then from (3.9)

$$E_t\left(\frac{\mathrm{d}P}{P}\right) = [\alpha\bar{W}_1 + \alpha_W\bar{W}_2 + r(1 - \bar{W}_1 - \bar{W}_2)]\mathrm{d}t = r(t)\mathrm{d}t \tag{3.12}$$

since the portfolio is riskless. Eqs. (3.11) and (3.12) yield the famous Black–Scholes–Merton equations:

$$\frac{\overline{W}_1}{\overline{W}_2} = -\frac{\sigma_W}{\sigma} \qquad (3.13)$$

and

$$r = \alpha \overline{W}_1 + \alpha_W \overline{W}_2 - r\overline{W}_1 - r\overline{W}_2 + r, \qquad (3.14)$$

which simplify to

$$\frac{\alpha - r}{\sigma} = \frac{\alpha_W - r}{\sigma_W}. \qquad (3.15)$$

Eq. (3.15) says that the *net rate of return per unit of risk* must be the same for the two assets.

For a further special case

$$\alpha(S, t) = \alpha_0; \quad \sigma(S, t) = \sigma_0; \quad r(t) = r_0, \qquad (3.16)$$

where $\alpha_0$, $\sigma_0$ and $r_0$ are constants that are independent of $(S, t)$, by using eq. (3.15) and making the necessary substitutions from (3.5) and (3.6) we obtain the partial differential equation

$$\tfrac{1}{2} \sigma_0^2 S^2 F_{SS}(S, t) + r_0 S F_S(S, t) - r_0 F(S, t) + F_t(S, t) = 0. \qquad (3.17)$$

Its boundary condition is determined by the specifications of the asset. For the case of an option which can be exercised only at the expiration date $T$ with exercise price $E$, the boundary condition is

$$F(0, \tau) = 0, \quad \tau = T - t,$$

$$F(S, T) = \max [0, S - E]. \qquad (3.18)$$

Let $W(S, \tau; E, r_0, \sigma_0^2)$ denote the solution $F$, subject to the boundary condition. This solution is given by Black–Scholes (1973) and Merton (1973a) as

$$W(S, \tau; E, r_0, \sigma_0^2) = S\phi(\mathrm{d}_1) - E \mathrm{e}^{-r_0 \tau} \phi(\mathrm{d}_2), \qquad (3.19)$$

where

$$\phi(y) \equiv \frac{1}{(2\pi)^{1/2}} \int_{-\infty}^{y} e^{-s^2/2}\,ds, \qquad (3.20)$$

i.e. the cumulative normal distribution, with

$$d_1 \equiv \left[\log\left(\frac{S}{E}\right) + \left(r_0 + \frac{\sigma_0^2}{2}\right)\tau\right] \cdot \frac{1}{\sigma\sqrt{\tau}}. \qquad (3.21)$$

$$d_2 \equiv d_1 - \sigma\sqrt{\tau}.$$

## 4. Consumption and portfolio rules

Another useful application of stochastic calculus techniques is found in Merton (1971). Assume that in an economy all assets are of a limited liability type, that there exist continuously trading perfect markets with no transaction costs for all assets, and that prices per share $P_i(t)$ are generated by Itô processes, i.e.

$$\frac{dP_i}{P_i} = \alpha_i(P, t)\,dt + \sigma_i(P, t)\,dz_i, \qquad (4.1)$$

where $\alpha_i$ is the instantaneous conditional expected percentage change in price per unit of time and $\sigma_i^2$ is the instantaneous conditional variance per unit of time. In the particular case where the geometric Brownian motion hypothesis is assumed to hold for asset prices, $\alpha_i$ and $\sigma_i$ will be constants and prices will be stationarily and log-normally distributed.

To derive the correct budget equation it is necessary to examine the discrete time formulation of the model and then to take limits to obtain the continuous time form. Consider a period model with period length $h$, where all income is generated by capital gains. Suppose that wealth, $W(t)$, and $P_i(t)$ are known at the beginning of period $t$. The notation used is,

$N_i(t)$ = number of shares of asset $i$ purchased during period $t$, i.e. between $t$ and $t + h$, where $h > 0$; and

$C(t)$ = amount of consumption per unit of time during $t$.

The model assumes that the individual comes into period $t$ with wealth invested in assets so that

$$W(t) = \sum_{1}^{n} N_i(t-h)P_i(t). \qquad (4.2)$$

Note that we write $N_i(t-h)$ because this is the number of shares purchased for the portfolio during the period $t-h$ to $t$, evaluated at current prices $P_i(t)$. The decisions about the amount of consumption for period $t$, $C(t)$, and the new portfolio, $N_i(t)$, are simultaneously made at known current prices:

$$- C(t) h = \sum_{1}^{n} [N_i(t) - N_i(t-h)] P_i(t). \tag{4.3}$$

Increment eqs. (4.2) and (4.3) by $h$ to eliminate backward differences and thus obtain

$$
\begin{aligned}
- C(t+h) h &= \sum_{1}^{n} [N_i(t+h) - N_i(t)] P_i(t+h) \\
&= \sum_{1}^{n} [N_i(t+h) - N_i(t)] [P_i(t+h) - P_i(t)] \\
&\quad + \sum_{1}^{n} [N_i(t+h) - N_i(t)] P_i(t)
\end{aligned}
\tag{4.4}
$$

and

$$W(t+h) = \sum_{1}^{n} N_i(t) P_i(t+h). \tag{4.5}$$

Take limits in (4.4) and (4.5) as $h \to 0$ to conclude that

$$- C(t)\, dt = \sum_{1}^{n} dN_i(t)\, dP_i(t) + \sum_{1}^{n} dN_i(t) P_i(t) \tag{4.6}$$

and, similarly,

$$W(t) = \sum_{1}^{n} N_i(t) P_i(t). \tag{4.7}$$

Use Itô's lemma to differentiate $W(t)$ to obtain

$$dW(t) = \sum_{1}^{n} N_i(t)\, dP_i(t) + \sum_{1}^{n} dN_i(t) P_i(t) + \sum_{1}^{n} dN_i(t)\, dP_i(t). \tag{4.8}$$

The last two terms, $\sum_{1}^{n} dN_i P_i + \sum_{1}^{n} dN_i\, dP_i$, in (4.8) are the net value of additions to wealth from sources other than capital gains. If $dy(t)$ denotes the instantaneous flow of noncapital gains, i.e. wage income, then

$$dy - C(t)\, dt = \sum_{1}^{n} dN_i P_i + \sum_{1}^{n} dN_i\, dP_i,$$

which yields the budget equation

$$dW = \sum_{1}^{n} N_i(t)\,dP_i + dy - C(t)\,dt. \tag{4.9}$$

Define the new variable $\omega_i(t) = N_i(t)P_i(t)/W(t)$ and use (4.1) to obtain,

$$dW = \sum_{1}^{n} \omega_i W\alpha_i\,dt - C\,dt + dy + \sum_{1}^{n} \omega_i W\sigma_i\,dz_i. \tag{4.10}$$

Merton (1971) assumes that $dy = 0$, i.e. all income is derived from capital gains and also that $\sigma_n = 0$, i.e. the $n$th asset is risk free. Therefore, letting $\alpha_n = r$,

$$dW = \sum_{1}^{n-1} \omega_i(\alpha_i - r)W\,dt + (rW - C)\,dt + \sum_{1}^{n-1} \omega_i\sigma_i W\,dz_i$$

becomes the budget constraint.

The problem of choosing optimal portfolio and consumption rules for the individual who lives $T$ years can now be formulated. It is the following

$$\max E_0 \left[ \int_{0}^{T} u(C(t), t)\,dt + B(W(T), T) \right] \tag{4.11}$$

subject to

$$W(t) \geqslant 0 \quad \text{for all } t \in [0, T] \text{ w.p.1}, \tag{4.12}$$

$$dW = \sum_{1}^{n-1} \omega_i(\alpha_i - r)W\,dt + (rW - C)\,dt + \sum_{1}^{n-1} \omega_i\sigma_i W\,dz_i. \tag{4.13}$$

Here $u$ and $B$ are strictly concave in $C$ and $W$.

To derive the optimal rules we use the technique of stochastic dynamic programming presented in Chapter 2. Define

$$J(W, P, t) = \max_{\{C,\omega\}} E_t \left[ \int_{t}^{T} u(C, s)\,ds + B(W(T), T) \right] \tag{4.14}$$

and also define

$$\phi(\omega, C, W, P, t) = u(C, t) + \mathscr{L}[J], \tag{4.15}$$

where

$$\mathscr{L} = \frac{\partial}{\partial t} + \left[ \sum_{1}^{n} \omega_i\alpha_i W - C \right] \frac{\partial}{\partial W} + \sum_{1}^{n} \alpha_i P_i \frac{\partial}{\partial P_i}$$

$$+ \tfrac{1}{2} \sum_1^n \sum_1^n \sigma_{ij}\,\omega_i \omega_j\, W^2\, \frac{\partial^2}{\partial W^2} + \tfrac{1}{2} \sum_1^n \sum_1^n P_i P_j \sigma_{ij}\, \frac{\partial^2}{\partial P_i \partial P_j}$$

$$+ \sum_1^n \sum_1^n P_i\, W \omega_j\, \sigma_{ij}\, \frac{\partial^2}{\partial P_i \partial W}\,. \tag{4.16}$$

Under the assumptions of the problem there exists a set of optimal rules $\omega^*$ and $C^*$ satisfying

$$0 = \max_{\{C,\omega\}} \{\phi\,(C, \omega;\, W, P, t)\}$$

$$= \phi\,(C^*, \omega^*;\, W, P, t) \quad \text{for } t \in [0, T]. \tag{4.17}$$

In the usual fashion of maximization under constraint we define the Lagrangian $L \equiv \phi + \lambda\,[1 - \Sigma_1^n\,\omega_i]$ and obtain the first-order conditions:

$$0 = L_C(C^*, \omega^*) = u_C(C^*, t) - J_W, \tag{4.18}$$

$$0 = L_{\omega_k}(C^*, \omega^*) = -\lambda + J_W\,\alpha_k\,W + J_{WW} \sum_1^n \sigma_{kj}\,\omega_j^*\,W^2$$

$$+ \sum_1^n J_{jW}\,\sigma_{kj}\,P_j W, \quad k = 1, \ldots, n, \tag{4.19}$$

$$0 = L_\lambda(C^*, \omega^*) = 1 - \sum_1^n \omega_i^*. \tag{4.20}$$

Merton solves for $C^*$ and $\omega^*$ and inserts the solutions to (4.17) and (4.15) to obtain a complicated partial differential equation. See Merton (1971, p. 383). If this partial differential equation is solved for $J$, then its solution, after appropriate substitutions, could yield the optimal consumption $C^*$ and portfolio rules $\omega^*$.

## 5. Hyperbolic absolute risk-aversion functions

In this application we specialize the analysis of Merton (1971) as presented in the previous section for the class of *hyperbolic absolute risk aversion* (HARA) utility functions of the form

$$u(C, t) = e^{-\rho t}\,v(C),$$

where

$$v(C) = \frac{1-\gamma}{\gamma} \left( \frac{\beta C}{1-\gamma} + \eta \right)^{\gamma}. \tag{5.1}$$

Note that the *absolute risk aversion* denoted by $A(C)$ and defined to be $A(C) = -(v''/v')$ is given by

$$A(C) = -\frac{v''}{v'} = \frac{1}{\dfrac{C}{1-\gamma} + \dfrac{\eta}{\beta}} > 0, \tag{5.2}$$

provided $\gamma \neq 1; \beta > 0; (\beta C/(1-\gamma)) + \eta > 0;$ and $\eta = 1$ if $\gamma = -\infty$.

This family of utility functions is rich because by suitable adjustments of the parameters one can have a utility function with absolute, or relative, risk aversion that is increasing, decreasing or constant.

Without loss of generality assume that there are two assets, one risk-free and one risky. The return of the risk-free asset is $r$ and the price of risky asset is lognormally distributed satisfying

$$\frac{\mathrm{d}P}{P} = \alpha \mathrm{d}t + \sigma \mathrm{d}z. \tag{5.3}$$

The optimality equation for this special case is

$$0 = \frac{(1-\gamma)^2}{\gamma} \, \mathrm{e}^{-\rho t} \left[ \frac{\mathrm{e}^{\rho t} J_W}{\beta} \right]^{\gamma/(\gamma-1)} + J_t$$

$$+ \left[ (1-\gamma) \frac{\eta}{\beta} + rW \right] J_W - \frac{J_W^2}{J_{WW}} \frac{(\alpha-r)^2}{2\alpha\sigma^2} \tag{5.4}$$

subject to $J(W, T) = 0$. For simplicity we assume that the individual has a zero bequest function.

A solution of the partial differential equation in (5.4) is given by Merton (1973c, p. 213)

$$J(W, t) = \frac{\delta}{\gamma} \beta^{\gamma} \mathrm{e}^{-\rho t} \left[ \frac{\delta \left\{ 1 - \exp\left[ -\left( \frac{\rho-\gamma\nu}{\delta} \right)(T-t) \right] \right\}}{\rho - \delta\nu} \right]^{\delta}$$

$$\left[ \frac{W}{\delta} + \frac{\eta}{\beta r} \left\{ 1 - \exp\left[ -r(T-t) \right] \right\} \right]^{\gamma}, \tag{5.5}$$

where $\delta = 1-\gamma$, $\nu = r + (\alpha-r)^2/2\delta\sigma^2$ and $\delta$ is assumed positive. If $\delta < 0$, and therefore $\gamma > 1$, the solution $J(W, t)$ in (5.5) will hold only when $0 \leqslant W(t) \leqslant (\gamma-1)\eta[1-\exp(-r(T-t))]/\beta r$.

The explicit solutions for optimal consumption and portfolio rules are given below:

$$C^*(t) = \frac{[\rho-\gamma\nu]\left[W(t) + \dfrac{\delta\eta}{\beta r}\{1-\exp[r(t-T)]\}\right]}{\delta\left\{1 - \exp\left[\dfrac{\rho-\gamma\nu}{\delta}(t-T)\right]\right\}} - \frac{\delta\eta}{\beta} \qquad (5.6)$$

and

$$\omega^*(t) = \frac{\alpha-r}{\delta\sigma^2} + \frac{1}{W(t)}\frac{\eta(\alpha-r)}{\beta r\sigma^2}\{1-\exp[r(t-T)]\}. \qquad (5.7)$$

The basic observation obtained from the above two equations is that the *demand functions are linear in wealth*. It can be shown that the HARA family is the only class of concave utility functions which imply linear solutions.

## 6. Portfolio jump processes

We follow Merton (1971) to discuss an application of the maximum principle for jump processes in a portfolio problem. Consider a two-asset case with a common stock whose price is log-normally distributed and a risky bond which pays an instantaneous rate of interest $r$ when not in default, but in the event of default the price becomes zero. The process which generates the bond's price is assumed to be given by

$$dP = rP dt - P dq, \qquad (6.1)$$

with $q(t)$ being an independent Poisson process. The new budget equation that replaces (4.13) is

$$dW = \{\omega W(\alpha-r) + rW - C\}dt + \omega\sigma W dz - (1-\omega)W dq. \qquad (6.2)$$

Note that (6.2) is an example of mixed Wiener and Poisson dynamics. An application of the generalized Itô formula and the maximum principle for jump processes in section 12 of Chapter 3 yields the optimality equation

$$0 = u(C^*, t) + J_t(W, t) + \lambda [J(\omega^* W, t) - J(W, t)]$$
$$+ J_W(W, t) [(\omega^* (\alpha - r) + r) W - C^*] + \tfrac{1}{2} J_{WW}(W, t) \sigma^2 \omega^{*2} W^2, \quad (6.3)$$

where the optimal assumption $C^*$ and portfolio $\omega^*$ rules are determined by the implicit equations

$$0 = u_C(C^*, t) - J_W(W, t) \tag{6.4}$$

and

$$0 = \lambda J_W(\omega^* W, t) + J_W(W, t)(\alpha - r) + J_{WW}(W, t) \sigma^2 \omega^* W. \tag{6.5}$$

There is one additional novelty in this Merton problem that is not present in the pure Brownian motion case. That is, for a HARA utility function you must not only conjecture a solution from $J(W, t) = g(t) W^a$ and solve for the exponent by the method of equating exponents, you must also conjecture a form for the demand function $\omega W = dW + e$, where the term $d$ is dependent of wealth. This is a natural conjecture to make for the form of the demand function for the risky asset given the *separation theorem* which says that for utility functions of the hyperbolic absolute risk-aversion class that the proportion of wealth held in the risky asset should be independent of the investor's wealth level and independent of his age. It is natural to conjecture the same sort of separation theorem for the Poisson case as well. At any rate, the philosophy is to try it and see if it works. Furthermore, conjecture that the term $e = 0$. This is natural because if wealth is zero there would be no demand for the risky asset.

Next equate exponents on wealth in the partial differential equation for the state valuation function, $J$, to solve for the unknown exponent on wealth. It will turn out to be the same exponent as that on consumption in the utility function. Cancel all terms involving wealth off the partial differential equation (6.3) for the state valuation function. That will give an ordinary differential equation for the unknown function $g(t)$. Furthermore, examine the necessary condition (6.5) to determine the unknown proportion $d$. Cancel off all terms involving wealth and all terms involving the unknown function $g(t)$ in the necessary condition (6.5) to get the relationship

$$d = \frac{\alpha - r}{\sigma^2 (1 - \gamma)} + \frac{\lambda}{\sigma^2 (1 - \gamma)} d^{\gamma - 1}. \tag{6.6}$$

This last is the same as Merton's (80') in Merton (1971, p. 397) with $\omega = d$. Thus, conjecturing a linear demand function for the risky asset worked.

The moral of this exercise is that for the Poisson case it was necessary to conjecture a form of the demand function for the risky asset in order to obtain a closed form solution. But the conjecture of the appropriate form for the demand function was motivated by the form of the demand function that was derived in the pure Brownian motion case. In other words, in the pure Brownian motion case when the utility function was hyperbolic absolute risk averse then the demand function for the risky asset was linear and the proportion of the investor's portfolio held in the risky asset turned out to be independent of the investor's wealth level and of his or her age. This independence is called the separation theorem. The name separation theorem derives from the fact that the consumption decision and the portfolio diversification decision turn out to be determined independently of each other in this particular case.

Merton's (1971) paper contains several other examples of closed form solution determinations for Poisson processes. Furthermore, it contains closed form solution determinations for more general processes as well.

## 7. The demand for index bonds

Consider, as Fischer (1975) does, a household with three assets in its portfolio: a real bond, a risky asset and a nominal bond. Assume that the portfolio can be adjusted instantaneously and costlessly. We also assume that the rate of inflation is stochastically describable by the process

$$\frac{\mathrm{d}P}{P} = \Pi \,\mathrm{d}t + s\,\mathrm{d}z. \tag{7.1}$$

The real bond pays a real return of $r_1$ and a nominal return of $r_1$ plus the realized rate of inflation. Note that

$$\frac{\mathrm{d}Q_1}{Q_1} = r_1 \,\mathrm{d}t + \frac{\mathrm{d}P}{P} = (r_1 + \Pi)\,\mathrm{d}t + s\,\mathrm{d}z \equiv R_1 \,\mathrm{d}t + s_1 \,\mathrm{d}z_1 \tag{7.2}$$

is the equation describing the nominal return on the index bond. The nominal return on equity is

$$\frac{\mathrm{d}Q_2}{Q_2} = R_2 \,\mathrm{d}t + s_2 \,\mathrm{d}z_2, \tag{7.3}$$

where $R_2$ is the expected nominal return on equity per unit of time and $s_2^2$ is the variance of the nominal return per unit time. Using the results of application 2, if we let $dQ_3/Q_3 = R_3\,dt$ describe the deterministic nominal return of the nominal bond, then the real return on the nominal bond is

$$\frac{d(Q_3/P)}{Q_3/P} = (R_3 - \Pi + s_1^2)\,dt - s_1\,dz \equiv r_3\,dt - s_1\,dz_1. \tag{7.4}$$

Let $\omega_1$, $\omega_2$ and $\omega_3$ be the proportions of the portfolio held in real bonds, equity and nominal bonds, respectively. Obviously, $\omega_1 + \omega_2 + \omega_3 = 1$. The flow budget constraint, giving the change in nominal wealth, $W$, is similar to (4.10):

$$dW = \sum_1^3 \omega_i R_i W dt - PC dt + \sum_1^2 \omega_i s_i W dz_i, \tag{7.5}$$

where $C$ is the rate of consumption. Uncertainty about the change in nominal wealth arises from holdings of real bonds and equity. Since $\omega_3 = 1 - \omega_1 - \omega_2$, we may rewrite eq. (7.5) as

$$dW = \sum_1^2 \omega_i (R_i - R_3) W dt + (R_3 W - PC) dt + \sum_1^2 \omega_i s_i W dz_i. \tag{7.6}$$

We are now in a position to formulate the household's choice problem:

$$\max_{\{C, \omega_i\}} \mathrm{E}_0 \int_0^\infty u[C(t), t]\,dt \tag{7.7}$$

subject to (7.6) and

$$W(0) = W_0,$$

where $u$ is a strictly concave utility function in $C$, and $\mathrm{E}_0$ is the expectation conditional on $P(0)$.

The first-order necessary conditions of optimality are:

$$0 = u_C(C, t) - PJ_W, \tag{7.8}$$

$$0 = J_W(R_1 - R_3) + J_{WW} W(\omega_1 s_1^2 + \omega_2 \rho s_1 s_2) + J_{WP} P s_1^2, \tag{7.9}$$

$$0 = J_W(R_2 - R_3) + J_{WW} W(\omega_1 s_1 s_2 \rho + \omega_2 s_2^2) + J_{WP} P \rho s_1 s_2, \tag{7.10}$$

where, as before,

$$J(W, P, t) = \max_{\{C, \omega_i\}} \mathrm{E}_t \int_t^\infty u(C, s)\,ds$$

and $\rho$ is the instantaneous coefficient of correlation between the Wiener processes $dz_1$ and $dz_2$ and $|\rho| < 1$. It is now possible to solve for asset demands from the two equations (7.9) and (7.10) and the fact that $\Sigma \omega_i = 1$, to obtain

$$\omega_1 = -\frac{J_W}{J_{WW}W}\left[\frac{R_1 - R_3}{s_1^2(1 - \rho^2)} - \frac{\rho(R_2 - R_3)}{s_1 s_2(1 - \rho^2)}\right] - \frac{J_{WP}P}{J_{WW}W}, \qquad (7.11)$$

$$\omega_2 = -\frac{J_W}{J_{WW}W}\left[\frac{R_2 - R_3}{s_2^2(1 - \rho^2)} - \frac{\rho(R_1 - R_3)}{s_1 s_2(1 - \rho^2)}\right], \qquad (7.12)$$

$$\omega_3 = \frac{J_W}{J_{WW}W}\left[\frac{(R_1 - R_3)(s_2 - \rho s_1)}{s_1^2 s_2(1 - \rho^2)} + \frac{(R_2 - R_3)(s_1 - \rho s_2)}{s_1 s_2^2(1 - \rho^2)} - 1\right]. \qquad (7.13)$$

From (7.11)–(7.13) Fischer (1975) studies the complete properties of the demand functions for the three assets. In particular consider the demand function for index bonds in (7.11). Observe that the coefficient $-J_W/J_{WW}W$ is the inverse of the degree of relative risk aversion of the household. If we make the simplifying assumption that $\rho = 0$ in (7.11), then (7.11) says that the demand for index bonds depends on (i) the degree of relative risk aversion, (ii) the difference between expected nominal returns on the two types of bonds, $R_1 - R_3$, and (iii) the variance of inflation, $s_1^2$. But how about the term $J_{WP}P/J_{WW}W$ in (7.11)? This term can be related to the degree of relative risk aversion as follows:

$$J_{WP}P/J_{WW}W = -J_W/J_{WW}W - 1. \qquad (7.14)$$

To obtain (7.14) differentiate (7.8) first with respect to $P$ to get

$$u_{CC}\frac{\partial C}{\partial P} = J_W + PJ_{WP} \qquad (7.15)$$

and with respect to $W$ to get

$$u_{CC}\frac{\partial C}{\partial W} = PJ_{WW}. \qquad (7.16)$$

Finally, note that since consumption is a function of real wealth

$$\frac{\partial C}{\partial P} = -\frac{W}{P}\frac{\partial C}{\partial W}. \qquad (7.17)$$

Combining $(7.15)-(7.17)$ we get $(7.14)$.

There are many other valuable insights that this analysis uncovers. We mention just one more. Consider the yield differentials in terms of real returns when $\omega_1 = 0$, i.e. when the household has no index bonds in its portfolio, given by

$$r_1 - r_3 = -\frac{J_{WW}W}{J_W}\left[(\rho s_1 s_2 - s_1^2)\omega_2 - s_1^2(1-\omega_2)\right]. \tag{7.18}$$

Suppose that $\omega_2 = 1$, i.e. the net quantities of real and nominal bonds are zero. If there is positive covariance between equity returns and inflation then from $(7.18)$ we obtain that $r_1 - r_3 > 0$, which means that index bonds will have to pay a higher return than the expected real yield on nominal bonds. In other words, if equity is a hedge against inflation, then index bonds do not command a premium over nominal bonds. Conversely, if equity is not a hedge against inflation then index bonds will command a premium over nominal bonds.

## 8. Term structure in an efficient market

Methods of stochastic calculus similar to the ones used by Black and Scholes (1973) and Merton (1971), which were presented in earlier sections, have been used by Vasicek (1977) to give an explicit characterization of the term structure of interest rates in an efficient market. Following Vasicek (1977) we describe this model below.

Let $P(t, s)$ denote the price at time $t$ of a discount bond maturing at time $s$, with $t \leqslant s$. The bond is assumed to have a maturity value, $P(s, s)$, of one unit, i.e.

$$P(s, s) = 1. \tag{8.1}$$

The yield to maturity, $R(t, T)$, is the internal rate of return at time $t$ on a bond with maturity date $s = t + T$, given by

$$R(t, T) = -\frac{1}{T}\log P(t, t + T), \quad T > 0. \tag{8.2}$$

From $(8.2)$ the rates $R(t, T)$ considered as a function of $T$ will be referred to as *the term structure at time t*. We use $(8.2)$ to define the *spot rate* as the instantaneous borrowing and lending rate, $r(t)$, given by

$$r(t) = R(t, 0) = \lim_{T \to 0} R(t, T). \tag{8.3}$$

It is assumed that $r(t)$ is a continuous function of time described by a stochastic differential equation of the form

$$dr = f(r, t)\, dt + \rho(r, t)\, dz, \tag{8.4}$$

where, as usual, $z(t)$ is a Wiener process with unit variance. It is assumed that the price of a discount bond, $P(t, s)$, is determined by the assessment, at time $t$, of the development of the spot rate process (8.4) over the term of the bond, and thus we write

$$P(t, s) \equiv P(t, s, r(t)). \tag{8.5}$$

Eq. (8.5) shows that the spot rate is the only state variable for the whole term structure, which implies that the instantaneous returns on bonds of different maturities are perfectly correlated. Finally, we assume that there are no transactions costs, information is available to all investors simultaneously, and that investors act rationally; that is to say, we assume that the market is efficient. This last assumption implies that no profitable riskless arbitrage is possible.

From eqs. (8.4) and (8.5) by using Itô's lemma we obtain the stochastic differential equation

$$dP = P\mu(t, s, r(t))\, dt - P\sigma(t, s, r(t))\, dz, \tag{8.6}$$

which describes the bond price changes. In (8.6) the functions $\mu$ and $\sigma$ are defined as follows:

$$\mu(t, s, r) \equiv \frac{1}{P(t, s, r)} \left[ \frac{\partial}{\partial t} + f \frac{\partial}{\partial r} + \tfrac{1}{2} \rho^2 \frac{\partial^2}{\partial r^2} \right] P(t, s, r), \tag{8.7}$$

$$\sigma(t, s, r) \equiv - \frac{1}{P(t, s, r)} \rho \frac{\partial}{\partial r} P(t, s, r). \tag{8.8}$$

Consider now the quantity $q(t, r(t))$ given by

$$q(t, r) = \frac{\mu(t, s, r) - r}{\sigma(t, s, r)}, \quad t \leqslant s, \tag{8.9}$$

which is called the *market price of risk* and which specifies the increase in expected instantaneous rate of return on a bond per an additional unit of risk. Substitute the expressions for $\mu$ and $\sigma$ from (8.7) and (8.8) into (8.9), make the necessary rearrangements and obtain the *term structure equation* given by

$$\frac{\partial P}{\partial t} + (f + \rho q)\,\frac{\partial P}{\partial r} + \tfrac{1}{2}\,\rho^2\,\frac{\partial^2 P}{\partial r^2} - rP = 0. \tag{8.10}$$

Observe that (8.10) is a partial differential equation whose solution $P$ may be obtained once the spot rate process $r(t)$ and the market price of risk $q(t, r)$ are specified. The boundary condition of (8.10) is

$$P(s, s, r) = 1. \tag{8.11}$$

Knowing $P(t, s, r)$ as a solution of (8.10) subject to (8.11) allows us to obtain the term structure from (8.2).

Vasicek (1977) uses techniques presented in Friedman (1975) to write a representation for the bond price as a solution to the term structure equation (8.10) subject to (8.11), given by

$$P(t, s) = \mathrm{E}_t \exp \left( - \int_t^s r(u)\,\mathrm{d}u - \tfrac{1}{2} \int_t^s q^2(u, r(u))\,\mathrm{d}u \right. $$
$$\left. + \int_s^t q(u, r(u))\,\mathrm{d}z(u) \right), \quad t \leqslant s. \tag{8.12}$$

To obtain some economic insight in eq. (8.12), construct a portfolio consisting of a bond whose maturity approaches infinity, called a *long bond*, and lending or borrowing at the spot rate, with proportions $\lambda(t)$ and $1 - \lambda(t)$, respectively. Here we define $\lambda(t)$ as

$$\lambda(t) = \frac{\mu(t, \infty) - r(t)}{\sigma^2(t, \infty)}, \tag{8.13}$$

i.e.

$$\lambda(t)\,\sigma(t, \infty) = q(t, r(t)). \tag{8.14}$$

The price $Q(t)$ of such a portfolio satisfies the equation

$$\mathrm{d}Q = \lambda Q(\mu(t, \infty)\,\mathrm{d}t - \sigma(t, \infty)\,\mathrm{d}z) + (1 - \lambda)Qr\,\mathrm{d}t. \tag{8.15}$$

Eq. (8.15) can be integrated by evaluating the differential of $\log Q$ and using (8.14) to yield

$$\begin{aligned}
\mathrm{d}(\log Q) &= \lambda\mu(t, \infty)\,\mathrm{d}t - \lambda\sigma(t, \infty)\,\mathrm{d}z + (1 - \lambda)r\,\mathrm{d}t - \tfrac{1}{2}\,\lambda^2\sigma^2(t, \infty)\,\mathrm{d}t \\
&= r\,\mathrm{d}t + \tfrac{1}{2}\,q^2\,\mathrm{d}t - q\,\mathrm{d}z,
\end{aligned}$$

from which we obtain

$$\frac{Q(t)}{Q(s)} = \exp\left(-\int_t^s r(u)\,du - \tfrac{1}{2}\int_t^s q^2(u,r(u))\,du + \int_t^s q(u,r(u))\,dz\right).$$

Using this last equation we may rewrite (8.12) as

$$P(t,s) = E_t\,\frac{Q(t)}{Q(s)}, \quad t \leqslant s,$$

which means that the price of any bond measured in units of the value of a portfolio $Q$ follows a martingale,

$$\frac{P(t,s)}{Q(t)} = E_t\,\frac{P(u,s)}{Q(u)} \quad \text{for } t \leqslant u \leqslant s.$$

Therefore, we conclude that if the bond price at time $t$ is a certain fraction of the value of the portfolio $Q$, then the same will hold in the future.

## 9.  Market risk adjustment in project valuation

In this application we follow Constantinides (1978) to develop a rule which reduces the problem of valuation under market risk to a problem of valuation when the price of risk is zero.

Let $V(x, t)$ denote the market value of a project where such a project can be chosen to be an investment, an option, a claim on a firm, etc. The project is assumed to generate a stream of cash flows and $V(x, t)$ represents the time and risk adjusted value of these cash flows. The market value function $V(x, t)$ is specified by the state variable $x$ and time $t$, where we assume that $x$ changes according to the stochastic differential equation

$$dx = \mu(x,t)\,dt + \sigma(x,t)\,dz$$

$$= \mu\,dt + \sigma\,dz, \tag{9.1}$$

where for notational convenience we write $\mu \equiv \mu(x, t)$ and $\sigma \equiv \sigma(x, t)$. In (9.1), as earlier, $z$ is a Wiener process with unit variance. The cash return generated by the project during the time interval $(t, t + dt)$ is assumed to be nonstochastic given by $c\,dt$, where $c = c(x, t)$.

Consider now the return on the project in the time interval $(t, t + \mathrm{d}t)$; it is the sum of capital appreciation $\mathrm{d}V(x, t)$ and cash return $c\,\mathrm{d}t$. To obtain $\mathrm{d}V(x, t)$ we use Itô's lemma to conclude

$$\mathrm{d}V(x, t) = \left( V_t + \mu V_x + \frac{\sigma^2}{2} V_{xx} \right) \mathrm{d}t + \sigma V_x \mathrm{d}z, \qquad (9.2)$$

assuming $V(x, t)$ is twice continuously differentiable with respect to $x$ and once continuously differentiable with respect to $t$. Using (9.2) we may write the rate of return on the project as

$$\frac{\mathrm{d}V(x, t) + c\,\mathrm{d}t}{V(x, t)} = \frac{1}{V} \left( c + V_t + \mu V_x + \frac{\sigma^2}{2} V_{xx} \right) \mathrm{d}t + \frac{\sigma V_x}{V} \mathrm{d}z. \qquad (9.3)$$

From (9.3) we write the expected value per unit of time, $\alpha_P$, and the covariance with the market per unit of time, $\sigma_{PM}$, as

$$\alpha_P = \left( c + V_t + \mu V_x + \frac{\sigma^2}{2} V_{xx} \right) \Big/ V \qquad (9.4)$$

and

$$\sigma_{PM} = \rho \sigma_M \sigma V_x / V, \qquad (9.5)$$

where $\rho = \rho(x, t)$ is the instantaneous correlation coefficient between $\mathrm{d}z$ and the market return. In (9.5) $\sigma_M$ denotes the positive square root of the variance of the market portfolio.

At this point Constantinides (1978) uses a result stated in Merton (1973b) and proved in Merton (1972). Before we state this result we indicate the notation. Let $\alpha_i$ denote the expected rate of return of security $i$ per unit of time and let $\sigma_{ij}$ denote the covariance of returns per unit of time. The riskless borrowing–lending rate is denoted by $r$ and the subscript M refers to the market portfolio. The result is this: under certain assumptions, which lead to Merton's (1973) intertemporal capital asset pricing model, the equilibrium security returns must satisfy the equation

$$\alpha_i - r = \beta_i (\alpha_M - r), \qquad (9.6)$$

where $\beta_i \equiv \sigma_{iM} / \sigma_M^2$. Note that (9.6) is the continuous time analogue of the security market line of the classical capital asset pricing model. For example, see Francis and Archer (1979, p. 158). For our purposes we rewrite (9.6) as

$$\alpha_i - r = \lambda \sigma_{iM}/\sigma_M, \tag{9.7}$$

where $\lambda \equiv (\alpha_M - r)/\sigma_M$. After this disgression we return to our model. Substitute (9.4) and (9.5) in (9.7) to get

$$c - rV + V_t + (\mu - \lambda\rho\sigma)V_x + \frac{\sigma^2}{2} V_{xx} = 0, \tag{9.8}$$

which is a partial differential equation for which, under certain boundary conditions, we may obtain a solution $V(x, t)$ giving the market value of the project. To proceed a step further with the analysis let

$$\mu^* \equiv \mu^*(x, t) \equiv \mu(x, t) - \lambda\rho(x, t)\sigma(x, t) \tag{9.9}$$

and rewrite (9.8) as

$$c - rV + V_t + \mu^* V_x + \frac{\sigma^2}{2} V_{xx} = 0. \tag{9.10}$$

Next, we want to compare (9.10) with a similar equation describing the value of the project in a capital market which pays no premium for market risk, i.e. $\alpha_M - r = 0$. Set $\alpha_M - r = 0$ in (9.6) and use (9.4) and (9.5) to obtain

$$c - r\hat{V} + \hat{V}_t + \mu\hat{V}_x + \frac{\sigma^2}{2} \hat{V}_{xx} = 0, \tag{9.11}$$

where $\hat{V}(x, t)$ denotes the value of the project in a capital market which pays no premium for risk. The boundary conditions of (9.11) are identical to those imposed on $V$ in (9.10) because these conditions are independent of the market risk premium.

We conclude this application by comparing eqs. (9.10) and (9.11). A comparison shows that $V(x, t)$ may be considered as the market value of the project in a capital market which pays no premium for market risk provided $\mu^*(x, t)$ replaces $\mu(x, t)$. This observation leads Constantinides (1978) to suggest the following rule for determining the market value of a project. First, replace the drift $\mu(x, t)$ by $\mu^*(x, t)$ as in (9.9). Secondly, discount expected cash flows at the riskless rate.

## 10. Demand for cash balances

The various models that have been developed to explain the demand for money can be classified into two categories. Some models, such as Baumol (1952) and

Tobin (1956), assume that transactions occur in a steady stream which is perfectly foreseen, while other models, such as Olivera (1971) and Miller and Orr (1966), assume that net cash flows are completely random. In this application we follow Frenkel and Jovanovic (1980) to incorporate various aspects of these two categories.

Assume that changes in money holdings follow an Itô stochastic differential equation:

$$dM(t) = -\mu dt + \sigma dz(t),$$

$$M(0) = M_0, \mu \geqslant 0,$$

(10.1)

where $z$ is a Wiener process with unit variance, $M_0$ is the optimal initial money holdings, and $\mu$ is the deterministic part of net expenditures. From (10.1) upon integration we obtain

$$M(t) = M_0 - \mu t + \sigma z(t),$$

(10.2)

where $M(t)$ is normally distributed with mean $M_0 - \mu t$ and variance $\sigma^2 t$. The optimal level of money holdings is determined by minimizing the cost of financial management. We distinguish two such sources of cost: first, forgone earnings which depend on the interest rate $r$ and on the money holdings $M(t)$, and secondly, the cost of adjustments which depends on the frequency of adjustment and on the fixed cost $C$ per adjustment. It is assumed that an adjustment of the money stock is necessary whenever money holdings reach a lower bound. This lower bound is assumed to be zero. Note that costs from each source are random because at each period $t$ money holdings as described by (10.2) are random. Therefore the optimal size of cash balances is determined by minimizing expected cost.

It is analytically convenient to separate the expected cost into two parts: first, the expected cost that was incurred prior to the period of the first adjustment and, secondly, the expected cost that is incurred thereafter. The period when holdings reach zero and adjustment is necessary is random. For the analysis of the first part we write at period $t$ the instantaneous forgone earnings as $rM(t)$ and their present value as $rM(t)e^{-rt}$. Denote by $h(M, t \mid M_0, 0)$ the probability that money holdings $M(t)$ which at period $t = 0$ were at the optimal level $M_0$, have not reached zero prior to period $t$ at which time money holdings are $M$. Thus, the present value of expected forgone earnings up to the first adjustment may be written as

$$J_1(M_0) = r \int\limits_0^\infty e^{-rt} \left[ \int\limits_{M=0}^\infty M h(M,t \mid M_0,0) \, dM \right] dt. \qquad (10.3)$$

Frenkel and Jovanovic (1980) show that (10.3) may be simplified to be written as

$$J_1(M_0) = M_0 - (1-\alpha)\frac{\mu}{r}, \qquad (10.4)$$

where

$$\alpha = \exp\left\{ -\frac{M_0}{\sigma^2} [(\mu^2 + 2r\sigma^2)^{1/2} - \mu] \right\}. \qquad (10.5)$$

Putting aside (10.4) for a moment we analyze next the expected cost which is incurred following the first adjustment. Denote by $G(M_0)$ the present value of total expected cost and by $f(M_0, t)$ the probability that cash holdings reach zero at time $t$, having been the optimal level $M_0$ at $t = 0$. Thus, the present value of the expected cost following the first adjustment is given by

$$J_2(M_0) = \int\limits_0^\infty e^{-rt} [C + G(M_0)] f(M_0, t) \, dt. \qquad (10.6)$$

Note that $G(M_0)$ excludes the current fixed cost of adjustment which explains why $C$ is added to $G(M_0)$ in (10.6). Frenkel and Jovanovic (1980) argue that (10.6) can be simplified to be written as

$$J_2(M_0) = \alpha [C + G(M_0)]. \qquad (10.7)$$

Using (10.4) and (10.7) we may write the present value of total expected cost as

$$G(M_0) = M_0 - (1-\alpha)\frac{\mu}{r} + \alpha[C + G(M_0)],$$

which upon rearrangement reduces to

$$G(M_0) = \frac{M_0 + \alpha C}{1-\alpha} - \frac{\mu}{r}. \qquad (10.8)$$

Minimizing the expected cost of financial management $G(M_0)$ with respect to the optimal level of cash balances $M_0$ we obtain the necessary condition

$$(1-\alpha) + (M_0 + C)\frac{\partial \alpha}{\partial M_0} = 0. \qquad (10.9)$$

Expand (10.9) in Taylor series around $M_0$ and after terms of third and higher order are ignored solve for $M_0$ to obtain

$$M_0 = \left( \frac{2C\sigma^2}{(\mu^2 + 2r\sigma^2)^{1/2} - \mu} \right)^{1/2}. \tag{10.10}$$

Eq. (10.10) satisfies the homogeneity postulate with a rise of a given proportion in $\sigma$, $C$ and $\mu$ resulting in an equiproportional rise in $M_0$.

Two special cases are of interest. In the first case if we assume $\sigma^2 = 0$, we obtain by expanding the bracketed term in the denominator of eq. (10.10) around $\sigma^2 = 0$:

$$(\mu^2 + 2r\sigma^2)^{1/2} = \mu + \frac{1}{2} \frac{2r\sigma^2}{\mu} + \mathcal{O}(\sigma^4). \tag{10.11}$$

In (10.11), $\mathcal{O}(\sigma^4)$ denotes terms of order $\sigma^4$ or higher. Putting (10.11) in (10.10) we get

$$M_0 = \left( \frac{2C\sigma^2}{(r\sigma^2/\mu) + \mathcal{O}(\sigma^4)} \right)^{1/2}. \tag{10.12}$$

Finally, taking the limit in (10.12) as $\sigma^2 \to 0$ we conclude that

$$\lim_{\sigma^2 \to 0} M_0 = \left( \frac{2C\mu}{r} \right)^{1/2},$$

which is the result obtained by the Baumol–Tobin formulation of optimal transaction balances.

In the second special case we let $\mu = 0$ and evaluate (10.10) to get

$$M_0 = \left( \frac{2C\sigma^2}{(2r\sigma^2)^{1/2}} \right)^{1/2}, \tag{10.13}$$

which is similar to the results of the Miller–Orr model.

Thus, using stochastic calculus techniques and following the Frenkel–Jovanovic (1980) model an extension of some of the existing models of the demand for money has been achieved. In such an extension the implications of the two special cases are clear. In the first case, corresponding to the Baumol–Tobin framework, it is assumed that the process governing net disbursements is deterministic, i.e. $\sigma^2 = 0$. In the second case, corresponding to the Miller–Orr framework, it is assumed that the process governing net disbursements is stochastic without any drift, i.e. $\mu = 0$.

## 11. The price of systematic risk

In 1976 Steve Ross produced a theory of capital asset pricing that showed that the assumption that all systematic risk free portfolios earn the risk free rate of return plus the assumption that asset returns are generated by a $K$-factor model leads to the existence of prices $\lambda_0$, $\lambda_1$, $\lambda_2$, ..., $\lambda_K$ on mean returns and on each of the $K$-factors (Ross, 1976). These prices satisfied the property that expected returns $E\tilde{Z}_i \equiv a_i$ on each asset $i$ was a linear function of the standard deviation of the returns on asset $i$ with respect to each factor $k$, i.e.

$$a_i = \lambda_0 + \sum_{k=1}^{K} \lambda_k b_{ki}, \quad i = 1, 2, ..., N, \tag{11.1}$$

where the original model of asset returns is given by

$$\tilde{Z}_i = a_i + \sum_{k=1}^{K} b_{ki}\tilde{\delta}_k + \tilde{\epsilon}_i, \quad i = 1, 2, ..., N. \tag{11.2}$$

Here $\tilde{Z}_i$ denotes random *ex ante* anticipated returns from holding the asset one unit of time, $\tilde{\delta}_k$ is systematic risk emanating from factor $k$, $\tilde{\epsilon}_i$ is unsystematic risk specific to asset $i$, and $a_i$ and $b_{ki}$ are constants. Assume that the means of $\tilde{\delta}_k$ and $\tilde{\epsilon}_i$ are zero for each $k$ and $i$, that $\tilde{\epsilon}_1$, ..., $\tilde{\epsilon}_N$ are independent, and that $\tilde{\delta}_k$ and $\tilde{\epsilon}_i$ are uncorrelated random variables with finite variances for each $k$ and $i$.

Ross proved that $\lambda_0$, $\lambda_1$, ..., $\lambda_K$ exist that satisfy (11.1) by forming portfolios $\eta \in R^N$ such that

$$\sum_{i=1}^{N} \eta_i = 0, \tag{11.3}$$

and constructing the $\eta_i$ such that the coefficients of each $\tilde{\delta}_k$ in the portfolio returns,

$$\sum_{i=1}^{N} \eta_i \tilde{Z}_i = \sum_{i=1}^{N} \eta_i \left[ a_i + \sum_{k=1}^{K} b_{ki}\tilde{\delta}_k + \tilde{\epsilon}_i \right] =$$

$$= \sum_{i=1}^{N} \eta_i a_i + \sum_{k=1}^{K} \left( \sum_{i=1}^{N} b_{ki}\eta_i \right) \tilde{\delta}_k + \sum_{i=1}^{N} \eta_i \tilde{\epsilon}_i, \tag{11.4}$$

are zero, and requiring that

$$\sum_{i=1}^{N} \eta_i a_i = 0 \tag{11.5}$$

for all such systematic risk free zero wealth portfolios.

Here (11.3) corresponds to the zero wealth condition. The condition

$$0 = \sum_{i=1}^{N} b_{ki}\eta_i, \quad k = 1, 2, ..., K, \tag{11.6}$$

corresponds to the systematic risk-free condition. Actually, Ross did not require that (11.3) hold for all zero wealth systematic risk-free portfolios but only for those that are well diversified in the sense that the $\eta_i$ are of comparable size so that he could use the assumption of independence of $\tilde{\epsilon}_1, ..., \tilde{\epsilon}_N$ to argue that the random variable $\Sigma_{i=1}^{N} \eta_i \tilde{\epsilon}_i$ was small and hence bears a small price in a world of investors who would pay a positive price only for the avoidance of risks that could not be diversified away.

Out of this analysis Ross argues that the condition: for all $\eta \in R^N$,

$$\sum_{i=1}^{N} \eta_i = 0; \quad \sum_{i=1}^{N} \eta_i b_{ki} = 0, \quad k = 1, 2, ..., K \tag{11.7}$$

implies that in *equilibrium*

$$\sum_{i=1}^{N} \eta_i a_i = 0, \tag{11.8}$$

should hold.

All that (11.7) and (11.8) say is that at zero wealth, zero systematic risk portfolios should earn a zero mean rate of return. The condition in (11.7) and (11.8) is economically compelling because in its absence rather obvious arbitrage opportunities appear to exist.

Whatever the case, (11.7) and (11.8) imply that there exists $\lambda_0, \lambda_1, \lambda_2, ..., \lambda_K$ such that (11.1) holds and the proof is just simple linear algebra. Notice that Ross made no assumptions about mean variance investor utility functions or normal distributions of asset returns common to the usual Sharpe–Lintner type of asset pricing theories which are so standard in the finance literature.

However, Ross's model, like the standard capital asset pricing models in finance, does not link the asset returns to underlying sources of uncertainty. Application 15 of Chapter 3 will be used as a module in the construction of an intertemporal general equilibrium asset pricing model where relationships of the form (11.2) are determined within the model and hence the $\lambda_0, \lambda_1, ..., \lambda_K$ will be determined within the model as well. Such a model of asset price determination will preserve the beauty and empirical tractability of the Ross–Sharpe–Lintner formulation but at the same time will give us a context where we can ask general equilibrium questions such as: What is the impact of an increase of the progressivity of the income tax on the demand for and supply of risky assets and the $\lambda_0, \lambda_1, ..., \lambda_K$?

Let us get on with relating the growth model of section 15 of Chapter 3 to (11.1). For simplicity assume all processes $i$ are active, i.e. (15.11) of Chapter 3 holds with equality. We record (15.11) here for convenience

$$u'(c_t) = \beta E_t \{u'(c_{t+1}) f_i'(x_{it}, r_t)\}. \tag{11.9}$$

Now (11.2) is a special hypothesis about asset returns. What kind of hypothesis about *technological* uncertainty corresponds to (11.2)? Well, as an example, put for each $i = 1, 2, ..., N$

$$f_i(x_{it}, r_t) \equiv (A_{it}^0 + A_{it}^1 \bar{\delta}_{1t} + A_{it}^2 \bar{\delta}_{2t} + ... + A_{it}^K \bar{\delta}_{Kt}) f_i(x_{it})$$

$$\equiv r_{it} f_i(x_{it}), \tag{11.10}$$

where $A_{it}^k$ are constants and $\{\bar{\delta}_{kt}\}_{t=1}^\infty$ are independent and identically distributed random variables for each $k$. For each $t$ the mean of $\bar{\delta}_{kt}$ is zero, the variance is finite, and $\bar{\delta}_{st}$ is independent of $\bar{\delta}_{kt}$ for each $s$, $k$ and $t$. Furthermore, assume that $f(\cdot)$ is concave, increasing, twice differentiable, $f'(0) = \infty$, $f'(\infty) = 0$ and that there is a bound $\epsilon_0$ such that $r_t > \epsilon_0$ with probability 1 for all $t$. These assumptions are stronger than necessary but will enable us to avoid concern with technical tangentialities. Define, for all $t$, $\bar{\delta}_{0t} \equiv 1$, so that we may sum from $k = 0$ to $K$ in (11.11) below.

Insert (11.10) into (11.9) to get for all $t$, $k$ and $i$

$$u'(c_t) = \beta E_t \left\{ u'(c_{t+1}) \left( \sum_{k=0}^K A_{it}^k \bar{\delta}_{kt} \right) f_i'(x_{it}) \right\}$$

$$= \sum_{k=0}^K ([A_{it}^k f_i'(x_{it})] E_t \{\beta u'(c_{t+1}) \bar{\delta}_{kt}\}). \tag{11.11}$$

Now set (11.11) aside for a moment and look at the marginal benefit of saving one unit of capital and assigning it to process $i$ at the beginning of period $t$. At the end of period $t$, $r_t$ is revealed and extra output

$$\bar{Z}_{it} \equiv A_{it}^0 f'(x_{it}) + \sum_{k=1}^K A_{it}^k f_i'(x_{it}) \bar{\delta}_{kt}, \tag{11.12}$$

emerges. Putting

$$a_i \equiv A_{it}^0 f_i'(x_{it}); \quad b_{ki} \equiv A_{it}^k f_i'(x_{it}); \quad \bar{\delta}_{kt} = \bar{\delta}_k, \tag{11.13}$$

eq. (11.12) is identical with Ross's (11.2) with $\bar{\epsilon}_i \equiv 0$. We proceed now to generate the analogue to (11.1). Turn to (11.11) and rewrite it, using (11.13), as

$$u'(c_t) = \sum_{k=1}^{K} b_{ki} \, E_t \{\beta u'(c_{t+1}) \, \tilde{\delta}_{kt}\} + a_i \, E_t \{\beta u'(c_{t+1})\}. \tag{11.14}$$

Hence,

$$a_i = \frac{u'(c_t)}{\beta E_t \{u'(c_{t+1})\}} - \sum_{k=1}^{K} b_{ki} \left( \frac{E_t \{u'(c_{t+1}) \, \tilde{\delta}_{kt}\}}{E_t \{u'(c_{t+1})\}} \right) \tag{11.15}$$

so that $\lambda_0, \lambda_1, ..., \lambda_K$, defined by

$$\lambda_0 \equiv \frac{u'(c_t)}{\beta E_t \{u'(c_{t+1})\}} \; ; \quad \lambda_k \equiv - \frac{E_t \{u'(c_{t+1}) \, \tilde{\delta}_{kt}\}}{E_t \{u'(c_{t+1})\}} \;, \tag{11.16}$$

yields

$$a_i = \lambda_0 + \sum_{k=1}^{K} b_{ki} \lambda_k. \tag{11.17}$$

Here the subscript $t$ is dropped to ease the notation. These results are extremely suggestive and show that the model studied may be quite rich in economic content. Although the model is normative, in the next section we shall turn it into an equilibrium asset pricing model so that the $\lambda_k$ become equilibrium risk prices. Let us explore the economic meanings of (11.16) in some detail.

Suppose that $K = 1$ and that there is a risk-free asset $N$ in the sense that

$$b_{N1} \equiv A_{Nt}^1 f'(x_{Nt}) = 0, \tag{11.18}$$

i.e.

$$A_{Nt}^1 = 0. \tag{11.19}$$

Then by (11.19) we obtain

$$a_N = \lambda_0; \quad a_i = a_N + b_{1i}\lambda_1, \tag{11.20}$$

so that for all $i, j \neq N$

$$(a_i - a_N)/b_{1i} = (a_j - a_N)/b_{1j}. \tag{11.21}$$

The second part of eq. (11.20) corresponds to the security market line which says that expected return and risk are linearly related in a one-factor model. Eq. (11.21) corresponds to the usual Sharpe–Lintner–Mossin capital asset pricing model result that in equilibrium the excess return per unit of risk must be equated across all assets.

The economic interpretation of $\lambda_0$ given in (11.16) is well known and needs no explanation here. Look at the formula for $\lambda_k$. The covariance of the marginal utility of consumption at time $t + 1$ with the zero mean finite variance shock $\tilde{\delta}_{kt}$ appears in the numerator. Since output increases when $\tilde{\delta}_{kt}$ increases and since $c_{t+1} = g(y_{t+1})$ does not decrease when $y_{t+1}$ increases, therefore this covariance is likely to be negative so that the sign of $\lambda_k$ is positive. We will look into the determinants of the magnitudes of $\lambda_0, \lambda_1, ..., \lambda_K$ in more detail later. Let us show how this model may be helpful in the empirical problem in estimating the $\lambda_0, \lambda_1, ..., \lambda_K$ from time series data.

First, how is one to close Ross's model (11.2) since the $\tilde{Z}_i$ are subjective? The most natural way to close the model in markets as well organized as U.S. securities markets would seem to be rational expectations: the subjective distribution of $\tilde{Z}_i$ is equal to the actual or objective distribution of $\tilde{Z}_i$. We shall show that our asset pricing model under rational expectations, which is developed below, generates the same solution as the normative model. Hence, the convergence theorem implies that $\{x_t, c_t, x_{1t}, x_{2t}, ..., x_{Nt}\}_{t=1}^{\infty}$ converges to a stationary stochastic process.

Hence, the *mean ergodic theorem* which says very loosely that the *time* average of any function $G$ of a stationary stochastic process equals the *average* of $G$ over the stationary distribution of that process, allows us to apply time series methods developed for stationary stochastic processes to estimate $\lambda_0, \lambda_1, ..., \lambda_K$. As is well known, time series data are useful for the estimation of $\lambda_0, \lambda_1, ..., \lambda_K$.

Let us next turn to the development of the asset pricing model.

## 12. An asset pricing model

In this section we reinterpret the model of section 15 of Chapter 3 and add to it a market for claims to pure rents so that it describes the evolution of equilibrium asset prices. In this way we will not only generate a general equilibrium context in which to discuss the martingale property of capital asset prices, but also the model will contain a nontrivial investment decision, a nontrivial market for claims to pure rents, i.e. a stock market, as well as a market for the pricing of the physical capital stock.

We believe that there is considerable benefit in showing how to turn optimal growth models into asset pricing models. This is so because there is a large literature on stochastic growth models which may be carried over to the asset pricing problem with little effort.

We develop an asset pricing model much like that of Lucas (1978). The model contains one representative consumer whose preferences are identical to the planner's preferences given in (15.1) of Chapter 3. The model contains $N$ different firms that rent capital from the consumption side at rate $R_{t+1}$ at each date so as to maximize

$$\pi_{i,t+1} \equiv f_i(x_{it}, r_t) - R_{i,t+1} x_{it}. \tag{12.1}$$

Notice that it is assumed that each firm $i$ makes its decision to hire $x_{it}$ after $r_t$ is revealed. Here $R_{i,t+1}$ denotes the rental rate on capital prevailing in industry $i$ at date $t+1$. It is to be determined within the model.

The model will introduce a stock market in such a way that the real quantity side of the model is the same as that of the growth model in equilibrium. Our model is closed under the assumption of rational expectations. The quantity side of the model is essentially an Arrow–Debreu model, as is the model of Lucas (1978). We introduce securities markets in such a way that there is a security for each state of the world. However, there is a separate market where claims to the rents in (12.1) are competitively traded. Recall that in an Arrow–Debreu economy the rents are redistributed in a lump-sum fashion.

The model is in the spirit of Lucas' (1978) model where each firm $i$ has outstanding one perfectly divisible equity share. Ownership of $a\%$ of the equity shares in firm $i$ at date $t$ entitles one to $a\%$ of profits of the firm $i$ at date $t+1$. Equilibrium asset prices and equilibrium consumption, capital and output are determined by optimization under the hypothesis of rational expectations much as in Lucas (1978). Let us describe the model. The representative consumer solves

$$\text{maximize } E_1 \sum_{t=1}^{\infty} \beta^{t-1} u(c_t) \tag{12.2}$$

subject to

$$c_t + x_t + P_t \cdot Z_t \leqslant \pi_t \cdot Z_{t-1} + P_t \cdot Z_{t-1} + \sum_{i=1}^{N} R_{it} x_{i,t-1} \equiv y_t, \tag{12.3}$$

$$c_t \geqslant 0; \quad x_t \geqslant 0; \quad Z_t \geqslant 0; \quad x_{it} \geqslant 0; \quad i = 1, 2, ..., N, \quad \text{all } t, \tag{12.4}$$

$$c_1 + x_1 + P_1 \cdot Z_0 = \pi_1 \cdot Z_0 + P_1 \cdot Z_0 + \sum_{i=1}^{N} R_{i1} x_{i0} \equiv y_1, \tag{12.5}$$

$$Z_0 \equiv 1; \quad R_{i1} \equiv f_i'(x_{i0}, r_0); \quad \Pi_{i1} \equiv f_i(x_{i0}, r_0) - f_i'(x_{i0}, r_0) x_{i0};$$

$$x_0, \{x_{i0}\}_{i=1}^{N} \text{ given.}$$

Note that $c_t$, $x_t$, $P_{it}$, $Z_{it}$, $\pi_{it}$ and $R_{it}$, all assumed measurable $\mathcal{F}_t$, denote consumption at date $t$, total capital stock owned at date $t$ by the consumer, price of one share of firm $i$ at date $t$, number of shares of firm $i$ owned by the individual at date $t$, profits of firm $i$ at date $t$, and rental factor, i.e. $R_{it}$ is principle plus interest obtained on a unit of capital leased to firm $i$. Here a dot denotes scalar product.

Firm $i$ is assumed to hire $x_{it}$ so as to maximize (12.1). The consumer is assumed to lease capital $x_{it}$ at date $t$ to firm $i$ before $r_t$ is revealed. Hence $R_{i,t+1}$ is uncertain at date $t$. The consumer, in order to solve his problem at date 1, must form expectations on $\{P_{it}\}_{t=1}^{\infty}$, $\{R_{it}\}_{t=1}^{\infty}$, $\{\pi_t\}_{t=1}^{\infty}$ and maximize (12.2) subject to (12.3) and (12.4). In this way notional demands for consumption goods and equities as well as notional supplies of capital stocks and capital services to each of the $N$ firms are drawn up by the consumer side of the economy. Similarly for the firm side. We close the model with this definition: The collection of stochastic processes $\mathcal{R} \equiv (\{\bar{P}_{it}\}_{t=1}^{\infty}, \{\bar{R}_{it}\}_{t=1}^{\infty}, \{\bar{\pi}_{it}\}_{t=1}^{\infty}, \{\bar{x}_{it}\}_{t=1}^{\infty}, \{\bar{Z}_{it}\}_{t=1}^{\infty}, i = 1, 2, ..., N, \{\bar{c}_t\}_{t=1}^{\infty}, \{\bar{x}_t\}_{t=1}^{\infty})$ is a *rational expectations equilibrium* (R.E.E.) if facing $\mathcal{P} \equiv (\{\bar{P}_{it}\}_{t=1}^{\infty}, \{\bar{R}_{it}\}_{t=1}^{\infty}, \{\bar{\pi}_{it}\}_{t=1}^{\infty})$ the consumer chooses

$$x_t = \bar{x}_t; \quad x_{it} = \bar{x}_{it}; \quad c_t = \bar{c}_t; \quad Z_{it} = \bar{Z}_{it}, \quad \text{a.e.,} \tag{12.6}$$

and the $i$th firm chooses

$$x_{it} = \bar{x}_{it} \tag{12.7}$$

and furthermore

(asset market clears)    $\bar{Z}_{it} \leqslant 1$ if $\bar{Z}_{it} < 1, \bar{P}_{it} = 0$, a.e., $\tag{12.8}$

(goods market clears)    $\bar{c}_t + \bar{x}_t = \sum_{i=1}^{N} f_i(\bar{x}_{i,t-1}, r_{t-1})$, a.e., $\tag{12.9}$

(capital market clears)    $\sum_{i=1}^{N} \bar{x}_{it} = \bar{x}_t$, a.e. $\tag{12.10}$

Here a.e. means almost everywhere. This ends the definition of R.E.E. that we will use in this section.

It is easy to write down first-order necessary conditions for an R.E.E. Let us start on the consumer side first. We drop upper bars to ease notation. At date $t$ if the consumer buys a share of firm $i$ the cost is $P_{it}$ units of consumption goods. The marginal cost at date $t$ in utils forgone is $u'(c_t)P_{it}$. At the end of period $t$, $r_t$ is revealed and $P_{i,t+1}$ and $\pi_{i,t+1}$ become known. Hence, the consumer obtains

$$u'(c_{t+1})(P_{i,t+1} + \pi_{i,t+1}) \tag{12.11}$$

extra utils at the beginning of $t + 1$ if he collects $\pi_{i,t+1}$ and sells the share ex-dividend at $P_{i,t+1}$. But these utils are uncertain and are received one period into the future. The expected present value of utility gained at $t + 1$ is

$$\beta E_t \{u'(c_{t+1})(P_{i,t+1} + \pi_{i,t+1})\}. \tag{12.12}$$

Consumer equilibrium in the market for asset $i$ requires that the marginal opportunity cost at date $t$ be greater than or equal to the present value of the marginal benefit of dividends and ex-dividend sale price at date $t + 1$:

$$P_{it}u'(c_t) \geqslant \beta E_t \{u'(c_{t+1})(\pi_{i,t+1} + P_{i,t+1})\}, \quad \text{a.e.,} \tag{12.13}$$

$$P_{it}u'(c_t)Z_{it} = \beta E_t \{u'(c_{t+1})(\pi_{i,t+1} + P_{i,t+1})\}Z_{it}, \quad \text{a.e.} \tag{12.14}$$

Similar reasoning in the rental market yields

$$u'(c_t) \geqslant \beta E_t \{u'(c_{t+1})(R_{i,t+1})\}, \quad \text{a.e.,} \tag{12.15}$$

$$u'(c_t)x_{it} = \beta E_t \{u'(c_{t+1})R_{i,t+1}\}x_{it}, \quad \text{a.e.} \tag{12.16}$$

It would be nice if the first-order necessary conditions (12.13)–(12.16) characterized consumer optima. But it is well known that a transversality condition at infinity is also needed to completely characterize optima. Recent work by Benveniste and Scheinkman (1977) allows us to prove

**Lemma 12.1.** Suppose A.1 of section 15 in Chapter 3 holds and assume that $\mathscr{P}$ is such that $W(y_t, t) \to 0$ as $t \to 0$, where $W(y_t, t)$ is defined by

$$W(y_t, t) = \text{maximum } E_1 \sum_{s=t}^{\infty} \beta^{s-1} u(c_s)$$

subject to (12.3)–(12.5) with $t$ replaced by $s$ and 1 replaced by $t$. Recall that $y_t$ denotes the r.h.s. of (12.3). Then, given $\{P_{it}\}_{t=1}^{\infty}, \{\pi_{it}\}_{t=1}^{\infty}, \{R_{it}\}_{t=1}^{\infty}, i = 1, 2, ..., N$, optimum solutions $\{Z_{it}\}_{t=1}^{\infty}, \{x_{it}\}_{t=1}^{\infty}, i = 1, 2, ..., N, \{c_t\}_{t=1}^{\infty}, \{x_t\}_{t=1}^{\infty}$ characterized by (12.13) and (12.14) and

$$\text{TVC}_{\infty} \text{ (equity market)} \quad \lim_{t \to \infty} E_1 \{\beta^{t-1}u'(c_t)P_t \cdot Z_t\} = 0, \tag{12.17}$$

$$\text{TVC}_{\infty} \text{ (capital market)} \quad \lim_{t \to \infty} E_1 \{\beta^{t-1}u'(c_t)x_t\} = 0. \tag{12.18}$$

*Proof.* Suppose $\{\overline{Z}_t\}$, $\{\overline{c}_t\}$ and $\{\overline{x}_t\}$ satisfy (12.13)–(12.17) and let $\{Z_t\}$, $\{c_t\}$ and $\{x_t\}$ be any other collection of stochastic processes satisfying the same initial conditions and (12.3)–(12.5). Compute for each $T$ an upper bound to the shortfall:

$$E_1 \left\{ \sum_{t=1}^{T} \beta^{t-1} u(c_t) - \sum_{t=1}^{T} \beta^{t-1} u(\overline{c}_t) \right\} \tag{12.19}$$

$$\leq E_1 \left\{ \sum_{t=1}^{T} \beta^{t-1} u'(\overline{c}_t)(c_t - \overline{c}_t) \right\} \tag{12.20}$$

$$= E_1 \left\{ \sum_{t=1}^{T} \beta^{t-1} u'(\overline{c}_t) \left[ \pi_t \cdot Z_{t-1} + P_t \cdot Z_{t-1} + \sum_{i=1}^{N} R_{it} x_{i,t-1} \right. \right.$$

$$\left. - P_t \cdot Z_t - x_t - \pi_t \cdot \overline{Z}_{t-1} - P_t \cdot \overline{Z}_{t-1} - \sum_{i=1}^{N} R_{it} \overline{x}_{i,t-1} \right.$$

$$\left. \left. + P_t \cdot \overline{Z}_t + \overline{x}_t \right] \right\} \tag{12.21}$$

$$= E_1 \{\beta^{T-1} u'(\overline{c}_T) [P_T \cdot (\overline{Z}_T - Z_T) + \overline{x}_T - x_T]\} \tag{12.22}$$

$$\leq E_1 \{\beta^{T-1} u'(\overline{c}_T) [P_T \cdot \overline{Z}_T + \overline{x}_T]\} \to 0 \quad \text{as } T \to \infty. \tag{12.23}$$

Here eqs. (12.13)–(12.16) were used to telescope out the middle terms in the series of the r.h.s. of (12.21).

The terms corresponding to date 1 cancel each other because the initial conditions are the same. Hence only the terms of the r.h.s. of (12.22) remain of all the terms of the r.h.s. of (12.20) and (12.21). That the r.h.s. of (12.22) has an asymptotic upper bound of zero follows from (12.17), (12.18) and the non-negativity of $Z_T$ and $x_T$. This shows that (12.13)–(12.18) imply optimality. Notice that no assumptions on $W(y_t, t)$ are needed to get this side of the proof.

Now let $\{\overline{Z}_t\}$, $\{\overline{c}_t\}$ and $\{\overline{x}_t\}$ be optimal given $\{P_t, R_t, \pi_t\}$. Since $u'(0) = \infty$ implies that $\overline{c}_t > 0$ a.e. and $W$ is differentiable at $\overline{y}_t$ we have by concavity of $W$ and $u \geq 0$ that

$$W(y_t, t) \geq W(y_t, t) - W(y_t/2, t) \geq W'(y_t, t) y_t/2 = \beta^{t-1} u'(c_t) y_t/2. \tag{12.24}$$

Hence,

$$E_1 W(y_t, t) \to 0, t \to \infty \text{ implies } E_1 \beta^{t-1} u'(c_t) y_t \to 0, t \to \infty. \tag{12.25}$$

But

$$y_t \equiv \pi_t \cdot Z_{t-1} + P_t \cdot Z_{t-1} + \sum_i R_{it} x_{i,t-1},$$ (12.26)

so that by the first-order necessary conditions

$$E_1 \beta^{t-1} u^1(c_t) \left[ (\pi_t + P_t) \cdot Z_{t-1} + \sum_i R_{it} x_{i,t-1} \right]$$

$$= E_1 \beta^{t-2} u'(c_{t-1}) P_{t-1} \cdot Z_{t-1} + E_1 \beta^{t-2} u'(c_{t-1}) x_{t-1}$$ (12.27)

because in more detail (12.13)–(12.16) imply

$$x_{i,t-1} u'(c_{t-1}) = \beta E_{t-1} [u'(c_t) R_{it}] x_{i,t-1},$$ (12.28)

$$\beta^{-1} x_{t-1} u'(c_{t-1}) = E_{t-1} \left[ u'(c_t) \left( \sum_i R_{it} x_{i,t-1} \right) \right],$$ (12.29)

$$P_{i,t-1} u'(c_{t-1}) Z_{i,t-1} = \beta E_{t-1} [u'(c_t) (\pi_{it} + P_{it}) Z_{i,t-1}],$$ (12.30)

$$\beta^{-1} u'(c_{t-1}) P_{t-1} \cdot Z_{t-1} = E_{t-1} [u'(c_t) (\pi_t \cdot Z_{t-1} + P_t \cdot Z_{t-1})].$$ (12.31)

Hence, because $P_{t-1} \geqslant 0, Z_{t-1} \geqslant 0$, (12.25) implies

$$E_1 \beta^{t-2} u'(c_{t-1}) P_{t-1} \cdot Z_{t-1} \to 0 \quad \text{as } t \to \infty$$ (12.32)

$$E_1 \beta^{t-2} u'(c_{t-1}) x_{t-1} \to 0, \qquad \text{as } t \to \infty,$$ (12.33)

as was to be shown.

The first part of this argument follows Malinvaud (1953) and the second part is taken from Benveniste and Scheinkman (1977). Lemma 12.1 is important because it characterizes consumer optima.

The assumption that $E_1 W(y_t, t) \to 0$, as $t \to \infty$, restrains $\mathscr{P}$. It requires that $\mathscr{P}$ be such that along any path in $\mathscr{P}$ utils cannot grow faster than $\beta^t$ on the average. A general sufficient condition on $\mathscr{P}$ for $E_1 W(y_t, t) \to 0$ can be given by what should be a straightforward extension of the methods of Brock and Gale (1969) and McFadden (1973) to our setup.

An obvious sufficient condition is that the utility function be bounded, i.e. there are numbers $\underline{B} < \bar{B}$ such that for all $c \geqslant 0, \underline{B} \leqslant u(c) \leqslant \bar{B}$.

We note that the method used here of introducing a stock market into this type of model where an investment decision is present was first developed by Scheinkman (1977) in the certainty case.

Next, we establish a basic lemma.

**Lemma 12.2.**

(i) Let $X = (\{\bar{c}_t\}_{t=1}^{\infty}, \{\bar{x}_{it}\}_{t=1}^{\infty}, \{\bar{x}_t\}_{t=1}^{\infty})$ solve the optimal growth problem (15.1) of Chapter 3 and define

$$\bar{R}_{it+1} \equiv f_i'(\bar{x}_{it}, r_t), \bar{\pi}_{i,t+1} \equiv f_i(\bar{x}_{it}, r_t) - f_i'(\bar{x}_{it}, r_t)\bar{x}_{it}. \tag{12.34}$$

Then let $\{\bar{P}_{it}\}_{t=1}^{\infty}, i = 1, 2, ..., N$, satisfy (12.30) and (12.32). Put

$$\bar{Z}_{it} = 1. \tag{12.35}$$

Then $(\{\bar{P}_{it}\}_{t=1}^{\infty}, \{\bar{R}_{it}\}_{t=1}^{\infty}, \{\bar{\pi}_{it}\}_{t=1}^{\infty}, \{\bar{x}_{it}\}_{t=1}^{\infty}, \{\bar{Z}_{it}\}_{t=1}^{\infty}, i = 1, 2, ..., N, \{\bar{c}_t\}_{t=1}^{\infty}, \{\bar{x}_t\}_{t=1}^{\infty}) \equiv \mathcal{R}$ is an R.E.E.

(ii) Let $\mathcal{R}$ be an R.E.E. Then $X$ solves the optimal growth problem (15.1) of Chapter 3.

*Proof.* Let $X$ solve the optimal growth problem (15.1). It is obvious that $\mathcal{R}$ satisfies the first-order necessary conditions for an R.E.E. by its very definition. What is at issue is the TVC$_\infty$ (12.17) and (12.18). Put

$$V(x_{t-1}, t-1) \equiv \text{maximum } E_1 \sum_{s=t}^{\infty} \beta^{s-1} u(c_s) \tag{12.36}$$

subject to

$$c_s + x_s = \sum_{j=1}^{N} f_j(x_{j,s-1}, r_{s-1}), \tag{12.37}$$

$$\sum_{j=1}^{N} x_{js} \equiv x_s; \quad x_{js} \geq 0, \quad j = 1, 2, ..., N, c_s \geq 0, x_s \geq 0,$$

$$s = t, t+1, ..., x_{t-1} \text{ given.} \tag{12.38}$$

Then following a similar argument as that in (12.24)–(12.33) we have, since $u$ is bounded, that for any $x_t \geq 0$, $V(x_t, t) \to 0$ as $t \to \infty$ and

$$V(x_t, t) \geq V(x_t, t) - V(x_t/2, t) \geq V'(x_t, t)x_t/2$$
$$= E_1[\beta^t u'(c_{t+1}) f_i'(x_{it}, r_t)x_t/2]$$
$$= E_1[\beta^{t-1} u'(c_t)x_t/2] \geq 0. \tag{12.39}$$

Since the l.h.s. of (12.39) must go to zero the r.h.s. must also. Hence

$$E_1[\beta^{t-1} u'(c_t)x_t] \to 0 \quad \text{as } t \to \infty \tag{12.40}$$

along any optimum program. This establishes (12.18).

What about (12.17)? Here the stochastic process $\{\bar{P}_{it}\}_{t=1}^{\infty}$ was assumed to have been constructed from the quantity side of the model by use of (12.30) so that the $\text{TVC}_\infty$ (12.32) was satisfied. Hence $\text{TVC}_\infty$ (12.17) is satisfied by the very construction of $\{\bar{P}_{it}\}_{t=1}^{\infty}$. This establishes that (i) implies (ii).

In showing that (ii) implies (i) it is clear that the first-order necessary conditions for the quantity side of an R.E.E. boil down to the first-order conditions for the optimal growth problem. What must be established is the $\text{TVC}_\infty$ (12.40). But this follows from (12.18) of lemma 12.1. This ends the proof of lemma 12.2.

Lemma 12.2 shows that the quantity side of any competitive equilibrium may be manufactured from solutions to the growth problem. This fact will enable us to identify the Ross prices. Furthermore, it will be used in the existence proof of an asset pricing function which is developed in the next section. We now return to the discussion of the relationship between the growth model of section 15 of Chapter 3 and the risk prices of Ross. This will facilitate the economic interpretation of an R.E.E. stochastic process $(\{\bar{R}_{it}\}_{t=1}^{\infty}, \{\bar{P}_{it}\}_{t=1}^{\infty}, \{\bar{\pi}_{it}\}_{t=1}^{\infty})$.

Drop the upper bars off equilibrium quantities from this point on in order to ease the notation. Assume that conditions are such that all asset prices are positive with probability 1 in equilibrium. Then $\bar{Z}_{it} = 1$ w.p.1 and from (12.30) we get for each $t$,

$$u'(c_t) = \beta E_t[u'(c_{t+1})Z_{it}]; \quad Z_{it} \equiv (P_{i,t+1} + \pi_{i,t+1})/P_{it}. \tag{12.41}$$

Next, profit maximization implies

$$f_i'(x_{it}, r_t) = R_{i,t+1}, \pi_{i,t+1} = f_i(x_{it}, r_t) - f_i'(x_{it}, r_t)x_{it}. \tag{12.42}$$

Turning to the rental market suppose that all processes are used w.p.1. Then (12.42) and (12.28) give us for each $i$ and $t$:

$$u'(c_t) = \beta E_t[u'(c_{t+1})f_i'(x_{it}, r_t)]. \tag{12.43}$$

Observe that we are not entitled to write the returns $Z_{it}$ defined by (12.41) in the linear Ross form (11.1) unless $P_i(y_{t+1})$ is linear in $y_{t+1}$. An example is presented in section 16 where $P_i(y_{t+1})$ turns out to be linear in $y_{t+1}$. But first we must show that an asset pricing function exists. This is done in the next section.

## 13. Existence of an asset pricing function

Since in equilibrium the quantity side of the asset pricing model of section 12 is the same as the $N$ process growth model of section 15 of Chapter 3, we therefore may use the facts collected in section 15 to prove the existence of an asset pricing function $P(y)$ in much the same way as in Lucas (1978). We begin by making an assumption.

**Assumption 1.** Assume for all $r \in R$:
(a) $f_i'(0, r) = +\infty, i = 1, 2, ..., N$.
(b) $\pi_i(x, r) \equiv f_i(x, r) - f_i'(x, r) x > 0$ for all $x > 0$.

Assumption 1(a) implies that (12.15) holds with equality in equilibrium. Also, assumption 1(b) implies (12.13) holds with equality in equilibrium. Let us search, as does Lucas (1978), for a bounded continuous function $P_i(y)$ such that in equilibrium the following holds:

$$P_{it} u'(c_t) = P_i(y_t) u'(c_t) = \beta E_t [u'(c_{t+1}) (\pi_{i, t+1} + P_i(y_{t+1}))]. \tag{13.1}$$

Convert the foregoing problem into a fixed point problem. Note first from section 15 of Chapter 3 that

$$u'(c_t) = U'(y_t), \quad t = 1, 2, ..., \tag{13.2}$$

$$\pi_{i, t+1} f_i(x_{it}, r_t) - f_i'(x_{it}, r_t) x_{it} \equiv \pi_i(x_{it}, r_t) = \pi_i(\eta_i(x_t) x_t, r_t)$$
$$= \pi_i[\eta_i(h(y_t)) h(y_t), r_t]$$
$$\equiv J_i(y_t, r_t), \tag{13.3}$$

$$y_{t+1} = \sum_{j=1}^{N} f_j(\eta_j(x_t) x_t, r_t)$$
$$= \sum_{j=1}^{N} f_j[\eta_j(h(y_t)) h(y_t), r_t] \equiv Y(y_t, r_t). \tag{13.4}$$

Put

$$G_i(y_t) \equiv \beta \int_R U'[Y(y_t, r)] J_i(y_t, r) \mu(dr),$$

$$F_i(y_t) \equiv P_i(y_t) U'(y_t), \tag{13.5}$$

$$(T_i F_i)(y_t) \equiv G_i(y_t) + \beta \int_R F_i[Y(y_t, r)] \mu(dr). \tag{13.6}$$

Then for each $i$, (13.1) may be written as

$$F_i(y_t) = (T_i F_i)(y_t).\tag{13.7}$$

Problem (13.7) is a fixed point problem in that we search for a function $F_i$ that remains fixed under operator $T_i$. In order to use the contraction mapping theorem to find a fixed point $F_i$ we must show first that $T_i$ sends the class of bounded continuous functions on $[0, \infty)$, call it $C[0, \infty)$, into itself. The results of section 15 established that all of the functions listed in (13.2)–(13.6) are continuous in $y_t$. We need

**Lemma 13.1.** *If $U(y)$ is bounded on $[0, \infty)$ then $G_i(y)$ is bounded.*

*Proof.* First by concavity of $U$ we have

$$U(y) - U(0) \geqslant U'(y)(y - 0) = U'(y)y.\tag{13.8}$$

Hence, there is $B$ such that

$$U'(y)y \leqslant B \quad \text{for all } y \in [0, \infty).\tag{13.9}$$

Secondly,

$$\int_R U'[Y(y,r)]J_i(y,r)\mu(\mathrm{d}r) = \int_R \{U'[Y(y,r)]Y(y,r)J_i(y,r)/Y(y,r)\}\mu(\mathrm{d}r)$$
$$\leqslant B\int[J_i(y,r)/Y(y,r)]\mu(\mathrm{d}r)\leqslant B,$$

since $f_i' \geqslant 0$ implies

$$f_i - f_i' x_{it} \equiv J_i \leqslant f_i; \quad Y \equiv \sum_{j=1}^N f_j; \quad J_i/Y \leqslant 1.$$

Thus, $G_i$ is bounded by $\beta B$. This ends the proof.

Next, we show that if

$$\|F_i\| \equiv \sup_{y \in [0, \infty)} |F_i(y)|\tag{13.10}$$

is chosen to be the norm on $C[0, \infty)$, then $T_i$ is a contraction with modulus $\beta$.

**Lemma 13.2.** $T_i: C[0, \infty) \to C[0, \infty)$ *is a contraction with modulus $\beta$.*

*Proof.* We must show that for any two elements, $F$ and $G$ in $C[0, \infty]$,

$$\| T_i F - T_i G \| \leqslant \beta \| F - G \|. \tag{13.11}$$

Now for $y \in [0, \infty)$ we have

$$\begin{aligned}
| T_i F(y) - T_i G(y) | &= \beta \left| \int (F[Y(y,r)] - G[Y(y,r)]) \mu(dr) \right| \\
&\leqslant \beta \int | F(y') - G(y') | \mu(dr) \\
&\leqslant \beta \int \sup_{y' \in [0, \infty)} | F(y') - G(y') | \mu(dr) \\
&= \beta \| F - G \|.
\end{aligned} \tag{13.12}$$

Take the supremum of the l.h.s. of (13.12) to get

$$\| T_i F - T_i G \| \leqslant \beta \| F - G \|. \tag{13.13}$$

This ends the proof.

**Theorem 13.1.** For each $i$ there exists exactly one asset pricing function of the form $P_i(y)$, where $P_i \in C[0, \infty)$.

*Proof.* Apply the contraction mapping theorem to produce a fixed point $F_i(y) \in C[0, \infty)$. Put

$$P_i(t) = \bar{F}_i(y)/U'(y). \tag{13.14}$$

It is clear that $P_i(y)$ satisfies (13.1). Furthermore, the very definition of $T_i$ and $P_i(y)$ that satisfies (13.1) is such that $P_i(\cdot) U'(\cdot) \equiv \bar{F}_i(\cdot)$ is a fixed point of $T_i$. This ends the proof.

Note that assumption 1(a) is not needed for the existence theorem. However, assumption 1(b) is needed in the theorem so that (13.1) holds with equality.

## 14. Certainty equivalence formulae

What we propose to do in this section is to use the asset pricing model of section 12 to construct a version of the price-equals-present-value-of-dividends ($P$ = PVD) formula for the pricing of common stocks. In equilibrium our formula must hold.

The formula will be derived from the following special case of the model of section 12:

$$f_i(x_{it}, r_t) \equiv \overline{f}_i(x_{it}) (A_i^0 + A_i^1 \, \tilde{\delta}_t), \quad A_i^0 > 0. \tag{14.1}$$

In eq. (14.1) observe that $\{\tilde{\delta}_t\}_{t=1}^{\infty}$ is an independent and identically distributed sequence of random variables with zero mean and finite variance $\sigma^2$. The numbers $A_i^0$ and $A_i^1$ and the random variables $\tilde{\delta}_t$ are assumed to satisfy: there is $\epsilon_0 > 0$ such that for each $t$

$$P[A_i^0 + A_i^1 \, \tilde{\delta}_t \geqslant \epsilon_0] = 1. \tag{14.2}$$

Optimum profits are given by

$$\pi_i(x_{it}, r_t) = \overline{f}_i(x_{it}) (A_i^0 + A_i^1 \, \tilde{\delta}_t) - \overline{f}_i'(x_{it}) x_{it} (A_i^0 + A_i^1 \, \tilde{\delta}_t)$$

$$= \overline{\pi}_i(x_{it}) (A_i^0 + A_i^1 \, \tilde{\delta}_t). \tag{14.3}$$

In order to shorten the notational burden in the calculations below, put

$$f_i(x_{it}, r_t) = \mu_{it} + \sigma_{it} \, \tilde{\delta}_t, \tag{14.4}$$

$$f_i'(x_{it}, r_t) = \mu_{it}' + \sigma_{it}' \, \tilde{\delta}_t, \tag{14.5}$$

$$\pi_i(x_{it}, r_t) = \overline{D}_{it} + V_{it} \, \tilde{\delta}_t, \tag{14.6}$$

where

$$\mu_{it} \equiv \overline{f}_i(x_{it}) A_i^0; \quad \sigma_{it} \equiv \overline{f}_i(x_{it}) A_i^1; \quad \mu_{it}' = \overline{f}_i'(x_{it}) A_i^0; \quad \sigma_{it}' = \overline{f}_i' A_i^1,$$

$$\overline{D}_{it} = \overline{\pi}_i(x_{it}) A_i^0; \quad V_{it} = \overline{\pi}_i(x_{it}) A_i^1. \tag{14.7}$$

All quantities will be evaluated at equilibrium levels unless otherwise noted. The notation is meant to be suggestive with $\overline{D}_{it}$ denoting average dividends or profits expected at date $t$, $V_{it}$ denoting the coefficient of variability of profits with respect to the process $\{\tilde{\delta}_t\}_{t=1}^{\infty}$, and so forth. For a specific parable think of the $\{\tilde{\delta}_t\}_{t=1}^{\infty}$ process as the market. Then production and profits in all industries $i = 1$, 2, ..., $N$ are affected by the market. High values of $\tilde{\delta}_t$ correspond to booms and low values correspond to slumps. Industries $i$ with $A_i^1 > 0$ are procyclical. Those with $A_i^1 < 0$ are countercyclical, and those with $A_i^1 = 0$ are acyclical.

**Assumption 1.** There is at least one industry, call it $N$, that is acyclical. The $N$th industry will be called *risk free*. For emphasis we will sometimes say that $N$ is *systematic risk free*.

In order that all industries be active in equilibrium and that output remain bounded we shall assume the following:

**Assumption 2.** (i) $\bar{f_i}'(0) = \infty$, $i = 1, 2, ..., N$. (ii) $\bar{f_i}'(\infty) = 0$, $i = 1, 2, ..., N$.

Assumption 2(i) guarantees that all $x_{it} > 0$ along an equilibrium. Assumption 2(ii) implies there is a bound $B$ such that $x_{it} \leqslant B$ w.p.1 for all $i$ and $t$.

Although concavity of $f(x)$ and $f(0) = 0$ imply that optimum profits are nonnegative we shall require that profits are positive for each $x > 0$, i.e.

**Assumption 3.** For all $x > 0$, $\pi_i(x) \equiv \bar{f_i}(x) - \bar{f_i}'(x) x > 0$.

Assumption 3 will be used to show that equity prices are positive in equilibrium.

By the first-order necessary conditions of equilibrium $(12.13)-(12.16)$, $(14.2)$, assumption 2(i) and assumption 3, it follows that

$$P_{it} u'(c_t) = \beta E_t[u'(c_{t+1}) (\pi_{i,t+1} + P_{i,t+1})], \quad \text{a.e.,} \tag{14.8}$$

$$u'(c_t) = \beta E_t[u'(c_{t+1}) R_{i,t+1}]$$

$$= \beta E_t u'(c_{t+1}) \mu'_{it} + \beta E_t[u'(c_{t+1}) \bar{\delta}_t] \sigma'_{it}, \quad \text{a.e.} \tag{14.9}$$

The r.h.s. of $(14.9)$ follows from $(14.5)$. It is clear from assumption $(3)$ that equity prices are positive since $\pi_{i,t+1}$ is positive w.p.1. Hence, both $(14.8)$ and $(14.9)$ are equalities and $Z_{it} = 1$.

The $P$ = PDV formula will be derived from $(14.8)$ and $(14.9)$ by recursion. Use $(14.3)$ and $(14.6)$ to get

$$\pi_{i,t+1} = \bar{D}_{it} + V_{it} \bar{\delta}_t. \tag{14.10}$$

In order to shorten notation put $u'(c_{t+1}) = u'_{t+1}$ for all $t$. From $(14.8)$ and $(14.10)$ we get

$$P_{it} u'(c_t) = \beta E_t u'_{t+1} \bar{D}_{it} + \beta E_t (u'_{t+1} \bar{\delta}_t) V_{it} + \beta E_t[u'_{t+1} P_{i,t+1}]. \tag{14.11}$$

Notice that $\mu'_{it}$, $\sigma'_{it}$, $\bar{D}_{it}$, $V_{it}$ are, in theory at least, observable. Hence, if we recurse (14.11) forward by replacing $t$ by $t + 1$ in (14.8) and insert the result into (14.11), we can use (14.9) to solve for

$$E_t m_t \equiv \frac{\beta E_t u'_{t+1}}{u'_t} \ ; \quad E_t n_t \equiv \frac{\beta E_t (u'_{t+1} \bar{\delta}_t)}{u'_t} \ ;$$

$$m_t \equiv \frac{\beta u'_{t+1}}{u'_t} \ ; \quad n_t \equiv \frac{\beta u'_{t+1} \bar{\delta}_t}{u'_t} \ , \tag{14.12}$$

in terms of $\mu'_{it}$, $\sigma'_{it}$ and build up a $P$ = PDV formula for $P_{it}$. To do so from (14.11) we get

$$\begin{aligned}
P_{it} &= E_t m_t \bar{D}_{it} + E_t n_t V_{it} + \beta E_t [u'_{t+1} P_{i,t+1}]/u'_t \\
&= E_t m_t \bar{D}_{it} + E_t n_t V_{it} + E_t \{m_t [E_{t+1} m_{t+1} \bar{D}_{i,t+1} + E_{t+1} n_{t+1} V_{i,t+1} \\
&\quad + \beta E_{t+1} (u'_{t+2} P_{i,t+2})/u'_{t+1}]\} \\
&= E_t m_t \bar{D}_{it} + E_t n_t V_{it} + E_t [m_t E_{t+1} m_{t+1} \bar{D}_{i,t+1} + m_t E_{t+1} n_{t+1} V_{i,t+1}] + \dots \\
&\quad + E_t [m_t E_{t+1} m_{t+1} \dots E_{t+T} (m_{t+T} \bar{D}_{i,t+T})] \\
&\quad + E_t [m_t E_{t+1} m_{t+1} \dots E_{t+T-1} m_{t+T-1} E_{t+T} n_{t+T} V_{i,t+T}] \tag{14.13} \\
&\quad + E_t [m_t E_{t+1} m_{t+1} \dots E_{t+T} (m_{t+T} P_{i,t+T+1})].
\end{aligned}$$

**Assumption 4.** The utility function $u(\cdot)$ is such that for all $\{P_{it}, \pi_{it}, R_{it}\}_{t=1}^{\infty}$, $i = 1, 2, \dots, N$, the $\text{TVC}_\infty$ is necessary for a consumer's maximum. Now the $\text{TVC}_\infty$ implies that

$$E_t [m_t E_{t+1} m_{t+1} \dots E_{t+T} (m_{t+T} P_{i,t+T+1})] \to 0 \quad \text{as} \quad T \to \infty. \tag{14.14}$$

By (14.9) we get for each $t$,

$$1 = E_t m_t \mu'_{it} + E_t n_t \sigma'_{it}. \tag{14.15}$$

Therefore if $\sigma'_{Nt} \equiv 0$, (14.15) implies

$$E_t m_t = 1/\mu'_{Nt}; \quad E_t n_t = [(\mu'_{Nt} - \mu'_{it})/\sigma'_{it}] (1/\mu'_{Nt}) \equiv -\Delta_t/\mu'_{Nt}. \tag{14.16}$$

Note here that $\Delta_t$ is the classical risk adjusted rate of return from portfolio theory.

Also, $\mu'_{Nt}$ is principal plus interest obtained by employing a marginal unit in process $N$. It is important to observe that $\Delta_t$ is independent of $i$. Furthermore, the Ross risk price $\lambda_t$ is determined by $\lambda_t = \beta \Delta_t$. This follows from (11.16) and (14.16). Turn now to the $K$-factor case.

In the $K$-factor case put

$$f_i(x_{it}, r_t) \equiv \bar{f}_i(x_{it})\left( \sum_{k=0}^{K} A_i^k \, \tilde{\delta}_{t+1}^k \right), \quad \tilde{\delta}_{t+1}^0 \equiv 0, A_i^0 > 0. \tag{14.17}$$

Put the same assumptions on the data as in section 13. Then, as in (14.4), (14.5) and (14.6), we may write

$$f_i(x_{it}, r_t) = \mu_{it} + \sum_{k=1}^{K} \sigma_{it}^k \tilde{\delta}_{t+1}^k, \tag{14.18}$$

$$f_i'(x_{it}, r_t) = \mu'_{it} + \sum_{k=1}^{K} \sigma_{it} \tilde{\delta}_{t+1}, \tag{14.19}$$

$$\pi_i(x_{it}, r_t) = \bar{D}_{it} + \sum_{k=1}^{K} V_{it}^k \tilde{\delta}_{t+1}^k \equiv \pi_{i,t+1}, \tag{14.20}$$

where the entities in (14.18)–(14.20) are defined as in (14.7). Keep the same assumptions as above. Then (14.8) and (14.9) become

$$P_{it}u'_t = \beta E_t \left[ u'_{t+1} \left( \bar{D}_{it} + \sum_{k=1}^{K} V_{it}^k \tilde{\delta}_{t+1}^k \right) + u'_{t+1} P_{i,t+1} \right], \tag{14.21}$$

$$u'_t = \beta E_t \left[ u'_{t+1} \left( \mu'_{it} + \sum_{k=1}^{K} \sigma_{it}^k \tilde{\delta}_{t+1}^k \right) \right]. \tag{14.22}$$

Define

$$m_t \equiv (\beta u'_{t+1})/u'_t, n_t^k \equiv (\beta u'_{t+1} \tilde{\delta}_{t+1}^k)/u'_t. \tag{14.23}$$

Then letting $N$ be a risk-free process, i.e. $\sigma'^k_{Nt} \equiv 0$ for all $k$ and $t$, we get

$$E_t m_t = 1/\mu'_{Nt} \tag{14.24}$$

and for each $i$

$$1 = (E_t m_t)\mu'_{it} + \sum_{k=1}^{K} \sigma'^k_{it} (E_t n_t^k) = \mu'_{it}/\mu'_{Nt} + \sum_{k=1}^{K} \sigma'^k_{it} (E_t n_t^k). \tag{14.25}$$

Hence, from (14.25) it follows that

$$(\mu'_{Nt} - \mu'_{it})/\mu'_{Nt} = \sum_{k=1}^{K} o'^k_{it}\, (E_t n^k_t), \quad i = 1, 2, ..., N. \tag{14.26}$$

Here it is assumed that a unique solution of (14.26) for $E_t n^k_t$ exists and is defined by

$$E_t n^k_t \equiv - \Delta^k_t/\mu'_{Nt}. \tag{14.27}$$

Note also that the Ross price of systematic risk $k$, $\lambda_{kt}$, satisfies $\lambda_{kt} = \beta \Delta^k_t$ which follows from (11.16) and (14.27). Let

$$\xi_{it} \equiv \bar{D}_{it} - V_{it}\,\Delta_t,$$

then (14.8) becomes

$$P_{it} = \frac{\xi_{it}}{\mu'_{Nt}} + E_t\left\{\left[\frac{\beta u'_{t+1}}{u'_t}\,\mu'_{Nt}\right] (P_{i,t+1}) \right\}\Big/\mu'_{Nt}$$

$$\equiv \{\xi_{it} + E_t \hat{P}_{i,t+1} \}/\mu'_{Nt}.$$

Hence,

$$E_t \hat{P}_{i,t+1} + \xi_{it} = \mu'_{Nt} P_{it}. \tag{14.28}$$

Eq. (14.28) says that investing $P_{it}$ in the stock market must give the same expected return after paying for the services of risk bearing as investing it in the risk-free process. It states that the stock market is a *fair game* taking into account the opportunity cost of funds and the cost of risk bearing. Clearly $\xi_{it} = 0$, and $\mu'_{Nt} = 1$ is necessary for a martingale. For specific preferences eq. (14.28) is testable and its violation would signal market inefficiency in our model world.

It is worth pointing out here that if the random variable

$$(\beta u'_{t+1}/u'_t)\mu'_{Nt} \equiv I_t \tag{14.29}$$

is independent of $P_{i,t+1}$ at date $t$, then (14.16) implies that (14.28) may be rewritten as

$$E_t P_{i,t+1} + \xi_{it} = \mu'_{Nt} P_{it}. \tag{14.30}$$

Eq. (14.30) is *directly testable* since it contains no subjective entities, unlike (14.28). The problem of deriving equations like (14.30) that contain no entities that are subjective and hence are directly testable remains open. The assumption of independence is way too strong. Perhaps weaker assumptions exist that will produce a formula analogous to (14.30) that holds in at least an approximate sense.

## 15. A testable formula

In what follows a formula is developed that is directly testable under the hypothesis of linearity of the asset pricing functions $P_i(y)$. An example where $P_i(y)$ is linear is given in the following section.

**Theorem 15.1.** Assume assumptions $1-4$ of section 14. Furthermore, assume that there are constants $K_i$ and $L_i$ such that

$$P_i(y) = K_i y + L_i, \quad i = 1, 2, ..., N. \tag{15.1}$$

Then, for each $t$ and $i$

$$\mu'_{Nt} P_{it} = E_t P_{i,t+1} - \sum_{k=1}^{K} \Delta_t^k S_{it}^k + E_t \pi_{i,t+1} - \sum_{k=1}^{K} \Delta_t^k V_{it}^k \tag{15.2}$$

must hold.

*Proof.* Here by (15.1) and (14.20) we may write

$$P_{i,t+1} \equiv P_i(y_{t+1}) \equiv \bar{P}_{it} + \sum_{k=1}^{K} S_{it}^k \bar{\delta}_{t+1}^k,$$

$$\bar{P}_{it} \equiv E_t P_{i,t+1}. \tag{15.3}$$

$$\pi_{i,t+1} \equiv \bar{D}_{it} + \sum_{k=1}^{K} V_{it}^k \bar{\delta}_{t+1}^k; \quad \bar{D}_{it} \equiv E_t \pi_{i,t+1}, \tag{15.4}$$

where $E_t P_{i,t+1}$, $E_t \pi_{i,t+1}$, $S_{it}^k$ and $V_{it}^k$ do not depend upon $y_{t+1}$ but depend on $(x_{it}, ..., x_{Nt})$ only.

In order to establish (15.2) it must be shown that (15.3) and (15.4) hold. By (14.17) and the definition of $y_{t+1}$ we have

$$y_{t+1} \equiv \sum_{j=1}^{N} f_j(x_{j+1}, r_t) = \sum_j \bar{f}_j(x_{jt}) \left( \sum_{k=0}^{K} A_j^k \bar{\delta}_{t+1}^k \right). \tag{15.5}$$

Hence, $y_{t+1}$ is a linear combination of the stocks $\tilde{\delta}^k_{t+1}$ with weights that depend only upon $(x_{it}, ..., x_{Nt})$. So also is $P_{i,t+1}$. Thus, (15.3) holds for appropriate $S^k_{it}$ since $P_{i,t+1}$ is linear in $y_{t+1}$. Eq. (14.20) is identical to (15.4).

Divide both sides of (14.21) by $u'_t$ to get

$$P_{it} = E_t \left[ m_t \left( \bar{D}_{it} + \sum_{k=1}^{K} V^k_{it} \tilde{\delta}^k_{t+1} \right) \right] + E_t[m_t P_{i,t+1}]. \tag{15.6}$$

Put, using (14.23),

$$\bar{m}_t \equiv E_t m_t, \bar{n}^k_t \equiv E_t n^k_t. \tag{15.7}$$

By (15.6) and (15.7) we have

$$P_{it} = \bar{m}_t \bar{D}_{it} + \sum_{k=1}^{K} V^k_{it} \bar{n}^k_t + E_t \left[ m_t \left( \bar{P}_{it} + \sum_{k=1}^{K} S^k_{it} \tilde{\delta}^k_{t+1} \right) \right]$$

$$= \bar{m}_t \bar{D}_{it} + \sum_{k=1}^{k} V^k_{it} \bar{n}^k_t + \bar{m}_t \bar{P}_{it} + \sum_{k=1}^{K} S^k_{it} \bar{n}^k_t.$$

But (14.24) and (14.27) imply

$$\mu'_{Nt} P_{it} = \bar{D}_{it} + \bar{P}_{it} - \sum_{k=1}^{K} \Delta^k_t (V^k_{it} + S^k_{it}).$$

This ends the proof.

It is worth pointing out that although (15.2) contains no subjective entities and, hence, is *directly testable* it was derived under the strong hypothesis of linearity of the asset pricing function $P_i(y_{t+1})$. The linearity hypothesis was needed to be able to write the one-period returns $Z_{it}$ to holding asset $i$ in the linear form (11.2) of Ross. The linear form of $Z_{it}$ was used, in turn, to derive (15.2). We suspect that strong conditions will be required on utility and technology to be able to write equilibrium asset returns in the form (11.2). Hence (15.2) is not general; it holds only as a linear approximation.

The economic content of (15.2) is compelling. It is a standard *no arbitrage profits* condition. The price of risk bearing over the time interval $[t, t+1]$ sells for $\Delta^k_t$ per unit of risk of type $k$. At date $t$ risk emerges from two sources: (i) $\pi_{i,t+1}$ and (ii) $P_{i,t+1}$. Profits contain $V^k_{it}$ units of risk of type $k$. The price of stock $i$ at date $t+1$ contains $S^k_{it}$ units of risk of type $k$. Hence, the total cost of risk bearing from all sources of risk for all types of risk is

$$\sum_{k=1}^{K} \Delta_t^k (V_{it}^k + S_{it}^k).$$

Thus, (15.2) just says that the risk-free earnings from an investment of $P_{it}$ must equal the sum of risk adjusted sale value of stock $i$ at date $t + 1$ and risk adjusted profits.

## 16. An example

In this section we present a solved example that illustrates the ideas presented in sections $11-15$. Let the data be given by

$$u(c) = \log c, \tag{16.1}$$

$$f_i(x_i, r) = A_i(r)x_i^\alpha, \quad i = 1, 2, ..., N, \ \ 0 < \alpha < 1. \tag{16.2}$$

We shall assume that for all $i$,

$$A_i(r) > 0 \quad \text{for all } r \in R,$$

and $A_i(1)$ is continuous in $r$. Since $R$ is compact each $A_i(r)$ has a positive lower bound $\underline{A}_i > 0$.

First-order necessary conditions become, for all $t$,

$$\frac{1}{c_t} \geqslant \beta\alpha E_t \left[ \frac{1}{c_{t+1}} A_i(r_t)x_{it}^{\alpha-1} \right], \tag{16.3}$$

$$\frac{1}{c_t}x_{it} = \beta\alpha E_t \left[ \frac{1}{c_{t+1}} A_i(r_t)x_{it}^{\alpha-1} \right] x_{it}, \quad i = 1, 2, ..., N, \tag{16.4}$$

$$\lim_{t\to\infty} E_1 \left[ \frac{\beta^{t-1}}{c_t} x_t \right] = 0. \tag{16.5}$$

Conjecture an optimum solution of the form

$$c_t = (1 - \lambda)y_t; \quad x_t = \lambda y_t; \quad x_{it} = \eta_i x_t; \quad \sum_{i=1}^{N} \eta_i = 1, \tag{16.6}$$

where

$$\lambda > 0; \quad \eta_i \geq 0, \quad i = 1, 2, ..., N. \tag{16.7}$$

Insert (16.6) into (16.3) and (16.4), solve for $\lambda$, $\{\eta_i\}_{i=1}^N$ and check that (16.5) is satisfied. Doing this we get

$$\frac{1}{(1-\lambda)y_t} \geq \beta \alpha E_t \left\{ \frac{1}{(1-\lambda)y_{t+1}} A_i(r_t) x_{it}^{\alpha-1} \right\}, \tag{16.8}$$

if and only if

$$\frac{1}{y_t} \geq \beta \alpha E_t \left\{ \frac{A_i(r_t)}{\sum\limits_{j=1}^N A_j(r_t) x_{jt}^{\alpha-1}} x_{it}^{\alpha-1} \right\},$$

if and only if

$$\frac{1}{y_t} \geq \beta \alpha E_t \left\{ \frac{\eta_i^{\alpha-1} x_t^{\alpha-1} A_i(r_t)}{\sum\limits_{j=1}^N A_j(r_t) \eta_j^{\alpha} x_t^{\alpha}} \right\},$$

if and only if

$$\frac{1}{y_t} \geq \beta \alpha E_t \left\{ \frac{A_i(r_t)}{\sum\limits_{j=1}^N A_j(r_t) \eta_j^{\alpha}} \right\} \frac{\eta_i^{\alpha-1}}{x_t} \equiv \frac{\beta \alpha \eta_i^{\alpha-1} \Gamma_i}{x_t},$$

if and only if

$$x_t \geq \beta \alpha \eta_i^{\alpha-1} \Gamma_i y_t. \tag{16.9}$$

Set (16.9) aside for the moment. From (16.4), following the same steps that we used to get (16.9), we obtain

$$\frac{x_{it}}{y_t} = \beta \alpha \eta_i^{\alpha} \Gamma_i, \tag{16.10}$$

if and only if

$$x_{it} = \beta \alpha \eta_i^{\alpha} \Gamma_i y_t = \eta_i x_t. \tag{16.11}$$

Hence (16.9) holds with equality for all $t$ and $i$. Since it is well known and is easy to see that for $N = 1$

$$\lambda = \beta\alpha,$$

it is natural to conjecture for $N \geqslant 1$ that

$$\lambda = \beta\alpha; \quad \eta_i^{\alpha-1} \Gamma_i = 1, \quad i = 1, 2, ..., N, \tag{16.12}$$

and test (16.5). If (16.12) satisfies (16.5) then we have found an optimum solution and hence the unique optimum solution.

Continuing, we have

$$\Gamma_i \equiv E\left\{ \frac{A_i(r)}{\displaystyle\sum_{j=1}^{N} A_j(r)\eta_j^\alpha} \right\} = \eta_i^{1-\alpha}; \quad \sum_{j=1}^{N} \eta_j = 1. \tag{16.13}$$

It can be shown that (16.13) has a unique solution $\{\bar{\eta}_i\}_{i=1}^{N}$.

It is easy to check that (16.6) with $\bar{\lambda} \equiv \alpha\beta$, $\eta_i \equiv \bar{\eta}_i$, $i = 1, 2, ..., N$, generates a solution that not only satisfies (16.3) and (16.4) by construction but also satisfies (16.5). We leave this to the reader.

Let us use the solution to calculate an example of an equilibrium asset price function from the work of section 12. From (12.35) and (12.32) we get

$$E_t[u'(c_{t+1})(P_{i,t+1} + \bar{\pi}_{i,t+1})\bar{P}_{it}] = E_t[\alpha A_i(r_t)\bar{x}_{it}^{\alpha-1} u'(c_{t+1})]$$

$$= E_t[\alpha A_i(r_t)\bar{\eta}_i^{\alpha-1}\bar{x}_t^{\alpha-1} u'(c_{t+1})], \tag{16.14}$$

$$\bar{\pi}_{i,t+1} = A_i(r_t)\bar{x}_{it}^\alpha - \alpha A_i(r_t)\bar{x}_{it}^{\alpha-1}\bar{x}_{it}$$

$$= (1-\alpha)A_i(r_t)\bar{x}_{it}^\alpha = (1-\alpha)A_i(r_t)\bar{\eta}_i^\alpha\bar{x}_t^\alpha. \tag{16.15}$$

Hence, the first-order necessary condition for an asset pricing function of the form $P_{it} = P_i(y_t)$ becomes for $u(c) = \log c$, using $c_t = (1-\bar{\lambda})y_t$:

$$P_i(y_t)/y_t = \beta E_t[(P_i(y_{t+1}) + \pi_{i,t+1})/y_{t+1}]. \tag{16.16}$$

Eqs. (16.15) and (16.16) give us

$$P_i(y_t)/y_t = \beta E_t\left\{ (1-\alpha)A_i(r_t)\bar{\eta}_i^\alpha\bar{x}_t^\alpha \bigg/ \left[ \sum_{j=1}^{N} A_i(r_t)\bar{\eta}_j^\alpha\bar{x}_t^\alpha \right] \right.$$

$$+ P_i(y_{t+1})/y_{t+1}\Big\}$$

$$\equiv \beta(1-\alpha)\,\bar{\eta}_i + \beta E_t[P_i(y_{t+1})/y_{t+1}], \quad i = 1, 2, ..., N. \quad (16.17)$$

Here by (16.13)

$$\bar{\eta}_i = E_t\left[A_i(r_t)\,\bar{\eta}_i^\alpha \Big/ \left(\sum_{j=1}^{N} A_j(r_t)\,\bar{\eta}_j^\alpha\right)\right]. \quad (16.18)$$

The system of equations (16.17) is in a particularly suitable form for the application of the contraction mapping theorem to produce a unique fixed point $\bar{P}(y) \equiv (\bar{P}_1(y), ..., \bar{P}_N(y))$ that solves (16.17). Rather than do this we just conjecture a solution of the form

$$\bar{P}_i(y) = \bar{K}_i y, \quad i = 1, 2, ..., N, \quad (16.19)$$

and find $\bar{K}_i$ from (16.19) by equating coefficients. Obviously from (16.19) $\bar{K}_i$ satisfies

$$\bar{K}_i = \beta(1-\alpha)\,\bar{\eta}_i + \beta\bar{K}_i, \quad i = 1, 2, ..., N, \quad (16.20)$$

so that

$$\bar{K}_i = (1-\beta)^{-1}\beta(1-\alpha)\,\bar{\eta}_i, \quad i = 1, 2, ..., N. \quad (16.21)$$

Since the r.h.s. of (16.19) is a contraction of modulus $\beta$ on the space of bounded continuous functions on $[0, \infty)$ with values in $R^N$, the solution (16.19) is the only solution such that each $P_i(y)/y$ is bounded and continuous on $[0, \infty)$.

We now have a solved example. It is interesting to examine the dependence of $P_i(y)$ on the problem data from (16.19) and (16.21).

First, in the one-asset case we find $\eta_N = 1$ from (16.18) so that

$$P(y) = \frac{\beta}{1-\beta}(1-\alpha)y. \quad (16.22)$$

Hence, (i) the asset price decreases as the elasticity of output with respect to capital input increases; (ii) the variance of output has no effect on the stock price; and (iii) the asset price increases when $\beta$ increases.

Result (i) follows because profit's share of national output is inversely related to $\alpha$. One would expect (ii) from the log utility function. One would expect (iii) because as $\beta$ increases the future is worth more relative to the present and hence savings should increase thereby forcing asset prices to rise.

Furthermore, (16.22) says that asset price increases as current available income $y$ increases.

Secondly, in the multi-asset deterministic case we have

$$P_i(y) = \frac{\beta}{1 - \beta} (1 - \alpha)\bar{\eta}_i y, \quad i = 1, 2, ..., N. \tag{16.23}$$

We can see that if the coefficient $A_i$ measures the productivity of firm $i$ using the common technology $x^\alpha$ so that output of $i$ is $A_i x^\alpha$, then firms that are relatively more productive bear higher relative prices for their stock. Absolute productivity does not affect relative prices. This is so because $\bar{\eta}_i$ is homogeneous of degree zero in $(A_1, ..., A_N)$.

This is again one of those results that looks intuitively clear after hindsight has been applied. The consumers in this economy have no other alternative but to lease capital or to invest in stock in the $N$ firms. Hence, if the productivity of all of them is halved the constellation of asset price relatives will not change although output will drop. This type of result is specific to the log utility and Cobb–Douglas production technologies.

## 17. Miscellaneous applications and exercises

(1) In the Black–Scholes application of section 3 suppose that the stock price $S$ upon which the option is written is given by

$$\frac{dS}{S} = \alpha dt + \sigma_1 dz_1 + \sigma_2 dz_2 \tag{17.1}$$

instead of eq. (3.7). Observe that in (17.1) we have two correlated noises with

$$E_t dz_1^2 = dt; \quad E_t dz_1 dz_2 = \rho dt; \quad E_t dz_2^2 = dt.$$

Derive the Black–Scholes equation in this case and conjecture what would happen to the option as $\rho \to 1$ or $\rho \to -1$.

(2) More on Black–Scholes: The Black–Scholes option pricing model of section 3 shows that the price of an option $F$ is a function of the stock price, $S$, upon which the option is written, the exercise price $E$, the time to maturity of the option $\tau$, the riskless rate of interest $r$ and the instantaneous variance rate on the stock price $\sigma^2$. Intuitively we would expect:

(a) $\partial F/\partial S > 0$, i.e. as the stock price rises so does the option price,

(b) $\partial F/\partial E < 0$, i.e. as the exercise price rises the option price falls,

(c) $\partial F/\partial \tau > 0$, i.e. as the time to maturity increases the price of the option rises,

(d) $\partial F/\partial r > 0$, i.e. as the riskless rate of interest rises so does the option price,

(e) $\partial F/\partial \sigma^2 > 0$, i.e. as the variance rate rises so does the price of the option. Check to see if (a)–(e) are true. See Merton (1973a) and Smith (1976).

(3) Merton's costate equation: Consider Merton's model of section 4 in the special case of one risky asset and one risk-free asset, where

$$\max E_0 \left\{ \int_0^T u(C,t)dt + B(W(T)) \right\}$$

$$dW = \omega_1(\alpha_1 - r)Wdt + (rW - C)dt + \omega_1 W\sigma_1 dz_1,$$

$$dP_1/P_1 = \alpha_1 dt + \sigma_1 dz_1, \quad W_0, P_1(0) \text{ given.}$$

Here $\alpha_1$ and $\sigma_1$ are independent of $P_1$ and $t$, and $\omega_1$ denotes the percentage of $W$ in the risky asset and the risk-free asset pays interest rate $r$. Let $p(t)$ denote the current value of the costate where the state is $W(t)$. Use the Hamiltonian and the costate equation from Chapter 3 plus the necessary conditions that the controls $C$ and $\omega_1$ must satisfy and write the costate in the form:

$$dp = \mu_p(p,t)dt + \sigma_p(p,t)dz_1. \tag{17.2}$$

Find the functions $\mu_p$ and $\sigma_p$. What do you get if $\alpha_1$ and $\sigma_1$ are dependent on time? Stochastically integrate (17.2). Can you find the solution for $p(t)$ in terms of $p_0$?

How is the value $p(T)$ determined? Does $p(t)$, given $p_0$, depend upon $U(C,t)$, $W_0$, $T$ and $B(W(T))$? Does homogeneity of beliefs on $\alpha_1, r$ and $\sigma_1$ imply homogeneity of behavior of the investors regardless of their tastes, age and initial wealth? How would you qualify this statement?

(4) Consider a specific case developed by Vasicek (1977) as an illustration of his general model presented in section 8. Assume that the market price of risk $q(t,r)$ is constant and independent of time and the level of the spot rate. We write $q(t,r) \equiv q$. Next assume that the spot rate $r(t)$ follows the process

$$dr = \alpha(\gamma - r)dt + \rho dz, \quad \alpha > 0, \tag{17.3}$$

as a specific case of (8.4). Eq. (17.3) is called the *Ornstein–Uhlenbeck process* and when $\alpha > 0$ it is called in the finance literature the *elastic random walk*. The great advantage of this process is that it has a stationary distribution. Under the above specific assumptions find the solution of the term structure equation (8.10) and study its economic content. See Vasicek (1977).

(5) A simple generalization of Constantinides' (1978) model presented in section 9 is as follows. Suppose that the cash flow generated by the project in the time interval $(t, t + dt)$ is stochastic given by $c\,dt + s\,dz_2$, with $c = c(x, t)$ and $s = s(x, t)$ and $z_2$ is a Wiener process. Rewrite (9.1) as

$$dx = \mu\,dt + \sigma\,dz_1$$

and assume that $z_1$ and $z_2$ are correlated. Under these assumptions find the new equations which replace (9.3) and (9.8) and interpret the resulting rule.

(6) Consider the following tableau of asset returns:

$$dx_i/x_i = \alpha_i\,dt + \sigma_i\,dz, \quad i = 1, 2, ..., N. \tag{17.4}$$

Here $dx_i/x_i$ is instantaneous return on stock $i$. Note that the shock $dz$ is common to all stocks. Use the hypothesis that all $dz$-risk-free portfolios earn the risk-free rate of return, $r$, to show

$$(\alpha_i - r)/\sigma_i = (\alpha_j - r)/\sigma_j \tag{17.5}$$

for all $i, j$. Assume that $\sigma_N \equiv 0$ and $\alpha_N \equiv r$.

Now introduce inflation. Let the price level, $P$, follow

$$dP/P = \pi\,dt + \sigma\,dw. \tag{17.6}$$

Let the returns in (17.4) above be nominal. Assume that $E_t dw\,dz = \rho\,dt$. Work out, using stochastic calculus, the real return

$$\frac{d(x_i/P)}{x_i/P} = (\ )dt + (\ )dz + (\ )dw.$$

$$\equiv A_i\,dt + B_i\,dz + C_i\,dw, \tag{17.7}$$

i.e. fill in the ( ).

Next, use a hedging and equilibrium argument analogous to that of Black–Scholes to derive a relationship between the coefficients $A_i$, $B_i$ and $C_i$ that must hold in equilibrium. Feel free to assume existence of d$z$ and d$w$ risk-free assets. It is usually hypothesized that stock returns are negatively related to anticipated inflation. Can you find a $\{\alpha_i, \sigma_i\}_{i=1}^{N}, \pi, \sigma, \rho$ structure consistent with this hypothesis in the equilibrium relationship among $A_i, B_i$ and $C_i$?

(7) Consider the following continuous time analogue of the model presented in section 15 of Chapter 3 and sections 11–16 of this chapter.

A representative consumer solves

$$\max \mathrm{E}_0 \int_0^{\infty} c^{-\beta t} u(c(t)) \mathrm{d}t$$

$$\sum_1^{N} x_i(t) = x(t) \tag{17.8}$$

subject to

$$c(t)\mathrm{d}t + \mathrm{d}x + \sum_1^{N} P_i \mathrm{d}E_i = \sum_1^{N} \mathrm{d}\pi_i E_i + \sum_1^{N} \mathrm{d}R_i x_i,$$

$$x(0) = x_0 \text{ given}; \quad E_i(0) = 1, \quad i = 1, 2, \ldots N,$$

where

$$\mathrm{d}R_i = A_i \mathrm{d}t + B_i \mathrm{d}z, \tag{17.9}$$

$$\mathrm{d}\pi_i = a_i \mathrm{d}t + b_i \mathrm{d}z, \quad i = 1, 2, \ldots, N. \tag{17.10}$$

The firm's problem is given by

$$\max \mathrm{d}\pi_i = \max \left[ f_i(x_i) - x_i A_i \right] \mathrm{d}t + \left[ \sigma_i(x_i) - x_i B_i \right] \mathrm{d}z \tag{17.11}$$

to yield

$$f_i'(x_i) = A_i; \quad \sigma_i'(x_i) = B_i, \tag{17.12}$$

$$a_i = f_i(x_i) - x_i f_i'(x_i); \quad b_i = \sigma_i(x_i) - x_i \sigma_i'(x_i). \tag{17.13}$$

Here $E_i(t)$ denotes the number of shares of firm $i$ at $t$, $x(t)$ is real output at time $t$ and $c(t)\mathrm{d}t$ is amount of consumption during $(t, t + \mathrm{d}t)$. Ownership of $E_i(t)$ shares at $t$ gives right to d$\pi_i$ units of output at $t + \mathrm{d}t$. Investment of $x_i(t)$ in process $i$ at $t$ gives right to d$R_i$ new output at $t + \mathrm{d}t$ as well as the original $x_i(t)$. At each date $t$, $x(t)$ amount of output is allocated across $i = 1, 2, \ldots, N$ firms with rentals $dR_i$ to be received at $t + dt$ conditioned upon what happens at $t + dt$. Equilibrium is defined as in section 12 of this chapter. Recall that in equilibrium the representative consumer faces $d\pi_i$, $dR_i$ and $P_i$ parametrically and solves (17.8) for $c, x, \{x_i\}_{i=1}^N$ and $\{E_i\}_{i=1}^N$.

Approximate problem (17.8) by its discrete time analogue, take limits and drop higher-order terms to derive the following conditions of optimality:

$$\beta - E_t\left(\frac{du'}{u'dt}\right) = E_t\left(\frac{dR_i}{dt}\right) + E_t\left(\frac{du'dR_i}{u'dt}\right), \tag{17.14}$$

$$-E_t\left(\frac{du'}{u'dt}\right) = E_t\left(\frac{Z_i}{dt}\right) + E_t\left(\frac{du_iZ_i}{u'dt}\right), \tag{17.15}$$

$$Z_i \equiv (d\pi_i + dP_i)/P.$$

In equilibrium, the instantaneous means and standard deviations of $dR_i$ and $d\pi_i$ can be written as time independent functions of $x(t)$ at each $t$. Explain why this is so. Similarly, explain why $P_i(t)$ may be written as $P_i(x(t))$. Suppose next that the $N$th process is risk-free, i.e. $\sigma_N(x_N) = 0$. Put $r(x) \equiv f_N'(x_N)$ in equilibrium. Use Itô's lemma on $P_i(x)$ to derive the Lintner–Sharpe certainty equivalence formula

$$r(x)P_i(x) = P_{ix}(x)\mu(x) + \tfrac{1}{2}P_{ixx}(x)\sigma^2(x) + a_i(x)$$

$$+ [P_{ix}(x)\sigma(x) + b_i(x)]E_t\left(\frac{du'dz}{u'dt}\right), \tag{17.16}$$

where $dx \equiv \mu(x)dt + \sigma(x)dz$. Explain why (17.16) makes economic sense.

## 18. Further remarks and references

In this chapter we have presented a number of applications to illustrate various stochastic techniques. The number of research papers in finance that use such stochastic techniques has increased significantly over the last two decades and there are several recent books of readings which have collected major contributions such as Szegö and Shell (1972), Ziemba and Vickson (1975), Levy and Sarnat (1977) and Bicksler (1979), among others. Papers from these books may be used to supplement the applications in this chapter.

The Black–Scholes theory of option pricing and the subsequent modifications of this theory by several authors have generated great interest with a rapid literature growth in this area. For an overview of the major results in option pricing we suggest the review articles by Smith (1976, 1979).

In the original Black–Scholes (1973) paper several assumptions were made such as: (1) there are no penalties for short sales; (2) there are no taxes and no transactions costs; (3) the market operates continuously and the stock price follows an Itô process; (4) the stock pays no dividends and the option can only be exercised at the terminal date of the contract; and (5) the riskless rate is known and constant. Under these assumptions Black and Scholes show that a riskless hedge can be formulated using proper proportions of call options and shares of the underlying stock. Such an instantaneously riskless hedge yields a rate of return equal to the known constant riskless rate. Since 1973 several authors have modified these assumptions and, more importantly, the Black–Scholes option pricing theory has found many applications in various areas of finance.

Some modifications of the original assumptions include the following. Merton (1973a) analyzes the option model with a stochastic interest rate. Ingersoll (1976) has included a differential tax structure. Cox and Ross (1976) have priced options for alternative stochastic processes. Merton (1976) has studied the problem of a discontinuous return structure. Smith (1976) discusses some additional modifications.

For purposes of illustration we follow Smith (1979) to illustrate some applications of the Black–Scholes theory of option pricing which was initially developed for a European call. Recall that a *European call* is an option to buy a share of a stock at the maturity date of the contract for a stated amount called the exercise price. Section 3 and exercise (2) of this chapter show that the solution of the price of a call option may be denoted as $F(X, \tau, E, r, \sigma^2)$. Merton (1973a) studies a European *put*, i.e. an option to sell a share of stock at the maturity date of the contract for a given exercise price. He finds that when borrowing and lending rates are equal, then the price of a European put is equal to the value of a portfolio of a European call with the same terms as the put, riskless bonds with a face value equal to the exercise price of options and a short position in the stock.

The Black–Scholes call pricing model can also be used in pricing the *debt* and *equity* of a firm. Assume that: (1) the firm issues pure discount bonds with no dividends until after the bonds mature at which time the bondholders are paid, if possible, and the residual, if available, is paid to the stockholders; (2) the total value of the firm is unaffected by capital structure, i.e. a Modigliani–Miller (1958) world applies; and (3) there are homogeneous expectations about the dynamic behavior of the value of the firm's assets with a log-normal distribution having a constant rate of return. Under these assumptions and a given constant riskless rate, the Black–Scholes theory provides the correct valuation of the firm's equity. In this case, issuing bonds is equivalent to the stockholders selling the assets of the firm to the bondholders for the proceeds of the issue plus a call option to repurchase these assets with an exercise price equal to the face value of the bonds. Thus, the equity of the firm is like a call option.

Next suppose that instead of having a bond contract calling for only one payment of principal plus interest at the maturity date, we consider convertible bonds. A *convertible bond* offers the bondholder at the maturity date the option to either receive the face value of the bond or new shares equal to a fraction of the value of the assets of the firm. Applying the Black–Scholes theory it has been shown that a convertible bond is equivalent to a nonconvertible bond plus a call option; see Ingersoll (1977) and his references.

The Black–Scholes theory can also be applied to the pricing of various other contingent claims, such as the pricing of underwriting contracts, the pricing of collateralized loans, the pricing of leases, and the pricing of insurance, among others. Smith (1979) summarizes the results of such pricings and provides appropriate bibliographical references.

A simplified approach to option pricing has been suggested recently by Cox and Ross (1979). They present a simple discrete time option pricing formula in which the fundamental economic principles of option valuation by arbitrage methods are clear and intuitive. Furthermore, their approach requires only elementary mathematics, yet it contains as a special limiting case the Black–Scholes model. Specifically, Cox and Ross (1979) suggest that whenever stock price movements conform to a discrete binomial process or to a limiting form of such a process, options can be prices solely on the basis of arbitrage methods. To price an option by arbitrage methods there must exist a portfolio of other assets which exactly replicates in every state of nature the payoff received by an optimally exercised option.

The basic result of Cox and Ross may be stated as follows. Suppose that markets are perfect, that changes in the interest rate are never random, and that changes in the stock price are always random. In a discrete time model a necessary and sufficient condition for pricing options of all maturities and exercise prices by arbitrage methods using only the stock and bonds in the portfolio is that in each period (1) the stock price can change from its beginning of period value to only ex-dividend values at the end of the period, and (2) the dividends and the size of each of the two possible changes are presently known functions depending at most on (i) current and past stock prices, (ii) current and past values of random variables whose changes in each period are perfectly correlated with the change in the stock price, and (iii) calendar time. For a verification of this result see Cox and Ross (1979).

Having briefly indicated some of the modifications and applications of the

Black–Scholes theory we inform the reader that two papers, by Harrison and Kreps (1979) and Kreps (1980), consider some foundational issues that arise in conjunction with the arbitrage theory of option pricing. The important point to consider is this: the ability to trade securities frequently can enable a few multiperiod securities to span many states of nature. In the Black–Scholes theory there are two securities and uncountably many states of nature, but because there are infinitely many trading opportunities and because uncertainty resolves nicely, markets are effectively complete. Thus, even though there are far fewer securities than states of nature, nonetheless markets are complete and risk is allocated efficiently. The question of what is important in determining the number of securities needed to have complete markets and an evaluation of the robustness of the Black–Scholes theory are presented in Kreps (1980).

Merton's (1971) continuous model discussed briefly in section 4 provides a general equilibrium framework for the analysis of consumption and investment decisions under uncertainty. Two earlier papers that studied a similar problem in discrete time are Fama (1970a) and Hakansson (1970). Some other important papers are Sharpe (1964), Lintner (1965a, 1965b) and Mossin (1966). The advantages of using continuous analysis are discussed in Merton (1975b) who claims that the continuous time solution is consistent with its discrete time counterpart when the trading interval is sufficiently small and that the assumptions required are descriptive of capital markets as they actually function. Furthermore, the continuous time analysis has all the advantages of simplicity and empirical tractability found in the classic mean–variance model but without its objectionable assumptions.

Merton (1973b) uses methods similar to Merton (1971) to develop an intertemporal capital asset pricing model. The model for the capital market is deduced from the portfolio selection behavior of an arbitrary number of investors who act to maximize the expected utility of lifetime consumption and who trade continuously in time. Explicit demand functions for assets are derived and it is shown that current demands are affected by the possibility of uncertain changes in future investment opportunities. The equilibrium relationships among expected returns are derived by aggregating demands and requiring the market to clear. It is shown that, contrary to the classical CAPM, expected returns on risky assets may differ from the riskless rate even when they have no systematic or market risk. Merton (1973b) has been extended by Breeden (1979). The extension is achieved by using the same continuous time framework as Merton (1973b). However, Breeden (1979) shows that Merton's multi-beta pricing equation can be collapsed into a single-beta equation, where the instantaneous expected excess return on any security is proportional to its beta, or covariance, with respect to aggregate consumption alone. The result is shown by Breeden to extend to a multi-good world with an asset's beta measured relative to aggregate real consumption.

Other extensions of Merton (1971, 1973b), just to mention a few, include Magill and Constantinides (1976) who introduce transaction costs in the portfolio selection problem, Constantinides (1980) who introduces personal tax consideration in the consumption investment problem, and Litzenberger and Ramaswamy (1979) who, in addition to personal taxes, also include dividends.

A useful survey paper that presents a general equilibrium approach for the analysis of consumption, investment and portfolio selection, is Merton (1981). In this paper Merton provides a comprehensive and unified survey of the consumption-saving choice and the portfolio selection choice under uncertainty. Both these choices are integral parts of the microeconomic theory of investment under uncertainty. This survey supplements our brief discussion in section 4 of this chapter. Note that the consistency of the pricing of equity by the Black– Scholes option theory with the continuous time capital asset pricing model is discussed in Smith (1979, pp. 89–90).

In section 7 we follow Fischer (1975) to illustrate the use of stochastic calculus techniques on the topic of indexed bonds. For a further, recent reference we indicate Liviatan and Levhari (1977). A brief history of the concept of indexation is presented in Humphrey (1974).

The term structure theory of interest rates has attracted much attention for a long time. For a general overview the books by Meiselman (1962), Malkiel (1966), Nelson (1972) and by Michaelsen (1973) can be found useful. Vasicek's (1977) paper is used in section 8 to illustrate the popularity of stochastic calculus techniques. A stochastic calculus approach is also used by Cox, Ingersoll and Ross (1978) to develop a general theory of the term structure of interest rates. Cox, Ingersoll and Ross use the continuous time CAPM framework developed by Merton (1973b) along with the rational expectations equilibrium model of Lucas (1978) to develop a complete intertemporal asset pricing model that is tractable and does not contradict individual expectations as the equilibrium moves through time. From such a model, Cox, Ingersoll and Ross develop a theory of the term structure of interest rates and derive closed form solutions for the term structure and the prices of bonds which are potentially testable. For details see Cox, Ingersoll and Ross (1978). Some other related papers are Richard (1978), Dothan (1978) and Brennan and Schwartz (1977).

Stochastic calculus methods have also found applications in the area of futures pricing. In section 6 of Chapter 1 we briefly introduced the notion of futures pricing. The reader who is interested in studying futures pricing should consult Black (1976) and Cox, Ingersoll and Ross (1980) and some of the references cited in these papers. Futures markets and forward markets have been treated by the academic literature as if they were synonymous. Cox, Ingersoll and Ross (1980) explain the fundamental differences and clarify the appropriate relationships between these two markets.

The demand for cash balances and the related issues on cash management are well established areas of research. Some references, other than the ones indicated in application (10), are Eppen and Fama (1968, 1969), Neave (1970), Vial (1972), Constantinides (1976) and Constantinides and Richard (1978). Related issues that might be of interest to some readers are discussed in Daellenbach and Archer (1969) and Crane (1971).

In sections 11–18 we follow Brock (1978) to develop an intertemporal general equilibrium theory of capital asset pricing inspired by Merton (1973b). We note, however, that Merton's intertemporal capital asset pricing model is not a general equilibrium theory in the sense of Arrow–Debreu because technological sources of uncertainty are not related to the equilibrium prices of risky assets. This is done in these sections by integrating ideas from modern finance and stochastic growth models. Basically what is done is to modify the stochastic growth model of Brock and Mirman (1972) in order to incorporate a nontrivial investment decision into the asset pricing model of Lucas (1978). This is done in such a way as to preserve the empirical tractability of the Merton formulation and at the same time determine the risk prices derived by Ross (1976) in his arbitrage theory of capital asset pricing. Ross's price of systematic risk $k$ at date $t$, denoted by $\lambda_{kt}$, which is induced by the source of systematic risk $\tilde{\delta}_{kt}$, is determined by the covariance of the marginal utility of consumption with $\delta_{kt}^N$. In this way Ross's $\lambda_{kt}$ are determined by the interaction of sources of production uncertainty and the demand for risky assets. Furthermore, this model provides a context in which conditions may be found on tastes and technology that are sufficient for equilibrium returns to be a linear function of the uncertainty in the economy. Linearity of returns is necessary for Ross's theory.

More advances should be expected along the lines of introducing imperfect information and inquiry into what rules firms should follow in order to maximize equilibrium welfare of the representative consumer when some contingency markets are absent. Finally, a more difficult and perhaps more interesting problem would be to introduce heterogeneous consumers so that borrowing on future income could be introduced and the impact of this on the price of risk could be investigated.

# SELECTED BIBLIOGRAPHY

Alchian, A. (1974), "Information, Martingales, and Prices", *Swedish Journal of Economics*, 76, 3–11.

Anderson, B. and J. Moore (1971), *Linear Optimal Control*, Prentice-Hall, Englewood Cliffs, New Jersey.

Antosiewicz, H. (1958), "A Survey of Liapunov's Second Method, Contributions to the Theory of Nonlinear Oscillations", *Annals of Mathematical Studies*, 4, 141–166.

Aoki, M. (1967), *Optimization of Stochastic Systems*, Academic Press, New York.

Aoki, M. (1976), *Optimal Control and System Theory in Dynamic Economic Analysis*, North-Holland Publishing Company, New York.

Araujo, A. and J. Scheinkman (1977), "Smoothness, Comparative Dynamics and the Turnpike Property", *Econometrica*, 45, 601–620.

Arnold, L. (1974), *Stochastic Differential Equations: Theory and Applications*, John Wiley & Sons, New York.

Arrow, K.J. (1971), *Essays in the Theory of Risk Bearing*, Markham Publications, Chicago.

Arrow, K.J., and M. Kurz (1970), *Public Investment, The Rate of Return, and Optimal Fiscal Policy*, The John Hopkins Press, Baltimore.

Ash, R.B. (1972), *Real Analysis and Probability*, Academic Press, Inc., New York.

Aström, K.J. (1970), *Introduction to Stochastic Control Theory*, Academic Press, Inc., New York.

Athans, M. and P.L. Falb (1966), *Optimal Control*, McGraw-Hill, New York.

Bachelier, L. (1900), "Théorie de la Spéculation", *Annales de l'Ecole Normale Superieure*, 17, 21–86. Translated in Cootner, P.H., Ed. (1964), *The Random Character of the Stock Market Prices*, MIT Press, Cambridge, Massachusetts, 17–75.

Balakrishnan, A.V. (1973), *Stochastic Differential Systems*, Springer-Verlag, New York.

Baron, D. (1970), "Price Uncertainty, Utility, and Industry Equilibrium in Pure Competition", *International Economic Review*, 11, 463–480.

Baron, D. (1971), "Demand Uncertainty in Imperfect Competition", *International Economic Review*, 12, 196–208.

Baron, D.P. and R. Forsythe (1979), "Models of the Firm and International Trade Under Uncertainty", *The American Economic Review*, 69, 565–574.

Batra, R.N. (1975), "Production Uncertainty and the Heckscher–Ohlin Theorem", *Review of Economic Studies*, 42, 259–268.

Baumol, W.J. (1952), "The Transactions Demand for Cash: An Inventory Theoretic Approach", *Quarterly Journal of Economics*, 66, 545–556.

Bellman, R. (1957), *Dynamic Programming*, Princeton University Press, Princeton, New Jersey.

Benes, U.E. (1971), "Existence of Optimal Stochastic Control Laws", *SIAM Journal on Control*, 9, 446–472.

Benveniste, L. and J. Scheinkman (1977), "Duality Theory for Dynamic Optimization Models of Economics", University of Chicago working paper, forthcoming in Journal of Economic Theory.

Benveniste, L.M. and J.A. Scheinkman (1979), "On the Differentiability of the Value Function in Dynamic Models of Economics", *Econometrica*, 47, 727–732.

Berkovitz, L.D. (1974), *Optimal Control Theory*, Springer-Verlag, New York.

Bertsekas, D.P. (1976), *Dynamic Programming* and *Stochastic Control*, Academic Press, Inc., New York.

Bertsekas, D.P. and S.E. Shreve (1978), *Stochastic Optimal Control*, Academic Press, New York.

Bewley, T. (1972), "Existence of Equilibria in Economies With Infinitely Many Commodities", *Journal of Economic Theory*, 4, 514-540.

Bewley, T. (1977), "The Permanent Income Hypothesis: A Theoretical Formulation", *Journal of Economic Theory*, 16, 252–292.

Bewley, T. (1978), "General Equilibrium with Market Frictions", Paper presented at the Econometric Society Meetings in Chicago.

Bharucha-Reid, A.T. (1960), *Elements of the Theory of Markov Processes and Their Applications*, McGraw-Hill Company, New York.

Bharucha-Reid, A.T., ed. (1970), *Probabilistic Methods in Applied Mathematics*, Volume 2 Academic Press, New York.

Bharucha-Reid, A.T. (1972), *Random Integral Equations*, Academic Press, New York.

Bicksler, J.L., ed. (1979), *Handbook of Financial Economics*, North-Holland Publishing Company, Amsterdam.

Billingsley, P. (1968), *Convergence of Probability Measures*, John Wiley & Sons, New York.

Billingsley, P. (1979), *Probability and Measure*, John Wiley & Sons, New York.

Bismut, J.M. (1973), "Conjugate Convex Functions in Optimal Stochastic Control", *Journal of Mathematical Analysis and Applications*, 44, 384–404.

Bismut, J.M. (1975), "Growth and Optimal Intertemporal Allocation of Risks", *Journal of Economic Theory*, 10, 239–257.

Bismut, J.M. (1976), "Theory Probabiliste du Controle des Diffusions", *Memoirs of the American Mathematical Society*, 4.

Black, F. and M. Scholes (1973), "The Pricing of Options and Corporate Liabilities", *Journal of Political Economy*, 81, 637–654.

Black, F. (1976), "The Pricing of Commodity Contracts", *Journal of Financial Economics*, 3, 167–179.

Borch, K.H. (1968), *The Economics of Uncertainty*, Princeton University Press, Princeton, New Yersey.

Borg, G. (1949), "On a Liapunov Criterion of Stability", *American Journal of Mathematics*, 71, 67–70.

Bourguignon, F. (1974), "A particular Class of Continuous-Time Stochastic Growth Models", *Journal of Economic Theory*, 9, 141–158.

Boyce, W.M. (1970), "Stopping Rules for Selling Bonds", *The Bell Journal of Economics and Management Science*, 1, 27–53.

Brainard, W. (1967), "Uncertainty and the Effectiveness of Policy", *American Economic Review*, 57, 411–425.

Breeden, D.T. (1979), "An Intertemporal Asset Pricing Model With Stochastic Consumption and Investment Opportunities", *Journal of Financial Economics*, 7, 265–296.

Breiman, L. (1964), "Stopping Rule Problems", in: E.F. Beckenbach, ed., *Applied Combinatorial Mathematics*, John Wiley & Sons, New York.

Brennan, M.J. and E.S. Schwartz (1977), "Savings Bonds, Retractable Bonds and Callable Bonds", *Journal of Financial Economics*, 5, 67–88.

Brock, W.A. (1977), "Global Asymptotic Stability of Optimal Control: A Survey of Recent Results", in: M. Intriligator, ed., *Frontiers of Quantitative Economics*, volume III, American Elsevier Publishers, New York, 207–237.

Brock, W.A. (1976), "Some Applications of Recent Results on the Asymptotic Stability of Optimal Control to the Problem of Comparing Long-Run Equilibria", Cornell University working paper.

Brock, W.A. (1978), "Asset Prices in a Production Economy", University of Chicago working paper.

Brock, W.A. (1979), "An Integration of Stochastic Growth Theory and the Theory of Finance—Part I: The Growth Model", in: J. Green and J. Scheinkman, eds., *Essays in Honor of L. McKenzie*, Academic Press, New York.

Brock, W.A. and D. Gale (1969), "Optimal Growth Under Factor Augmenting Progress", *Journal of Economic Theory*, 1, 229–243.

Brock, W.A. and M.J.P. Magill (1979), "Dynamics Under Uncertainty", *Econometrica*, 47, 843–868.

Brock, W.A. and M. Majumdar (1978), "Global Asymptotic Stability Results for Multisector Models of Optimal Growth Under Uncertainty when Future Utilities are Discounted", *Journal of Economic Theory*, 18, 225–243.

Brock, W.A. and L. Mirman (1972), "Optimal Economic Growth and Uncertainty: The Discounted Case", *Journal of Economic Theory*, 4, 479–513.

Brock, W.A. and L. Mirman (1973), "Optimal Economic Growth and Uncertainty: The No Discounting Case", *International Economic Review*, 14, 560–573.

Brock, W.A., M. Rothschild and J. Stiglitz (1979), "Notes on Stochastic Capital Theory", University of Wisconsin working paper.

Brock, W.A. and J. Scheinkman (1977), "Some Results on Global Asymptotic Stability of Control Systems", in: J.D. Pitchford and S.J. Turnovsky, eds., *Applications of Control Theory To Economic Analysis*, North-Holland, Amsterdam.

Brock, W.A. and J. Scheinkman (1976), "Global Asymptotic Stability of Optimal Control Systems with Applications to the Theory of Economic Growth", *Journal of Economic Theory*, 12, 164–190.

Bryson, A.E. and Y. Ho (1979), *Applied Optimal Control*, Halsted Publishing Press, Waltham, Massachusetts.

Bucy, R.S. (1965), "Stability and Positive Supermartingales", *Journal of Differential Equations*, 1, 151–155.

Burmeister, E. and A.R. Dobell (1970), *Mathematical Theories of Economic Growth*, Macmillan, New York.

Cass, D. and J.E. Stiglitz (1970), "The structure of Investor Preferences and Asset Returns, and Separability in Portfolio Allocation: A Contribution to the Pure Theory of Mutual Funds", *Journal of Economic Theory*, 2, 122–160.

Cass, D. and J.E. Stiglitz (1972), "Risk Aversion and Wealth Effects on Portfolios With Many Assets", *Review of Economic Studies*, 39, 331–354.

Cass, D. and K. Shell (1976), "The Structure and Stability of Competitive Dynamical Systems", *Journal of Economic Theory*, 12, 31–70.

Cesari, L. (1963), *Asymptotic Behavior and Stability Problems*, 2nd edn., Academic Press, New York.

Chow, G.C. (1970), "Optimal Stochastic Control of Linear Economic Systems", *Journal of Money, Credit and Banking*, 2, 291–302.

Chow, G.C. (1973), "Effect of Uncertainty on Optimal Control Policies", *International Economic Review*, 14, 632–645.

Chow, G.C. (1975), *Analysis and Control of Dynamic Economic Systems*, John Wiley & Sons, New York.

Chow, Y.S. and H. Robbins (1967), "A Class of Optimal Stopping Problems", in: *Proc. Fifth Berkeley Symposium Math. Statist. Prob.*, vol. 1, University of California Press, Los Angeles, pp. 419–426.

Chow, Y.S., H. Robbins and D. Siegmund (1971), *Great Expectations: The Theory of Optimal Stopping*, Houghton Mifflin Company, Boston.

Chung, K.L. (1974), *A Course in Probability Theory*, Academic Press, New York.

Çinlar, E. (1975), *Introduction to Stochastic Processes*, Prentice-Hall, Inc., Englewood Cliffs, New Jersey.

Constantinides, G.M. (1976), "Stochastic Cash Management With Fixed and Proportional Transaction Costs", *Management Science*, 22, 1320–1331.

Constantinides, G.M. (1978), "Market Risk Adjustment in Project Valuation", *Journal of Finance*, 33, 603–616.

Constantinides, G.M. and S.F. Richard (1978), "Existence of Optimal Simple Policies for Discounted-Cost Inventory and Cash Management in Continuous Time", *Operations Research*, 26, 620–636.

Constantinides, G.M. (1980), "Capital Market Equilibrium with Personal Tax", University of Chicago, Graduate School of Business working paper.

Cootner, P.H., ed. (1964), *The Random Character of Stock Market Prices*, MIT Press, Cambridge, Massachusetts.

Cornell, B. (1977), "Spot Rates, Forward Rates and Exchange Market Efficiency", *Journal of Financial Economics*, 5, 55–65.

Cox, D.R. and H.D. Miller (1965), *The Theory of Stochastic Processes*, John Wiley & Sons, New York.

Cox, J.C. and S.A. Ross (1976), "The valuation of Options for Alternative Stochastic Processes", *Journal of Financial Economics*, 3, 145–166.

Cox, J. and S.A. Ross (1979), "Option Pricing: A Simplified Approach", *Journal of Financial Economics*, 7, 229–263.

Cox, J.C., J.E. Ingersoll and S.A. Ross (1978), "A Theory of The Term Structure of Interest Rates", Stanford University, Graduate School of Business working paper.

Cox, J.C. and J.E. Ingersoll, Jr. and S.A. Ross (1980), "The Relationship Between Forward Prices and Futures Prices", Stanford University working paper.

Crane, D.B. (1971), "A Stochastic Programming Model for Commercial Bank Bond Portfolio Management", *Journal of Financial and Quantitative Analysis*, 6, 955–976.

Daellenbach, H.G. and S.H. Archer (1969), "The Optimal Bank Liquidity: A Multi-period Stochastic Model", *Journal of Financial and Quantitative Analysis*, 4, 329–343.

Danthine, J.P. (1977), "Martingale, Market Efficiency and Commodity Prices", *European Economic Review*, 10, 1–17.

Danthine, J.P. (1978), "Information, Future Prices, and Stabilizing Speculation", *Journal of Economic Theory*, 17, 79–98.

Davis, M.H.A. (1973), "On The Existence of Optimal Policies in Stochastic Control", *SIAM Journal on Control*, 11, 587–594.

DeGroot, M. (1968), "Some Problems of Optimal Stopping", *Royal Statistical Society Bulletin*, 30, 108–122.

DeGroot, M.H. (1970), *Optimal Statistical Decisions*, McGraw-Hill Company, New York.

Dellacherie, C. (1974), *Integrales Stochastiques Par Rapport Aux Processus de Wiener et de Poisson*, Lecture Notes in Mathematics, no. 381, Springer-Verlag, New York.

Dellacherie, C. and P.A. Meyer (1978), *Probabilities and Potential*, North-Holland, New York.

Diamond, P.A. and M.E. Yaari (1972), "Implications of the Theory of Rationing for Consumer Choice Under Uncertainty", *American Economic Review*, 62, 333–343.

Diamond, P.A. and J.E. Stiglitz (1974), "Increases In Risk and In Risk Aversion", *Journal of Economic Theory*, 8, 337–360.

Diamond, P. and M. Rothschild, eds. (1978), *Uncertainty in Economics: Readings and Exercises*, Academic Press, Inc. New York.

Doob, J.L. (1953), *Stochastic Processes*, John Wiley & Sons, New York.

Doob, J.L. (1971), "What Is A Martingale?", *The American Mathematical Monthly*, 78, 451–462.

Doob, J.L. (1966), "Wiener's Work in Probability Theory", *Bulletin of the American Mathematical Society*, 72, 69–72.

Dothan, L.U. (1978), "On the Term Structure of Interest Rates", *Journal of Financial Economics*, 6, 59–69.

Dreyfus, S.E. (1965), *Dynamic Programming and the Calculus of Variations*, Academic Press, New York.

Dvoretzky, A. (1967), "Existence and Properties of Certain Optimal Stopping Rules", in: *Proc. Fifth Berkeley Symposium Math. Statist. Prob.*, University of California Press, Los Angeles, pp. 441–452.

Dynkin, E.B. (1965), *Markov Processes*, Springer-Verlag, New York.

Dynkin, E.B. and A.A. Yushkevich (1969), *Markov Processes: Theorems and Problems*, Plenum Press, New York.

Eppen, G.D. and E.F. Fama (1968), "Solutions for Cash-Balance and Simple Dynamic-Portfolio Problems", *Journal of Business*, 41, 94–112.

Eppen, G.D. and E.F. Fama (1969), "Cash Balance and Simple Dynamic Portfolio Problems with Proportional Costs", *International Economic Review*, 10, 119–133.

Fama, E. (1970a), "Multiperiod Consumption-Investment Decisions", *The American Economic Review*, 60, 163–174.

Fama, E.F. (1970b), "Efficient Capital Markets: A Review of Theory and Empirical Work", *The Journal of Finance*, 25, 383–417.

Fama, E.F. (1972), "Perfect Competition and Optimal Production Decisions Under Uncertainty", *Bell Journal of Economics and Management Science*, 3, 509–530.

Fama, E.F. (1975), "Short-Term Interest Rates As Predictors of Inflation", *American Economic Review*, 65, 269–282.

Fama, E. (1976), *Foundations of Finance*, Basic Books, New York.

Fama, E.F. and M. Miller (1972), *Theory of Finance*, Holt, Rinehart and Winston, New York.

Feller, W. (1954), "Diffusion Processes in One Dimension", *Transactions of the American Mathematical Society*, 97, 1–31.

Feller, W. (1968), *An Introduction to Probability Theory and its Applications*, vol. 1, John Wiley & Sons, New York.

Feller, W. (1971), *An Introduction to Probability Theory and its Applications*, vol. 2, John Wiley & Sons, New York.

Fischer, S. (1972), "Assets, Contingent Commodities and The Slutsky Equations", *Econometrica*, 40, 371–385.

Fischer, S. (1975), "The Demand for Index Bonds", *Journal of Political Economy*, 83, 509–534.

Fleming, W.H. (1969), "Optimal Continuous-Parameter Stochastic Control", *SIAM Review*, 11, 470–509.

Fleming, W.H. (1971), "Stochastic Control for Small Noise Intensities", *SIAM Journal on Control*, 9, 473–517.

Fleming, W.H. and R.W. Richel (1975), *Deterministic and Stochastic Optimal Control*, Springer-Verlag, New York.

Foldes, L. (1978a), "Optimal Saving and Risk in Continuous Time", *Review of Economic Studies*, 45, 39–65.

Foldes, L (1978b), "Martingale Conditions for Optimal Saving–Discrete Time", *Journal of Mathematical Economics*, 5, 83–96.

Francis, J.C. and S.H. Archer (1979), *Portfolio Analysis*, 2nd edn., Prentice-Hall, Inc., New Jersey.

Frenkel, J.A. and B. Jovanovic (1980), "On Transactions and Precausionary Demand for Money", *Quarterly Journal of Economics*, 95, 25–43.

Friedman, A. (1975), *Stochastic Differential Equations and Applications*, Academic Press, New York.

Friedman, M. and L.J. Savage (1948), "The Utility Analysis of Choices Involving Risk", *The Journal of Political Economy*, 56, 279–304.

Gal'perin, E.A. and N.N. Krasovskii (1963), "Cn the Stabilization of Stationary Motions in Nonlinear Control Systems", *Journal of Applied Mathematics and Mechanics*, 27, 1521–1546.

Gertler, M. (1979), "Money, Prices, and Inflation in Macroeconomic Models with Rational Inflationary Expectations", *Journal of Economic Theory*, 21, 222–234.

Gihman, I. and A.V. Skorohod (1969), *Introduction to the Theory of Random Processes*, Saunders, Philadelphia.

Gihman, I. and A.V. Skorohod (1972), *Stochastic Differential Equations*, Springer-Verlag, Inc., New York.

Gihman, I. and A.V. Skorohod (1974), *The Theory of Stochastic Processes*, Springer-Verlag, Inc., New York.

Gihman, I. and A.V. Skorohod (1975), *The Theory of Stochastic Processes*, Vol. 1, Springer-Verlag, Inc., New York.

Gihman, I. and A.V. Skorohod (1979), *The Theory of Stochastic Processes*, Vol. 2, Springer-Verlag, Inc., New York.

Gould, J.P. (1968), "Adjustment Costs in the Theory of Investment of the Firm", *Review of Economic Studies*, 35, 47–55.

Greenwood, P.H. and C.A. Ingene (1978), "Uncertain Externalities, Liability Rules and Resource Allocation", *The American Economic Review*, 68, 300–310.

Grossman, S. (1976), "On the Efficiency of Competitive Stock Markets where Traders have Diverse Information", *Journal of Finance*, 31, 573–585.

Grossman, S. and J.E. Stiglitz (1976), "Information and Competitive Price Systems", *American Economic Review*, 66, 246–253.

Grossman, S.J. and J.E. Stiglitz (1980), "The Impossibility of Informationally Efficient Markets", *The American Economic Review*, 70, 393–408.

Hadjimichalakis, M. (1971), "Equilibrium and Disequilibrium Growth with Money: The Tobin Models", *Review of Economic Studies*, 38, 457–479.

Hadley, G. (1964), *Nonlinear and Dynamic Programming*, Addison-Wesley publishing Co., Reading, Massachusetts.

Hahn, F.H. (1970), "Savings and Uncertainty", *Review of Economic Studies*, 37, 21–24.

Hahn, W. (1963), *Theory and Application of Liapunov's Direct Method*, Prentice-Hall, New York.

Hahn, W. (1967), *Stability of Motion*, Springer-Verlag, New York.

Hakansson, N.H. (1970), "Optimal Investment and Consumption Strategies Under Risk for a Class of Utility Functions", *Econometrica*, 38, 587–607.

Hale, J. (1969), "Dynamical Systems and Stability", *Journal of Mathematical Analysis and Applications*, 26, 39–59.

Hall, R.E. (1978), "Stochastic Implications of the Life Cycle–Permanent Income Hypothesis: Theory and Evidence", *Journal of Political Economy*, 86, 971–987.

Harrison, J.M. and D.M. Kreps (1979), "Martingales and Arbitrage in Multiperiod Securities Markets", *Journal of Economic Theory*, 20, 381–408.

Harrison, J.M. and S.R. Pliska (1981), "Martingales and Stochastic Integrals in the Theory of Continuous Trading", *Stochastic Processes and their Applications*, 11.

Hartman, P. (1961), "On the Stability in the Large for Systems of Ordinary Differential Equations", *Canadian Journal of Mathematics*, 13, 480–492.

Hartman, P. (1964), *Ordinary Differential Equations*, John Wiley & Sons, New York.

Hartman, P. and C. Olech (1962), "On Global Asymptotic Stability of Solutions of Ordinary Differential Equations", *Transactions of The American Mathematical Society*, 104, 154–178.

Helpman, E. and A. Razin (1979), *A Theory of International Trade Under Uncertainty*, Academic Press, New York.

Hestenes, M. (1966), *Calculus of Variations and Optimal Control Theory*, John Wiley & Sons, New York.

Hoel, P.G., S.C. Port and C.J. Stone (1972), *Introduction to Stochastic Processes*, Houghton Mifflin Company, Boston.

Hirshleifer, J. (1965), "Investment Decision Under Uncertainty: Choice-Theoretic Approaches", *Quarterly Journal of Economics*, 79, 509–536.

Hirshleifer, J. (1966), "Investment Decision Under Uncertainty: Applications of the State-Preference Approach", *Quarterly Journal of Economics*, 80, 252–277.

Hirshleifer, J. (1973), "Where Are We In The Theory of Information?", *American Economic Review*, 63, 31–39.

Houthakker, H. (1961), "Systematic and Random Elements in Short-Term Price Movements", *American Economic Review*, 51, 164–172.

Humphrey, T.M. (1974), "The Concept of Indexation in the History of Economic Thought", *The Economic Review of the Federal Reserve Bank of Richmond*, 60, 3–16.

Ingersoll, J. (1976), "A Theoretical and Empirical Investigation of the Dual-Purpose Funds: An Application of Contingent Claims Analysis", *Journal of Financial Economics*, 3, 83–123.

Ingersoll, J.E. (1977), "A Contingent-Claims Valuation of Convertible Securities", *Journal of Financial Economics*, 4, 289–322.

Intriligator, M.D. (1978), *Econometric Models, Techniques and Applications*, Prentice-Hall, Inc., New Jersey.

Ishii, Y. (1977), "On The Theory of The Competitive Firm Under Price Uncertainty: Note", *The American Economic Review*, 67, 768–769.

Itô, K. (1944), "Stochastic Integral", *Proceedings of the Imperial Academy*, Tokyo, 20, 519–524.

Itô, K. (1946), "On A Stochastic Integral Equation", *Proceedings of the Imperial Academy*, 22, 32–35.

Itô, K. (1950), "Stochastic Differential Equations In A Differentiable Manifold", *Nagoya Mathematics Journal*, 1, 35–47.

Itô, K. (1951a), "On A Formula Concerning Stochastic Differentials", *Nagoya Mathematics Journal*, 3, 55–65.

Itô, K. (1951b), "On Stochastic Differential Equations", *Memoirs of the American Mathematical Society, The American Mathematical Society*, Rhode Island.

Itô, K. (1961), *Lectures on Stochastic Processes*, Tata Institute of Fundamental Research, India.

Itô, K. and H.P. McKean, Jr. (1974), *Diffusion Processes and Their Sample Paths*, Second Printing, Springer-Verlag, New York.

Jensen, M.C. (1969), "Risk, The Pricing of Capital Assets, and The Evaluation of Investment Portfolios", *Journal of Business*, 42, 167–247.

Jensen, M.C., ed. (1972), "Capital Markets: Theory and Evidence", *Bell Journal of Economics and Management Science*, 3, 357–398.

Jensen, M.C. and J.B. Long, Jr. (1972), "Corporate Investment Under Uncertainty and Pareto Optimality In The Capital Markets", *Bell Journal of Economics and Management Science*, 3, 151–174.

Jovanovic, B, (1979a), "Job Matching and the Theory of Turnover", *Journal of Political Economy*, 87, 972–990.

Jovanovic, B. (1979b), "Firm-Specific Capital and Turnover", *Journal of Political Economy*, 87, 1246–1260.

Kantor, B. (1979), "Rational Expectations and Economic Thought", *Journal of Economic Literature*, 17, 1422–1441.

Karlin, S. and H.M. Taylor (1975), *A First Course in Stochastic Processes*, 2nd edn., Academic Press, New York.

Kats, I.I. and N.N. Krasovskii (1960), "On The Stability of Systems with Random Disturbances", *Journal of Applied Mathematics and Mechanics*, 24, 1225–1245.

Kihlstrom, R.E. and L.J. Mirman (1974), "Risk Aversion with Many Commodities", *Journal of Economic Theory*, 8, 361–388.

Khas'minskii, R.Z. (1962), "On the Stability of the Trajectory of Markov Processes", *Applied Mathematics and Mechanics (USSR)*, 26, 1554–1565.

Kolmogorov, A.N. (1950), *Foundations of the Theory of Probability*, Chelsea Publishing Company, New York.

Kozin, F. (1972), "Stability of the Linear Stochastic System", in: R. Curtain, ed., *Stability of Stochastic Dynamical Systems*, Springer-Verlag, New York.

Kozin, F. and S. Prodromou (1971), "Necessary and Sufficient Conditions for Almost Sure Sample Stability of Linear Itô Equations", *SIAM Journal of Applied Mathematics*, 21, 413–424.

Krasovskii, N.N. (1965), *Stability of Motion*, Stanford University Press, California.

Kreps, D.M. (1980), "Multiperiod Securities and the Efficient Allocation of Risk: A Comment on the Black–Scholes Option Pricing Model", Technical Report 306, Institute for Mathematical Studies in the Social Sciences, Stanford University, California.

Kunita, H. (1970), "Stochastic Integrals Based on Martingales Taking Values in Hilbert Spaces", *Nagoya Mathematics Journal*, 38, 41–52.

Kunita, H. and S. Watanabe (1967), "On Square Integrable Martingales", *Nagoya Mathematics Journal*, 30, 209–245.

Kushner, H.J. (1965), "On the Stochastic Maximum Principle: Fixed Time of Control", *Journal of Mathematical Analysis and Applications*, 11, 78–92.

Kushner, H.J. (1967a), *Stochastic Stability and Control*, Academic Press, Inc., New York.

Kushner, H.J. (1967b), "Optimal Discounted Stochastic Control for Diffusion Processes", *SIAM Journal On Control*, 5, 520–531.

Kushner, H.J. (1971), *Introduction to Stochastic Control*, Holt, Rinehart and Winston, New York.

Kushner, H.J. (1972), "Stochastic Stability", in: R. Curtain, ed., *Stability of Stochastic Dynamical Systems*, Springer-Verlag, New York.

Kushner, H.J. (1975), "Existence Results for Optimal Stochastic Controls", *Journal of Optimization Theory and Applications*, 15, 347–360.

Kussmaul, A.V. (1977), *Stochastic Integration and Generalized Martingales*, Pitman, London.

Kwakernaak, H. and R. Sivan (1972), *Linear Optimal Control Systems*, Wiley–Interscience, New York.

Ladde, G.S. and V. Lakshmikantham (1980), *Random Differential Inequalities*, Academic Press, New York.

LaSalle, J.P. (1964), "Recent Advances in Liapunov Stability Theory", *SIAM Review*, 6, 1–11.

LaSalle, J.P. and S. Lefschetz (1961), *Stability by Liapunov's Direct Method*, Academic Press, New York.

Lau, M. (1977), "The Behavior of the Exchange Rate in a Two-Country Monetary Model", University of Chicago, Department of Economics, unpublished paper.

Lee, E.B. and L. Markus (1967), *Foundations of Optimal Control Theory*, John Wiley & Sons, New York.

Lefschetz, S. (1965), *Stability of Nonlinear Control Systems*, Academic Press, New York.

Leland, H.E. (1972a), "On Turnpike Portfolios", in: G. Szego and K. Shell, eds., *Mathematical Methods in Investment and Finance*, North-Holland Publishing, Amsterdam, pp. 24–33.

Leland, H.E. (1972b), "Theory of the Firm Facing Uncertain Demand", *The American Economic Review*, 62, 278–291.

Leland, H.E. (1974), "Production Theory and the Stock Market", *Bell Journal of Economics and Management Science*, 5, 125–144.

Leonardz, B. (1974), *To Stop or Not to Stop: Some Elementary Optimal Stopping Problems with Economic Interpretations*, Halsted Press, New York.

LeRoy, S. (1973), "Risk Aversion and the Martingale Property of Stock Prices", *International Economic Review*, 14, 436–446.

Levhari, D. (1972), "Optimal Savings and Portfolio Choice under Uncertainty", in: G. Szego and K. Shell, eds., *Mathematical Models in Investment and Finance*, North-Holland Publishing, Amsterdam, pp. 34–48.

Levhari, D. and N. Liviatan (1972), "On Stability in the Saddle-point Sense", *Journal of Economic Theory*, 4, 88–93.

Levhari, D. and T.N. Srinivasin (1969), "Optimal Savings under Uncertainty", *Review of Economic Studies*, 36, 153–163.

Levy, H. and M. Sarnat, eds. (1977), *Financial Decision Making under Uncertainty*, Academic Press, New York.

Liapunov, A. (1949), "Probleme General de la Stabilite du Mouvement", *Annals of Mathematical Studies*, 17, Princeton University Press, Princeton, New Jersey.

Lippman, S.A. and J.J. McCall (1976a), "The Economics of Job Search: A Survey, Part I", *Economic Inquiry*, 14, 155–189.

Lippman, S.A. and J.J. McCall (1976b), "The Economics of Job Search: A Survey, Part II", *Economic Inquiry*, 14, 347–368.

Lippman, S.A. and J.J. McCall (1976c), "Job Search in a Dynamic Economy", *Journal of Economic Theory*, 12, 365–390.

Lippman, S.A. and J.J. McCall, eds. (1979), *Studies in Economics of Search*, Elsevier North-Holland, Inc., New York.

Lintner, J. (1965a), "The Valuations of Risk Assets and the Selection of Risky Investments in Stock Portfolios and Capital Budgets", *Review of Economics and Statistics*, 47, 13–37.

Lintner, J. (1965b), "Security Prices, Risk, and Maximal Gains from Diversification", *The Journal of Finance*, 20, 587–615.

Litzenberger, R.H. and K. Ramaswamy (1979), "The Effect of Personal Taxes and Dividends on Capital Asset Prices", *Journal of Financial Economics*, 7, 163–195.

Liviatan, N. and D. Levhari (1977), "Risk and the Theory of Indexed Bonds", *American Economic Review*, 67, 366–375.

Loève, M.M. (1977), *Probability Theory*, 4th edn., Springer-Verlag, New York.

Lucas, R.E. (1967a), "Optimal Investment Policy and the Flexible Accelerator", *International Economic Review*, 8, 78–85.

Lucas, R.E. (1967b), "Adjustment Costs and the Theory of Supply", *Journal of Political Economy*, 75, 321–334.

Lucas, R.E. (1975), "An Equilibrium Model of the Business Cycle", *Journal of Political Economy*, 83, 1113–1144.

Lucas, R.E. (1978), "Asset Prices in an Exchange Economy", *Econometrica*, 46, 1429–1445.

Lucas, R.E. and E. Prescott (1971), "Investment under Uncertainty", *Econometrica*, 39, 659–681.

Lucas, R.E. and E.C. Prescott (1974), "Equilibrium Search and Unemployment", *Journal of Economic Theory*, 7, 188–209.

MaCurdy, T.E. (1978), "Two Essays on the Life Cycle", University of Chicago, Department of Economics, unpublished Ph.D. thesis.

Magill, M.J.P. (1977a), "A Local Analysis of *N*-Sector Capital Accumulation under Uncertainty", *Journal of Economic Theory*, 15, 211–219.

Magill, M.J.P. (1977b), "Some New Results on the Local Stability of the Process of Capital Accumulation", *Journal of Economic Theory*, 15, 174–210.

Magill, M.J.P. and G.M. Constantinides (1976), "Portfolio Selection with Transactions Costs", *Journal of Economic Theory*, 13, 245–263.

Majumdar, M. (1975), "Equilibrium in Stochastic Economic Models", in: R. Day and T. Groves, eds., *Adaptive Economic Models*, Academic Press, New York, 479–498.

Malinvaud, E. (1953), "Capital Accumulation and Efficient Allocation of Resources", *Econometrica*, 21, 233–268.

Malkiel, B.G. (1966), *Term Structure of Interest Rates: Expectations and Behavior Patterns*, Princeton University Press, Princeton, New Jersey.

Malliaris, A.G. (1978), "The Neoclassical Economic Growth Model under Uncertainty", paper presented at the Econometric Society Meetings in Chicago.

Malliaris, A.G. (1981), "Martingale Methods in Financial Decision Making", *SIAM Review*, forthcoming.

Mandelbrot, B.B. (1966), "Forecasts of Future Prices, Unbiased Markets and 'Martingale' Models", *Journal of Business*, 39, 242–255.

Mandelbrot, B.B. (1971), "When can Price be Arbitraged Efficiently? A Limit to the Validity of the Random Walk and Martingale Models", *Review of Economics and Statistics*, 53, 225–236.

Mandl, P. (1968), *Analytical Treatment of One-Dimensional Markov Processes*, Springer-Verlag, New York.

Mangasarian, O.L. (1963), "Stability Criteria for Non-Linear Ordinary Differential Equations", *SIAM Journal On Control*, 1, 311–318.

Mangasarian, O.L. (1966), "Sufficient Conditions for the Optimal Control of Nonlinear Systems", *SIAM Journal On Control*, 4, 139–152.

Mangasarian, O. (1969), *Nonlinear Programming*, McGraw-Hill, New York.

Mao, J.C. (1969), *Quantitative Analysis of Financial Decisions*, Macmillan Company, New York.

Markowitz, H.M. (1959), *Portfolio Selection: Efficient Diversification of Investments*, John Wiley & Sons, New York.

Markus, L. and H. Yamabe (1960), "Global Stability Criteria for Differential Systems", *Osaka Mathematical Journal*, 12, 305–317.

Massera, J.L. (1949), "On Liapunov's Conditions of Stability", *Annals of Mathematics*, 57, 705–721.

Massera, J.L. (1956), "Contributions to Stability Theory", *Annals of Mathematics*, 64, 183–206.

Mayer, W. (1976), "The Rybczynski, Stolper–Samuelson, and Factor-Price Equilization Theorems under Price Uncertainty", *The American Economic Review*, 66, 797–808.

McCall, J.J. (1967), "Competitive Production for Constant Risk Utility Functions", *Review of Economic Studies*, 34, 417–420.

McCall, J.J. (1971), "Probabilistic Microeconomics", *The Bell Journal of Economics and Management Science*, 2, 403–433.

McFadden, D. (1973), "On the Existence of Optimal Development Programmes in Infinite Horizon Economics", in: J. Mirrlees and N.H. Stern, eds., *Models of Economic Growth*, Halsted Press, Chicago.

McKean, H.P., Jr. (1969), *Stochastic Integrals*, Academic Press, New York.

McKenzie, L. (1976), "Turnpike Theory", *Econometrica*, 44, 841–865.

McNees, K. (1978), "The 'Rationality' of Economic Forecasts", *American Economic Review*, 68, 301–305.

McShane, E.J. (1974), *Stochastic Calculus and Stochastic Models*, Academic Press, New York.

Meiselman, D. (1962), *The Term Structure of Interest Rates*, Prentice-Hall, New Jersey.

Merton, R.C. (1971), "Optimum Consumption and Portfolio Rules in a Continuous-Time Model", *Journal of Economic Theory*, 3, 373–413.

Merton, R.C. (1972), "An Analytic Derivation of the Efficient Portfolio Frontier", *Journal of Financial and Quantitative Analysis*, 7, 1851–1872.

Merton, R. (1973a), "Theory of Rational Option Pricing", *Bell Journal of Economics and Management Science*, 4, 141–183.

Merton, R. (1973b), "An Intertemporal Capital Asset Pricing Model", *Econometrica*, 41, 867–887.

Merton, R. (1973c), "Erratum", *Journal of Economic Theory*, 6, 213–214.

Merton, R. (1974), "On the Pricing of Corporate Debt: The Risk Structure of Interest Rates", *Journal of Finance*, 29, 449–470.

Merton, R. (1975a), "An Asymptotic Theory of Growth under Uncertainty", *Review of Economic Studies*, 42, 375–393.

Merton, R.C. (1975b), "Theory of Finance from the Perspective of Continuous Time", *Journal of Financial and Quantitative Analysis*, 10, 659–674.

Merton, R.C. (1976), "Option Pricing when Underlying Stock Returns are Discontinuous", *Journal of Financial Economics*, 3, 125–144.

Merton, R.C. (1978), "On the Mathematics and Economic Assumptions of Continuous-Time Models", M.I.T. working paper 981-78.

Merton, R.C. (1981), "On the Microeconomic Theory of Investment under Uncertainty", in: K.J. Arrow and M.D. Intriligator, eds., *Handbook of Mathematical Economics*, Vol. 2, North-Holland Publishing Company, Amsterdam.

Metivier, M. and J. Pellaumail (1980), *Stochastic Integration*, Academic Press, New York.

Meyer, P.A. (1966), *Probability and Potentials*, Blaisdell Publishing Company, Waltham, Massachusetts.

Meyer, P.A. (1967), *Intégrales Stochastiques*, Lecture Notes in Mathematics, no. 39, Springer-Verlag, New York.

Meyer, P.A. (1972), *Martingales and Stochastic Integrals*, vol. 1, Springer-Verlag, Berlin.

Meyer, P.A. (1976), *Un Cours sur les Integrales Stochastiques*, Lecture Notes in Mathematics, no. 511, Springer-Verlag, New York.

Michaelsen, J.B. (1973), *The Term Structure of Interest Rates*, Intext Educational Publishers, New York.

Miller, M.H. and F. Modigliani (1958), "The Cost of Capital, Corporation Finance, and the Theory of Investment", *American Economic Review*, 48, 261–297.

Miller, M.H. and D. Orr (1966), "A Model of the Demand for Money by Firms", *Quarterly Journal of Economics*, 80, 413–435.

Miller, R.A. and K. Voltaire (1980), "A Sequential Stochastic Free Problem", University of Chicago working paper, forthcoming in *Economic Letters*.

Mills, E.S. (1959), "Uncertainty and Price Theory", *Quarterly Journal of Economics*, 73, 116–130.

Mirman, L.J. (1971), "Uncertainty and Optimal Consumption Decisions", *Econometrica*, 39, 179–185.

Mirman, L. (1973), "Steady State Behavior of One Class of One Sector Growth Models with Uncertain Technology", *Journal of Economic Theory*, 6, 219–242.

Mirman, L.J. and W.R. Porter (1974), "A Microeconomic Model of the Labor Market under Uncertainty", *Economic Inquiry*, 12, 135–145.

Mirman, L. and I. Zilcha (1975), "On Optimal Growth under Uncertainty", *Journal of Economic Theory*, 11, 329–339.

Mirman, L. and I. Zilcha (1976), "Unbounded Shadow Prices for Optimal Stochastic Growth Models", *International Economic Review*, 17, 121–132.

Mirman, L. and I. Zilcha (1977), "Characterizing Optimal Policies in a One-Sector Model of Economic Growth under Uncertainty", *Journal of Economic Theory*, 14, 389–401.

Miroshnichenko, T.P. (1975), "Optimal Stopping of the Integral of a Wiener Process", *Theory of Probability and Its Applications*, 20, 387–391.

Modigliani, F. and M.H. Miller (1958), "The Cost of Capital, Corporation Finance, and the Theory of Investment", *American Economic Review*, 48, 261–297.

Modigliani, F. and R. Shiller (1973), "Inflation, Rational Expectations and the Term Structure of Interest Rates", *Economica*, 40, 12–43.

Mortensen, D.T. (1973), "Generalized Costs of Adjustment and Dynamic Factor Demand Theory", *Econometrica*, 41, 657–665.

Mossin, J. (1966), "Equilibrium in a Capital Asset Market", *Econometrica*, 34, 768–783.

Mullineaux, D. (1978), "On Testing for Rationality: Another Look at the Livingston Price Expectations Data", *Journal of Political Economy*, 86, 329–336.

Muth, J.F. (1961), "Rational Expectations and the Theory of Price Movements", *Econometrica*, 29, 315–335.

Nagel, E. (1969), *Principles of the Theory of Probability*, 12th printing, The University of Chicago Press, Chicago.

Neave, E.H. (1970), "The Stochastic Cash-Balance Problem with Fixed Costs for Increases and Decreases", *Management Science*, 16, 472–490.

Nelson, C.R. (1972), *Term Structure of Interest Rates*, Basic Books, New York.

Nelson, R. (1961), "Uncertainty, Prediction, and Competitive Equilibrium", *Quarterly Journal of Economics*, 75, 41–62.

Neveu, J. (1965), *Mathematical Foundations of the Calculus of Probability*, Holden-Day, San Francisco.

Neveu, J. (1975), *Discrete-Parameter Martingales*, North-Holland Publishing Company Ltd., New York.

Ohlson, J.A. (1977), "Risk Aversion and the Martingale Property of Stock Prices: Comments", *International Economic Review*, 18, 229–234.

Olivera, J.H.G. (1971), "The Square-Root Law of Precautionary Reserves", *Journal of Political Economy*, 79, 1095–1104.

O'Neill, D.E. (1978), "Sources of Macroeconomic Martingales and Tests of the Martingale Property", paper presented at the American Economic Association Meetings in Chicago.

Papoulis, A. (1965), *Probability, Random Variables, and Stochastic Processes*, McGraw-Hill, New York.

Perrakis, S. (1980), "Factor-Price Uncertainty with Variable Proportions: Note", *The American Economic Review*, 70, 1083–1088.

Pontryagin, L.S. *et al.*, (1962), *The Mathematical Theory of Optimal Processes*, Wiley–Interscience, New York.

Poole, W. (1970), "Optimal Choice of Monetary Policy Instruments in a Simple Stochastic Macro Model", *Quarterly Journal of Economics*, 84, 197–216.

Prabhu, N.U. (1965), *Stochastic Processes*, Macmillan Company, New York.

Pratt, J.W. (1964), "Risk Aversion in the Small and in The Large", *Econometrica*, 32, 122–136.

Radner, R. (1970), "Problems in the Theory of Markets under Uncertainty", *The American Economic Review*, 60, 454–460.

Radner, R. (1974), "Market Equilibrium and Uncertainty: Concepts and Problems", in: M.D. Intriligator and D. Kendrick, eds., *Frontiers of Quantitative Economics*, North-Holland Publishing, Amsterdam, pp. 43–90.

Richard, S.F. (1978), "An Arbitrage Model of the Term Structure of Interest Rates", *Journal of Financial Economics*, 6, 33–57.

Rishel, R.W. (1970), "Necessary and Sufficient Dynamic Programming Conditions for Continuous Time Stochastic Optimal Control", *SIAM Journal on Control*, 8, 559–571.

Robbins, H. (1970), "Optimal Stopping", *American Mathematical Monthly*, 77, 333–343.

Rockafellar, R.T. (1970), *Convex Analysis*, Princeton University Press, Princeton, New Jersey.

Rockafellar, R.T. (1973), "Saddle Points of Hamiltonian Systems in Convex Problems of Lagrange", *Journal of Optimization Theory and Applications*, 12, 367–590.

Rockafellar, R.T. (1976), "Saddle Points of Hamiltonian Systems in Convex Lagrange Problems having a Nonzero Discount Rate", *Journal of Economic Theory*, 12, 71–113.

Roll, R. (1970), *The Behavior of Interest Rates: An Application of the Efficient Market Model to U.S. Treasury Bills*, Basic Books, New York.

Ross, S.A. (1975), "Uncertainty and the Heterogeneous Capital Goods Model", *Review of Economic Studies*, 42, 133–146.

Ross, S.A. (1976), "The Arbitrage Theory of Capital Asset Pricing", *Journal of Economic Theory*, 13, 341–360.

Rothschild, M. (1973), "Models of Market Organization with Imperfect Information: A Survey", *Journal of Political Economy*, 81, 1283–1308.

Rothschild, M. and J.E. Stiglitz (1970), "Increasing Risk: I. A Definition", *Journal of Economic Theory*, 2, 225–243.

Rothschild, M. (1974), "Searching for the Lowest Price when the Distribution of Prices is Unknown", *Journal of Political Economy*, 82, 689–711.

Roxin, E. (1965), "Stability in General Control Systems", *Journal of Differential Equations*, 1, 115–150.

Roxin, E. (1966), "On Stability in Control Systems", *SIAM Journal on Control*, 4, 357–372.

Ruiz-Moncayo, A. (1968), "Optimal Stopping for Functions of Markov Chains", *Annals of Mathematical Statistics*, 39, 1905–1912.

Samuelson, P. (1965), "Proof that Properly Anticipated Prices Fluctuate Randomly", *Industrial Management Review*, 6, 41–49.

Samuelson, P.A. (1969), "Lifetime Portfolio Selection by Dynamic Stochastic Programming", *Review of Economics and Statistics*, 51, 239–246.

Samuelson, P.A. (1972), "The General Saddle Point Property of Optimal-Control Motions", *Journal of Economic Theory*, 5, 102–120.

Samuelson, P.A. (1973), "Proof that Properly Discounted Present Values of Assets Vibrate Randomly", *Bell Journal of Economics and Management Science*, 4, 369–374.

Samuelson, P.A. and H.P. McKean, Jr. (1965), "Rational Theory of Warrant Pricing", and "Appendix: A Free Boundary Problem for the Heat Equation Arising from a Problem in Mathematical Economics", respectively, *Industrial Management Review*. Reprinted in R.C. Merton, ed. (1972), *The Collected Scientific Papers of Paul A. Samuelson*, M.I.T. Press, pp. 791–871.

Sandmo, A. (1970), "The Effect of Uncertainty on Saving Decisions", *Review of Economic Studies*, 37, 353–360.

Sandmo, A. (1971), "On the Theory of the Competitive Firm under Price Uncertainty", *The American Economic Review*, 61, 65–73.

Sargent, T. (1972), "Rational Expectations and the Term Structure of Interest Rates", *Journal of Money, Credit and Banking*, 4, 74–97.

Sargent, T.J. (1976), "A Classical Macroeconometric Model for the United States", *Journal of Political Economy*, 84, 207– 37.

Sargent, T.J. (1979), *Macroeconomic Theory*, Academic Press, Inc., New York.

Scheinkman, J.A. (1976), "On Optimal Steady States of *N*-Sector Growth Models when Utility is Discounted", *Journal of Economic Theory*, 12, 11–30.

Scheinkman, J.A. (1977), "Notes on Asset Pricing", The University of Chicago, Department of Economics working paper.

Scheinkman, J.A. (1978), "Stability of Separable Hamiltonians and Investment Theory", *Review of Economic Studies*, 45, 559–570.

Schuss, Z. (1980), *Theory and Applications of Stochastic Differential Equations*, Wiley-Interscience, Somerset, New Jersey.

Sharpe, W.F. (1964), "Capital Assets Prices: A Theory of Market Equilibrium under Conditions of Risk", *The Journal of Finance*, 19, 425–442.

Shell, K. (1972), "Selected Elementary Topics in the Theory of Economic Decision Making under Uncertainty", in: G. Szegö and K. Shell, eds., *Mathematical Methods in Investment and Finance*, North-Holland Publishing, Amsterdam, pp. 65–75.

Shiller, R.J. (1978), "Rational Expectations and the Dynamic Structure of Macroeconomic Models: A Critical Review", *Journal of Monetary Economics*, 4, 1–44.

Shiryayev, A.N. (1978), *Optimal Stopping Rules*, Springer-Verlag, Inc., New York.

Siegmund, D. (1967), "Some Problems in the Theory of Optimal Stopping Rules", *Annals of Mathematical Statistics*, 38, 1627–1640.

Simon, H.A. (1977), *Models of Discovery*, D. Reidel Publishing Company, Pallas Paperbacks, Boston.

Slutsky, E. (1937), "The Summation of Random Causes as the Source of Cyclic Processes", *Econometrica*, 5, 105–146.

Smith, C.W., Jr. (1976), "Option Pricing: A Review", *Journal of Financial Economics*, 3, 3–51.

Smith, C.W., Jr. (1979), "Applications of Option Pricing Analysis", in J.L. Bicksler, ed.

(1979) *Handbook of Financial Economics*, North-Holland Publishing Company, Amsterdam.

Smith, K. (1969), "The Effect of Uncertainty on Monopoly Price, Capital Stock, and Utilization of Capital", *Journal of Economic Theory*, 1, 48–59.

Solow, R.M. (1956), "A Contribution to the Theory of Economic Growth", *Quarterly Journal of Economics*, 70, 65–94.

Soong, T.T. (1973), *Random Differential Equations in Science and Engineering*, Academic Press, New York.

Stigum, B.P. (1969a), "Entrepreneurial Choice over Time under Conditions of Uncertainty", *International Economic Review*, 10, 426–442.

Stigum, B.P. (1969b), "Competitive Equilibria under Uncertainty", *Quarterly Journal of Economics*, 83, 533–561.

Stigum, B. (1972), "Balanced Growth under Uncertainty", *Journal of Economic Theory*, 5, 42–68.

Stigler, G.J. (1961), "The Economics of Information", *Journal of Political Economy*, 69, 213–225.

Stiglitz, J.E. (1969), "A Re-Examination of the Modigliani-Miller Theorem", *American Economic Review*, 59, 784–793.

Stiglitz, J.D. (1972), "Portfolio Allocation with Risky Assets", in: G. Szegö and K. Shell, eds., *Mathematical Methods in Investment and Finance*, North-Holland, Amsterdam, pp. 76–125.

Stratonovich, R.L. (1966), "A New Representation for Stochastic Integrals and Equations", *SIAM Journal of Control*, 4, 362–371.

Strauss, A. (1968), *An Introduction to Optimal Control Theory*, Springer-Verlag, New York.

Syski, R. (1967), "Stochastic Differential Equations", in: T.L. Saaty, ed., *Modern Nonlinear Equations*, McGraw-Hill, New York.

Szegö, G. and K. Shell, eds. (1972), *Mathematical Methods in Investment and Finance*, North-Holland Publishing Company, Amsterdam.

Tanaka, H. (1957), "On Limiting Distributions for One-Dimensional Diffusion Processes", *Bulletin of Mathematical Statistics*, 7, 84–91.

Tinbergen, J. (1952), *On the Theory of Economic Policy*, 2nd edn., North-Holland Publishing Company, Amsterdam.

Tobin, J. (1956), "The Interest-Elasticity of Transactions Demand for Cash", *Review of Economics and Statistics*, 38, 241–247.

Tobin, J. (1965), "Money and Economic Growth", *Econometrica*, 33, 671–684.

Treadway, A.B. (1969), "On Rational Entrepreneurial Behavior and the Demand for Investment", *Review of Economic Studies*, 36, 227–239.

Treadway, A.B. (1970), "Adjustment Costs and Variable Inputs in the Theory of the Competitive Firm", *Journal of Economic Theory*, 2, 329–347.

Treadway, A.B. (1971), "The Rational Multivariate Flexible Accelerator", *Econometrica*, 39, 845–855.

Tsokos, C.P. and W.J. Padgett (1974), *Random Integral Equations with Applications to Life Sciences and Engineering*, Academic Press, New York.

Tucker, H. (1967), *A Graduate Course in Probability*, Academic Press, Inc., New York.

Turnovsky, S.J. (1973), "Optimal Stabilization Policies for Deterministic and Stochastic Linear Economic Systems", *Review of Economic Studies*, 40, 79–95.

Turnovsky, S.J. (1977), *Macroeconomic Analysis and Stabilization Policy*, Cambridge University Press, New York.

Turnovsky, Stephen J. and E.R. Weintraub (1971), "Stochastic Stability of a General Equilibrium System under Adaptive Expectations", *International Economic Review*, 12, 71–86.

Van Moerbeke, P. (1974), "Optimal Stopping and Free Boundary Problems", *Rocky Mountain Journal of Mathematics*, 4, 539–577.

Vasicek, O. (1977), "An Equilibrium Characterization of the Term Structure", *Journal of Financial Economics*, 5, 177–188.

Vial, J.P. (1972), "A Continuous Time Model for the Cash Balance Problem", in: G.P. Szegö and K. Shell, eds., *Mathematical Methods in Investment and Finance*, North-Holland Publishing Company, Amsterdam, pp. 244–291.

Ville, J. (1939), *Etude Critique de la Notion de Collectif*, Paris, France.

Wald, A. (1944), "On Cumulative Sums of Random Variables", *Annals of Mathematical Statistics*, 15, 283–296.

Wong, E. and M. Zakai (1965), "On the Convergence of Ordinary Integrals to Stochastic Integrals", *Annals of Mathematical Statistics*, 36, 1560–1564.

Wonham, W.M. (1966a), "Liapunov Criteria for Weak Stochastic Stability", *Journal of Differential Equations*, 2, 195–207.

Wonham, W.M. (1966b), "A Liapunov Method for the Estimation of Statistical Averages", *Journal of Differential Equations*, 2, 365–377.

Wonham, W.M. (1970), "Random Differential Equations in Control Theory", in: A.T. Bharucha-Reid, ed. (1970), *Probabilistic Methods in Applied Mathematics*, Volume 2, Academic Press, New York, 131–212.

Wu, S.Y. (1979), "An Essay on Monopoly Power and Stable Price Policy", *The American Economic Review*, 69, 60–72.

Yahav, J.A. (1966), "On Optimal Stopping", *Annals of Mathematical Statistics*, 37, 30–35.

Yoshizawa, T. (1966), *Stability Theory of Liapunov's Second Method*, Mathematical Society of Japan, Tokyo, Japan.

Zabel, E. (1970), "Monopoly and Uncertainty", *Review of Economic Studies*, 37, 205–219.

Ziemba, W.T. and R.G. Vickson, Eds. (1975), *Stochastic Optimization Models in Finance*, Academic Press, New York.

# AUTHOR INDEX*

Alchian, A., 61
Anderson, B., 139
Antosiewicz, H., 136
Aoki, M., 132, 139
Araujo, A., 139
Archer, S.H., 237, 277
Arnold, L., 70, 72, 74, 75, 77, 79, 80, 81, 96, 99, 100, 102, 133
Arrow, K.J., 108, 138, 217
Ash, R.B., 3, 10, 11, 12, 29, 32, 60, 61
Åström, K.J., 66, 67, 70, 132, 133, 139
Athans, M., 139

Bachelier, L., 61
Balakrishman, A.V., 132
Baron, D., 214
Batra, R.N., 214
Baumol, W.J., 238
Bellman, R., 108, 139
Benes, U.E., 139
Benveniste, L., 139, 186, 249, 251
Berkowitz, L.D., 139
Bernoulli, D., 62
Bertsekas, D.P., 132, 139
Bewley, T., 184, 215
Bharucha-Reid, A.T., 62, 70, 72
Bicksler, J.L., 272
Billingsley, P., 5, 13, 15, 16, 19, 29, 32, 34, 37, 58, 60, 61, 62
Bismut, J.M., 115, 118, 119, 120, 139, 150, 160, 163, 214, 215
Black, F., 220, 221, 222, 233, 273, 276
Borg, G., 136
Bourguignon, F., 106, 141, 214
Boyce, W.M., 216
Brainard, W., 192, 216
Breeden, D.T., 275

Breiman, L., 63
Brennan, J.J., 276
Brock, W.A., 53, 138, 139, 140, 167, 168, 182, 185, 187, 188, 205, 214, 215, 216, 251, 277
Bryson, A.E., 139
Bucy, R.S., 136
Burmeister, E., 142

Cass, D., 139
Cesari, L., 136
Chow, G.C., 67, 132
Chow, Y.S., 45, 51, 52, 53, 63, 216
Chung, K.L., 60
Çinlar, E., 62, 63
Constantinides, G.M., 236, 237, 238, 270, 276, 277
Cornell, B., 62
Cox, D.R., 62, 200
Cox, J.C., 273, 274, 276, 277
Crane, D.B., 277

Daellenbach, H.G., 277
Danthine, J.P., 62
Davis, M.H.A., 139
DeGroot, M., 45, 47, 48, 49, 51, 63
Dellacherie, C., 61, 139
Diamond, P.A., 214
Dobell, A.R., 142
Doob, J.L., 19, 29, 61, 62, 69, 133, 139
Dothan, L.U., 276
Dreyfus, S.E., 139
Dvoretzky, A., 63
Dynkin, E.B., 62, 63

Eppen, G.D., 277

* Also see Selected Bibliography.

# SUBJECT INDEX